Water Science and Technology

Small Water and Wastewater Systems VII

Small Water and Wastewater Systems VII

Selected Papers from the 7th IWA International Specialty Conference on Small Water and Wastewater Systems, held in Mexico City on 7 – 10 March 2006

Issue Editor: S. Gonzalez Martinez

National University of Mexico, Institute of Engineering, 04510 Mexico DF, Mexico

Publishing

International Programme Committee

Organising Committee

Organized and Sponsored by

International Water Association
UNAM

British Library Cataloguing in Publication Data
A CIP catalogue record for this book is available from the British Library
ISBN: 1 84339 582 7
ISBN: 978-1-84339-582-9

Contents

Preface

The conferences series on Small Wastewater Treatment Plants started in 1989 in recognition of the importance of small systems. Two years later the IWA Specialist Group on Small Wastewater Treatment Plants was established. The discussion held in this group was characterized by great enthusiasm of the scientific and technical communities. Influenced by the general debate on sustainable development and driven by first proposals for an integrated water resources management, the name was changed to "Small Water and Wastewater Systems". Decentralised systems for water supply and wastewater treatment and their association with receiving bodies and industrial wastewater systems came into focus.

The Specialist Group made several efforts to define "Small". Eventually, an agreement was reached according to which small systems are those serving from a few persons up to 4000 population equivalent. Within this range, there are very small wastewater systems (one to ten families), medium small (50 to 500 inhabitants) and large small (500 to 4000 inhabitant equivalent) systems. Because of the size, many industrial water and wastewater systems can be included as "small systems".

Associated with the size and the resources for operation and maintenance, there is in many countries a lack of qualified personnel to provide the expertise for the optimal operation conditions. This is why small systems represent a challenge for the small communities or industry to achieve the effluent values imposed by the regulations. Therefore the "Small" systems must deal with a wide variety of knowledge topics from basic concepts of physics, chemistry, biochemistry and microbiology through the fundamentals of process engineering, to design practices and operational matters. Training of personnel is one of the most important keys to successful operation of the systems.

The 7th Specialty Conference was held in Mexico City on 7–10 March 2006, with the topics decentralised wastewater collection and treatment systems, ecological sanitation, industrial wastewater treatment, anaerobic treatment systems, upgrading wastewater treatment systems, environmental management systems, environmental microbiology and population dynamics, operation and monitoring wastewater treatment systems, wastewater reclamation and reuse, low tech/appropriate systems, sludge management, biofilm reactors, sequencing batch reactors, nutrient removal and recovery, wetland systems and stabilisation ponds. 147 papers were submitted to the conference secretariat, 78 were selected for oral presentation and 19 as posters. After the conference 32 papers were selected to be printed in this issue of *Water Science and Technology*. The participants were honored by the plenary presentations of Professors Peter Wilderer, Goen Ho and François Brissaud.

Simón González
Conference Chairman

Water Science & Technology **Vol 55 No 7 pp 1–9** © IWA Publishing 2007

Low technology systems for wastewater treatment: perspectives

F. Brissaud

Hydrosciences, MSE, Université Montpellier II, 34095 Montpellier Cedex 05, France
(E-mail: *brissaud@msem.univ-montp2.fr*)

Abstract Low technology systems for the treatment of wastewater are sometimes presented as remnants of the past, nowadays supposedly only meant to serve developing countries and remote rural areas. However, considering their advantages and disadvantages together with enhanced treatment requirements and recent research and technological developments, the future of these systems still appears promising. Successful applications of low technology systems require that more care is taken of their design and operation than often observed. Correlatively, more efforts should be made to decipher the treatment mechanisms and determine the related reaction parameters, so as to provide more deterministic approaches of the natural wastewater treatment systems and better predict their performance.

Keywords Extensive systems; low technology; natural systems; wastewater treatment

Introduction

Low technology systems, also called extensive or natural systems, are based on the imitation or adaptation of processes that occur naturally in soils and water bodies. They include mainly stabilization ponds, intermittent soil filtration and constructed wetlands. They incorporate few if any electro-mechanical equipment and are low energy consumers.

While big cities in developed countries unquestionably are to be served by sewage treatment plants based on conventional intensive technologies (physical–chemical treatment, activated sludge, etc.), conventional systems are in competition with low technology systems to treat the wastewater of medium and small-size communities. Indeed, with the exception of desert zones, large areas of land where extensive wastewater treatment facilities could be set up are seldom available in the vicinity of big cities where authorities are used to relying on sophisticated technologies and plants operated by highly skilled personnel to abide by discharge regulations and prevent failures that could damage the environment. Large towns can afford high treatment expenses, which is not the case for smaller communities.

On the contrary, though intensive techniques should not be discarded, there are strong arguments to claim that extensive sewage treatments are more appropriate for medium and small-size municipalities than conventional systems, the most popular of which are based on the activated sludge process. The perspectives offered to the development of natural wastewater treatment systems depend on (i) the balance between their advantages and disadvantages, (ii) the development of the treatment regulations, (iii) the continuation and success of ongoing researches and recent technological advances aimed at increasing the performance and reliability of these technologies and (iv) the efforts made to educate designers and operators who, unfortunately, are not always aware of the treatment processes, of the performance that can be expected and of the operational conditions required to make these systems efficient and sustainable.

doi: 10.2166/wst.2007.120

Advantages/disadvantages

Advantages

Though investment costs are not much lower than those of intensive techniques, extensive systems offer the decisive advantage of very low operation and maintenance (O&M) costs. Actually, no (or very limited) energy supply is required and as electro-mechanical equipment is reduced to a minimum, maintenance is relatively inexpensive. Power supply, which is an important part of the operation cost of conventional systems, is rapidly increasing due to the soaring cost of energy. As extensive techniques incorporate low technology equipment, O&M does not require highly skilled personnel and is not very time consuming. These advantages deserve the highest consideration when planning wastewater treatment in rural areas and developing countries. Indeed, as reported by Kengne *et al.* (2005), construction and O&M costs of conventional sewage treatment plants require large amounts of money that countries facing structural and financial adjustment cannot afford. The result is, for instance, that most of the conventional systems, especially activated sludge, present in the Sub-Sahara Africa, lack maintenance, are overloaded or simply out of order. In rural zones of developed countries, small community budgets, poor access to equipment, supplies and repair facilities, together with the difficulty linked to the O&M of many small plants scattered over large areas often preclude proper O&M of conventional systems and result in recurrent violation of discharge permits (US EPA, 2000).

All extensive and natural techniques allow significant to high microbial decontamination. Four to 5 log units removal of *Escherichia coli* can be obtained in stabilisation ponds (Brissaud *et al.*, 2005; Pearson *et al.*, 2005), more than 6 log units in shallow ponds (Von Sperling and Mascarenhas, 2005), up to 5 log units in batch operated reservoirs (Indelicato *et al.*, 1996) and 3 to 4 log units with infiltration percolation. One to more than 3 log units has been reported in vertical flow and horizontal flow constructed wetlands (Masi *et al.*, 2004; Baeder-Bederski *et al.*, 2004). Though decontamination performances are dependent on a number of factors, i.e. climatic conditions, the season of the year, the design and load and management of the plants, the potential to decontaminate wastewater is a very strong argument in favour of extensive techniques in many circumstances. Microbial decontamination is – or should be - the priority of sanitation strategies in most developing countries, particularly in tropical and arid regions where water reuse is on the agenda. Protection of bathing waters is of the utmost importance in tourist areas, such as the Mediterranean. In karstic areas, disinfection of wastewater is essential for the protection of groundwater quality. In water scarce countries, the most rewarding water reuse applications require pathogen removal.

In natural systems, sludge production is limited to the settling stage. As settling is performed in Imhoff tanks, septic tanks and ponds, most of the accumulated sludge is digested, the excess being removed twice a year from Imhoff tanks, 3–5 years from septic tanks and up to one year out of 15 for facultative ponds as demonstrated by Picot *et al.* (2005). Many vertical flow constructed wetlands are fed with raw wastewater that has just been degritted; sludge accumulates on the filters rather slowly. Molle *et al.* (2005) report sludge heights of 13 and 25 cm in the primary vertical filter of a plant designed for 1,600 p.e., after 9 and 15 years operation, respectively. SS mineralization rate was estimated to be more than 60% (Boutin *et al.*, 1997). Because the dry matter content was greater than 20%, sludge deposits were easily removed without damaging the reed rhizomes. Reducing both desludging frequency and the amount of sludge to be handled is a major advantage, not only in terms of labour saving but also because the regulations imposed on sludge disposal have been made so severe in several countries that the cost has dramatically increased.

Extensive techniques are reliable. Treatment units are gravity fed in many plants and therefore are not dependent on power supply; then, the main risks of failure events are linked to potential clogging of the infiltration systems and to possible submersion of the plants during storm events. When pumping is necessary, pumps, often doubled and easy to replace, are the only electro-mechanical tools required to assure a proper functioning.

Systems with a high detention time, such as stabilization ponds and reservoirs can buffer load variations, thus preventing the discharge of poorly treated water in the environment.

Extensive or natural techniques are sometimes classified as "green" techniques. Ponds and vegetated constructed wetlands and infiltration systems can be designed in order to enhance the aesthetic value of the plant settings, including "natural" shapes for the treatment units, islands, plant diversity, providing wildlife habitat, etc. However, wildlife access should be controlled in order to limit the presence of nuisance species and to prevent the detrimental impact of overpopulation of waterfowl, insects such as mosquitoes, floating macrophytes and even fishes.

Disadvantages

The main shortcoming of most extensive techniques is their large and often excessive footprints. They cannot, therefore, be applied in densely populated areas where land prices are often soaring. This is illustrated by the fate of many waste stabilization pond systems commissioned from the 1960s to the 1980s. As the population has increased and towns have sprawled into the neighbouring countryside, old waste stabilization ponds became overloaded and closer to the new settlements. In most cases, the capacity of the plant has been increased through the construction of a conventional biological treatment unit, the ponds – when they have not been decommissioned – being converted to a polishing stage. However other solutions that can be classified as natural systems or close to them are available. The most popular is the aeration of the primary ponds (Copin *et al.*, 2004). The subdivision of large ponds into smaller units is also very successful (Archer and Donaldson, 2003).

Another disadvantage is that, in many cases, low technology systems cannot meet all treatment objectives. If high organic matter removal is achieved in soil treatments and vertical flow constructed wetlands, COD abatement is hampered in ponds, reservoirs and other free water surface wetlands by algal cell content in the final treatment units. Though this particularity has been taken into account in several wastewater disposal regulations, e.g. E.U. Directive 91/271 EEC (EU, 1991), more stringent requirements are likely to be imposed in the near future. Fortunately, this drawback can be overcome by adding rock filters or constructed wetlands to polish the effluent (Tanner *et al.*, 2005). Nitrogen removal is temperature dependent. While even performances can be expected all the year long in hot climates, nitrogen reduction in stabilisation ponds and free water surface wetlands depends highly on the season in temperate and northern climates; low removal, if any, occurs during the cold seasons (Archer and O'Brien, 2005; Andersson *et al.*, 2005; Houweling *et al.*, 2005). In soil treatment systems and vertical flow constructed wetlands, ammonium nitrogen is effectively nitrified but without the possibility to remove nitrogen through denitrification. This is the reason why Cooper *et al.* (1999) recommend hybrid systems which combine vertical and horizontal flow constructed wetlands. Phosphorus may precipitate in ponds, reservoirs and free water surface wetlands provided the temperature is high enough, but is hardly eliminated in soil filtration systems and constructed wetlands. Several attempts to enhance phosphorus reduction through the utilisation of media capable of binding P – calcareous crushed rocks, calcite, apatite, ochre, iron slags – have not all been very successful (Arias and Brix, 2005; Shilton *et al.*, 2005).

Very stringent microbial criteria, i.e. ≤10 faecal coliforms per 100 mL, are not likely to be always met when using only low technology systems. Typically, free water surface systems cannot consistently supply effluents of less than 100 faecal coliforms per 100 mL. Indeed, recontamination by birds, rodents, amphibians and insects always occurs in natural environments. Though in these cases the meaning of current microbial criteria in terms of public health remains unclear, natural settings limit the quality of the effluents. Moreover, in temperate and cold climates, microbial decontamination in free water surface systems depends on the season, with lower performance in winter. Seasonal variations are much less marked in soil treatments. Low loads and deep vadose zones lead to very high microbial decontamination levels.

The efficiency of free water surface extensive technologies, particularly ponds and reservoirs, are dependent on meteorological conditions. Rainfall and evaporation result in differences between inlet and outlet flow rates. Tracer tests associated with records of climate parameters pointed to the key influence of temperature and wind on the hydrodynamic behaviour of ponds and the distribution of water detention times (Brissaud *et al.*, 2003). On the other hand, micro-organism die-off kinetics depends on temperature and solar radiation. Engineers are not accustomed to dealing with these contingencies; under-estimation of climate effects and poor understanding of hydrodynamics have contributed to the reputation of unreliability of the disinfection performances of extensive open systems.

Evaporation results in water losses inversely proportional to the water depth and proportional to the water residence time in free water surface bodies. Evaporation rate may reach values as high as 8.5 mm.day^{-1} and even more in arid regions. Therefore, it should be taken into account in pond design, particularly in those countries where evaporation is high and when treated water is to be reused in agriculture. Water losses can reach 20–30%, which results in a significant increase of salinity and tends to hamper water reuse in agriculture. Evaporation can be diminished through a reduction of the water detention time and water depth augmentation.

Potential adverse environmental impacts should not be underestimated when planning extensive wastewater treatment facilities. Odour nuisances and proliferation of offensive insects, such as mosquitoes and midges, may ruin valuable projects. In soil treatment systems, odour nuisance may occur when the applied wastewater remains at the soil surface for a long time, i.e. more than a few hours, after the application. This situation, which usually results from overloading or poor clogging control and is detrimental for the oxygen supply to the filter and the treatment efficiency, should be prevented through proper operation and management. In the case of underground distribution systems, odour nuisance can be due to aeration defects. Odour nuisances are likely to occur in ponds when sludge that has accumulated during the cold season starts being digested thanks to the Spring temperature increase. Offensive smells also result from temporary excess of organic load or from sludge accumulation. Several options can be taken to reduce the risk of offensive smells emitted by stabilisation pond systems (Copin *et al.*, 2004; Truppel *et al.*, 2005). One solution is to recirculate effluents from the second or third ponds to the inlet of the primary pond. An alternative solution consists in the mechanical aeration of the primary pond. Another option is to consider near desludging of the primary pond. Midge and mosquito proliferation can be fought against with either ecological methods (Kengne *et al.*, 2005) or pesticide spreading (Craggs *et al.*, 2005; Sweeney *et al.*, 2005).

Technology advances

Low technology systems face nowadays several challenges among which footprint reduction, enhanced discharged water quality requirements and better performance reliability and predictability seem the most important. Technology advances and recent research developments allow or are expected to allow us to face successfully these new objectives and assure the continuing development of natural wastewater treatment systems in various socio-economical contexts.

Footprint reduction

As the population is rapidly increasing in urban and sub-urban areas, land available at reasonable costs becomes every day scarcer. Therefore in these areas, plant footprint reduction will be a major asset in the competition between low technology and conventional systems. Several technological solutions have been proposed during the last decades. Thanks to small to moderate electric power consumption, facultative ponds can be replaced by high rate algal ponds – that can be part of Advanced Pond Systems (Craggs *et al.*, 2003) – or mechanically aerated ponds, thus reducing considerably the land surface required for pond systems. Introducing anaerobic ponds as a primary treatment stage has the same effect but does not require power supply; however, offensive smells are frequent in the neighbourhood of anaerobic ponds. As illustrated by Archer and Donaldson (2003), subdividing large ponds into smaller units may significantly improve the performance and save land surface. Another solution consisting in substituting vertical flow constructed wetlands for facultative ponds looks very promising.

Several attempts have been made to limit the land surface required by buried filters of on site domestic wastewater treatment systems. The recommended hydraulic load applied on sand filters should not exceed $0.03\,m^3\,m^{-2}$ per day for individual facilities and $0.05\,m^3\,m^{-2}\,d^{-1}$ for collective filters. This rather low load stems from three difficulties, the first one being to achieve a uniform distribution of wastewater at the surface of buried filters, the second the supply of oxygen and the third the management of clogging in confined and wet atmosphere. Several approaches allow increase of the hydraulic load. The first one consists in designing septic effluent distribution systems to provide a rather uniform spreading all over the filtrating beds; for this purpose it is highly recommended to feed the filter intermittently by pumping or using an automatic dosing batch, the flow rate being high enough to fill the distribution pipe network rapidly. Inserting distribution lines at the top of buried half pipes, as was proposed by the US EPA manual (2002) or similar systems lead to a more efficient and uniform use of the filters than the distribution through pipes inserted in gravel beds. Capillary forces in polyurethane sponge-like structures or geo-textile structures also improve water distribution over the filters. Half pipe leaching chambers and polyurethane and geo-textile structures provide for the circulation and renewal of the air over the filter, thus improving the oxygen supply, facilitating surface clogging management and allowing augmentation of the hydraulic load up to $0.13\,m^3\,m^{-2}\,d^{-1}$. The second approach consists in increasing the water content of the filter without diminishing the gas phase volume, which results in longer water detention time and increased treatment capacities. Different porous media have been tested successfully; it seems that hydraulic loads of $0.25\,m^3\,m^{-2}\,d^{-1}$ might be acceptable.

Enhanced requirements

Regulations and guidelines on the quality of treated wastewater either discharged in the environment or reused in irrigation or for other beneficial purposes, tend to become more stringent in many countries. More emphasis is put on the removal of nitrogen and phosphorus and on microbial decontamination. Treatment performance of low technology systems can be enhanced through either technological advances or combination of natural systems.

Though being already old concepts, baffles and rock filters should lead to considerable enhancement of pond system performance. Baffles can be designed on the basis of CFD modelling, which should improve their efficiency in the reduction of preferential pathways (Shilton and Mara, 2005). Rock filters are not that common in Europe. They have been used for years in the United States (Middlebrooks, 1995) and are part of current pond design in New Zealand (Archer and Donaldson, 2003). They are also used in Israel. A rock filter – the rock size being 40–80 mm – is a porous rock bed through which pond effluent travels, causing algae to settle out on the rock surfaces. Accumulated algae are then biologically degraded by bacteria growing on the biofilm wrapping on the rocks. In addition, significant ammonia removal and microbial decontamination can be achieved.

When microbial decontamination is at stake, innovative solutions such as the shallow ponds of Von Sperling and Mascarenhas (2005) and the batch operation of ponds and reservoirs (Juanicó and Dor, 1999) can provide high performance within relatively short detention times

Combining different natural systems, namely soil treatment and free water surface systems, offers the possibility to benefit from the advantages of each component of the treatment train. Though all possibilities have not yet been investigated, combining ponds and vertical flow constructed wetlands, infiltration percolation and horizontal flow constructed wetlands, vertical and horizontal flow constructed wetlands have proven to be effective.

Research

When the reliability of the natural systems stems essentially from their independency or low dependency on power supply, the simplicity of the electro-mechanical equipment and their buffer capacity – in the case of free water surface systems – performance prediction from the very design still needs to be improved. Low technology systems were originally based on empirical approaches. As these systems have successfully met the sanitation objectives of the communities they served, little need for scientific investigation was felt. Improving the reliability and predictability of extensive techniques as well as their performances requires not only technological innovation but also in-depth knowledge of the processes involved in the treatment and how they interact. Predicting the performance of a natural wastewater treatment systems demands that the relationships between the environmental and operation conditions and the quality of the final water are known. These relationships are expressed in the form of empirical or deterministic models. Models are often hybrid, partly empirical, partly deterministic. Though very helpful to the engineers, empirical models have several limitations; they cannot take all processes into account and they perform badly outside their calibration interval and when the implicit assumptions on which they have been elaborated are not verified. Deterministic models need the identification and mathematical expressions of the dominant mechanisms involved. This approach has been taken by Fritz *et al.* (1979) for the removal of nitrogen. However, while theory has been used since Marais (1974) and later developed by Mara and Pearson (1998) and others for the design of stabilisation ponds, most projects of soil treatment, constructed wetlands and even ponds are still designed on a purely empirical basis.

Stabilisation pond models combine theoretical water detention times with first order pollutant degradation kinetics. In fact, water detention times are highly dependent on meteorological conditions and therefore are not steady. Efforts are being made to calculate actual detention times as a function of the meteorological conditions through 3D CFD models. Expressing reaction kinetics in a deterministic form is hampered by the complexity of interactive processes occurring in free water surface systems; this is particularly true for microbial decontamination. For instance, light penetration in ponds, a key factor of algal photosynthesis, oxygen production and micro-organism inactivation, has not yet been completely elucidated, despite the results of recent research (Davies-Colley *et al.*, 2005; Heaven *et al.*, 2005).

Deterministic models of organic matter and nitrogen oxidation in soil treatment are available (Bancolé *et al.*, 2003). However, more calibration tests should be made, particularly as regards biomass development; thus a precious tool for clogging control would be available. Progress has been made in the understanding of the influence of the microbial ecosystem on bacterial indicator inactivation. Deterministic modelling of the transfer of micro-organisms, considered as colloids, in unsaturated porous media submitted to intermittent infiltration, is expected to allow more accurate predictions of microbial decontamination (Auset *et al.*, 2005).

Up to now, the approach of wastewater treatment in constructed wetlands has been mainly empirical. This might be due to the fact that a deterministic modeling has looked hardly feasible due to the introduction of plants rooted in the filters. However, more recently efforts are being made to characterize and model the hydraulics of these systems and, later on, the pollution degradation processes (Dittmer *et al.*, 2005; Schmid *et al.*, 2005).

Conclusion

Taking into account the economic situation of many countries, the development of economically sound sanitation strategies ensuring the protection of water resources, relies on (a) the availability of wastewater treatment techniques that meet disposal standards at low investment and O&M costs; (b) require low O&M skills and (c) are reliable. Thus, extensive technologies are likely to be preferred to intensive processes, particularly for those facilities that are to serve small and medium-sized communities. Investment costs of extensive wastewater treatment techniques are not much lower than those related to intensive technologies. Most low technology systems require large areas of available land and cannot be applied in areas that are too densely populated. However, extensive techniques do offer decisive advantages: very low O&M costs, easy O&M and high microbial decontamination potential.

Though technological innovation and transfers of recent research findings to engineering practice have enhanced the potential performances of these techniques, very stringent standards can hardly be met with high reliability. This is the case for health related criteria and also for nutrient removal. When the need for imposing very restrictive water quality criteria is demonstrated, combining low technology and conventional systems can be an option, particularly when a high microbial decontamination is required. As an example, combining infiltration percolation and ultra-violet irradiation is very effective and at a relatively low cost (Salgot *et al.*, 2002).

As many countries are now considering the equipment of small and middle-size communities, the future of natural systems looks promising, provided more care is taken of their design and operation than often observed. The current design of extensive treatment plants – consultants often use nothing more than a rule of thumb – should be improved to provide more predictable performances and reliable disinfection. These techniques should be taken into more consideration in the educational programmes of engineers and technicians.

References

Andersson, J.L., Kaliner Bastviken, S. and Tonderski, K.S. (2005). Free water surface wetlands for wastewater treatment in Sweden: nitrogen and phosphorus removal. *Wat. Sci. Tech.*, **51**(9), 39–46.

Archer, H.E. and Donaldson, S.A. (2003). Waste stabilization ponds upgrading at Blenheim and Seddon, New Zealand – case studies. *Wat. Sci. Tech.*, **48**(2), 17–23.

Archer, H.E. and O'Brien, B.M. (2005). Improving nitrogen reduction in waste stabilization ponds. *Wat. Sci. Tech.*, **51**(12), 133–138.

Arias, C.A. and Brix, H. (2005). Phosphorus removal in constructed wetlands: can suitable alternative media be identified? *Wat. Sci. Tech.*, **51**(9), 267–273.

Auset, M., Keller, A.A., Brissaud, F. and Lazarova, V. (2005). Intermittent filtration of bacteria and colloids in porous media. *Wat. Resour. Res.*, **41** W 09408, doi:10.1029/2004 WR003611.

Baeder-Bederski, O., Dürr, M., Borneff-Lipp, M., Netter, R., Kuschk, P., Mosig, P., Daeschein, G. and Müller, R.A. (2004). Reduction of micro-organisms in municipal sewage by means of planted and unplanted soil filters. *9th Int. IWA Conf. on Wetland Systems for Water Pollution Control. Avignon, 26–30 Sept.* pp. 435–441.

Bancolé, A., Brissaud, F. and Gnagne, T. (2003). Oxidation processes and clogging in intermittent unsaturated infiltration. *Wat. Sci. Tech.*, **48**(11/12), 139–146.

Boutin, C., Liénard, A. and Esser, D. (1997). Development of a new generation of reed-bed filters in France: First results. *Wat. Sci. Tech.*, **35**(5), 315–322.

Brissaud, F., Tournoud, M.G., Drakides, C. and Lazarova, V. (2003). Mixing and its impact on faecal coliform removal in a stabilisation pond. *Wat. Sci. Tech.*, **48**(2), 75–80.

Brissaud, F., Andrianarison, T., Brouillet, J.L. and Picot, B. (2005). Twenty years monitoring of Mèze stabilisation ponds. II – removal of faecal indicators. *Wat. Sci. Tech.*, **51**(12), 33–41.

Cooper, P., Griffin, P., Humphries, S. and Pound, A. (1999). Design of a hybrid reed bed system to achieve complete nitrification and denitrification of domestic sewage. *Wat. Sci. Tech.*, **40**(3), 283–289.

Copin, Y., Brouillet, J.L., Rivière, Y. and Brissaud, F. (2004). Waste stabilisation ponds in the wastewater management strategy of a French department. *6th Int IWA Conf. on Waste Stabilisation Ponds, Avignon, 27 Sept – 1st Oct.*, pp. 29–35.

Craggs, R., Davies-Colley, R.J., Tanner, C.C. and Sukias, J.P. (2003). Advanced pond system: performance with high rate ponds of differents depths and areas. *Wat. Sci. Tech.*, **48**(2), 259–267.

Craggs, R., Golding, L., Clearwater, S., Susarla, L. and Donovan, W. (2005). Control of chironomid midge larvae in wastewater stabilisation ponds: comparison of five compounds. *Wat. Sci. Tech.*, **51**(12), 191–199.

Davies-Colley, R.J., Craggs, R.J., Park, J. and Nagels, J.W. (2005). Optical characteristics of waste stabilisation ponds; recommendations for monitoring. *Wat. Sci. Tech.*, **51**(12), 153–161.

Dittmer, U., Meyer, D. and Langergraber, G. (2005). Simulation of a subsurface vertical flow constructed wetland for CSO treatment. *Wat. Sci. Tech.*, **51**(9), 225–232.

European Union (1991). *Council Directive Concerning Urban Wastewater Treatment*. 91/271 EEC of May 21, 1991, OJ NO L135/40 of May 30, 1991.

Fritz, J.J., Middleton, A.C. and Meredith, D.D. (1979). Dynamc process modeling of wastewater stabilization ponds. *JWPCF*, **51**(11), 2724–2743.

Heaven, S., Banks, C.J. and Zotova, E.A. (2005). Light attenuation parameters for waste stabilization ponds. *Wat. Sci. Tech.*, **51**(12), 143–152.

Houweling, C.D., Kharoune, L., Escalas, A. and Comeau, Y. (2005). Modeling ammonia removal in aerated facultative lagoons. *Wat. Sci. Tech.*, **51**(12), 139–142.

Indelicato, S., Barbagallo, S., Cirelli, G.L. and Zimbone, S.M. (1996). Re-use of municipal wastewater for irrigation in Italy. *Proc. of 7th International Conf. on Water and Irrigation*, Tel Aviv, Israel, 13–16 May 1996.

Juanicó, M. and Dor, I. (1999). *Reservoirs for Wastewater Storage and Reuse*, Berlin, Springer-Verlag.

Kengne Noumsi, I.M., Brissaud, F., Akoa, A., Atangan Eteme, R., Nya, J., Ndikefor, A. and Fonkou, T. (2005). Microphyte and macrophyte based lagooning in tropical regions. *Wat. Sci. Tech.*, **51**(12), 267–274.

Kengne Noumsi, I.M., Akoa, A., Atangan Eteme, R., Nya, J., Ngniado, A., Fonkou, T. and Brissaud, F. (2012). Mosquito development and biological control in a microphyte-based wastewater treatment plant. *Wat. Sci. Tech.*, **51**(12), 201–204.

Mara, D.D. and Pearson, H.W. (1998). *Design Manual for Waste Stabilisation Ponds in Mediterranean Countries*, Lagoon Technology International, Leeds, England.

Marais, G.v.R. (1974). Faecal bacteria kinetics in stabilization ponds. *J. Envir. Eng. Div., A.S.C.E., EE1*, 119–139.

Masi, F., Conte, G., Lepri, L., Martellini, T. and Del Bubba, M (2004). Endocrine disrupting chemicals (EDCs) and pathogens removal in anhybrid CW system for a tourist facility wastewater treatment and reuse. *9th Int. IWA Conf. on Wetland Systems for Water Pollution Control. Avignon*, 26–30 Sept. pp. 461–468.

Middlebrooks, E.J. (1995). Upgrading pond effluents: an overview. *Wat. Sci. Tech.*, **31**(12), 353–368.

Molle, P., Liénard, A., Boutin, C., Merlin, G. and Iwema, A. (2005). How to treat raw sewage with constructed wetlands: an overview of the French systems. *Wat. Sci. Tech.*, **51**(9), 11–21.

Pearson, H.W., Silva Athayde, S.T., Athayde, S.T., Jr and Silva, S.A. (2005). Implications for physical design: the effect of depth on the performance of waste stabilisation ponds. *Wat. Sci. Tech.*, **51**(12), 69–74.

Picot, B., Sambuco, J.P., Brouillet, J.L. and Rivière, Y. (2005). Waste stabilisation ponds: sludge accumulation technical and financial study on desludging and sludge disposal, case studies in France. *Wat. Sci. Tech.*, **51**(12), 227–234.

Salgot, M., Folch, M., Huertas, E., Tapias, J., Avellaneda, D., Girós, G., Brissaud, F., Vergés, C., Molina, J. and Pigem, J. (2002). Comparison of different advanced disinfection systems for wastewater reclamation. *Wat. Sci. Tech.: Wat. Supply*, **2**(3), 213–218.

Shilton, A. and Mara, D.D. (2005). CFD (computational luid dynamics) modelling of baffles for optimizing tropical waste stabilisation pond systems. *Wat. Sci. Tech.*, **51**(12), 103–106.

Shilton, A., Pratt, S., Drizo, A., Mahmood, B., Banker, S., Billings, L., Glenny, S. and Luo, D. (2005). 'Active' filters for upgrading phosphorus removal from pond systems. *Wat. Sci. Tech.*, **51**(12), 111–116.

Schmid, B.H., Stephan, U. and Hengl, M.A. (2005). Sediment deposition in constructed wetland ponds with emergent vegetation: laboratory study and mathematical model. *Wat. Sci. Tech.*, **51**(9), 307–314.

Sweeney, D.G., O'Brien, M.J., Cromar, N.J. and Fallowfield, H.J. (2005). Changes in waste stabilisation pond performance resulting from the retrofit of activated sludge treatment upstream:part II – management and operation issues. *Wat. Sci. Tech.*, **51**(12), 17–22.

Tanner, C.C., Craggs, R.J., Sukias, J.P.S. and Park, J.B.K. (2005). Comparison of maturation ponds and constructed wetlands as the final stage of an advanced pond system. *Wat. Sci. Tech.*, **51**(12), 307–314.

Truppel, A., Camargos, J.L.M., da Costa, R.H.R. and Belli Filho, P. (2005). Reduction of odors from a facultative pond using two different operation practices. *Wat. Sci. Tech.*, **51**(12), 205–211.

U.S. Environmental Protection Agency (2000). *Manual: Constructed wetlands. Treatment of municipal wastewater*. EPA/625/R-99/010. Office of Research and Development, Cincinnati, OH.

U.S. Environmental Protection Agency (2002). *On site wastewater treatment systems manual*. EPA/625/R-00/008.

Von Sperling, M. and Mascarenhas, L.C.A.M. (2005). Performance of very shallow ponds treating effluents from UASB reactors. *Wat. Sci. Tech.*, **51**(12), 83–90.

Membrane technology revolutionizes water treatment

P.A. Wilderer* and S. Paris**

*Institute of Advanced Studies on Sustainability, European Academy of Sciences and Arts, Amalienstr. 75, 80799 Munich, Germany (E-mail: *peter@wilderer.de*)

**Huber Technology, Industriepark Erasbach A1, 92334 Berching, Germany (E-mail: *stefania.paris@huber.de*)

Abstract Membranes play a crucial role in living cells, plants and animals. They not only serve as barriers between the inside and outside world of cells and organs. More importantly, they are means of selective transport of materials and host for biochemical conversion. Natural membrane systems have demonstrated efficiency and reliability for millions of years and it is remarkable that most of these systems are small, efficient and highly reliable even under rapidly changing ambient conditions. Thus, it appears to be advisable for technology developers to keep a close eye on Mother Nature. By doing so it is most likely that ideas for novel technical solutions are born. Following the concept of natural systems it is hypothesized that the Millennium Development Goals can be best met when counting on small water and wastewater treatment systems. The core of such systems could be membranes in which chemical reactions are integrated allowing recovery and direct utilization of valuable substances.

Keywords Active transport; anaerobic conversion; biofilm; biomimetics; cell membrane; embedded biochemical reactions; green refinery; Millennium Development Goals; natural membrane systems

Introduction

When browsing through the literature of the past 15 years one gets the impression that membrane technology constitutes the most important invention man has ever made. Application of membranes for the purpose of water and wastewater treatment is certainly a respectable demonstration of human ingenuity, however, membrane processes are not at all inventions of man. An even superficial investigation of natural biological systems reveals that membranes are widely used in nature for a multitude of purposes and have been for billions of years. As a matter of fact, membrane processes play an essential role in cell metabolism as well as in the metabolism of animals and plants. It can be hypothesized that membrane systems are the main reason why small systems like a cell, the lung, the kidney or the rumen of a cow function so effectively. It may be a good idea, therefore, to study the function of membranes as they are integrated in living systems and use the information gained for further development of membrane technology.

Learning from nature is the concept of a relatively new branch of science called "biomimetics" or "bionics". Attempts to use natural systems as a model have already led to a variety of innovative technical solutions. In this context, it should be noticed that living systems – a cell, an organ, a plant, an animal – are relatively small in size. Nevertheless, these systems are apparently highly efficient, flexible and adaptable to changing ambient conditions. Taking this into account, it is fair to assume that membrane technology derived from natural systems provides specific advantages to small systems. We could even go a step further by hypothesizing that membrane technology allows development and implementation of a new generation of small on site water supply and sanitation systems with re-use of the valuable substances, the purified wastewater for instance, as an integral part – often referred to as "Decentralized Sanitation and Re-use" (DeSa/R).

doi: 10.2166/wst.2007.121

Water Science & Technology **Vol 55 No 7 pp 11–20** © IWA Publishing 2007

DeSa/R sytems are considered highly important in the process of meeting the Millennium Development Goals (MDGs) because they allow rapid while stepwise installation of water infrastructure especially in newly built residential areas of the fast growing cities, where the demand of water supply and sanitation is of utmost importance for social stability. The key component in this concept is most probably membrane technology.

In the following, membrane processes as they occur in living systems are briefly described, followed by a chapter in which current applications of membrane technology are summarized. Finally, some ideas are presented on how to translate natural membrane systems into advanced membrane technology.

Membrane systems on duty in nature

It is worth noticing that membranes (Latin: membrana = cuticle) play a very important, even crucial role in natural systems. In lungs, for instance, membranes serve as an interface between a gaseous phase (air) and a liquid phase (blood). They separate both phases from each other, but – importantly – they allow transfer to and dissolution of oxygen in the blood flowing along the membrane, and transfer of carbon dioxide from the blood stream into the atmosphere. Similarly, membranes exposed to the anaerobic atmosphere in the rumen of a cow allow transfer to and dissolution of hydrolysis products (e.g. fatty acids and amino-acids) in the blood. Cell membranes are not only barriers and transfer units but provide a structure within which chemical processes are executed, electron transfer processes for instance (respiration chain), and enzymatically catalyzed binding and transport processes.

Cell membranes

Membranes are vital for all living entities. They separate the cell from the outside world, and they also separate compartments inside the cell to protect important processes and events. In addition to their separation function, membranes have diverse other functions in the different regions and organelles of a cell (Alberts *et al.*, 1994). In particular, they allow communication of the cells with the outside world by controlling selective transfer of materials into and from the cell. Some of the most important physical, chemical and physiological functions of cells occur in or on the membrane.

Membranes contain proteins and lipids. However, the proportion of each varies depending on the membrane. Myelin, for instance, insulating nerve fibers, contains only 18% protein and 76% lipid. Mitochondrial inner membranes contain 76% protein and only 24% lipid. Plasma membranes of human red blood cells and mouse liver contain nearly equal amounts of proteins (44 and 49% respectively) and lipids (43 and 52% respectively).

The lipids are arranged in the form of a lipid bilayer. Proteins are embedded in this structure. They pass through the bilayer and are called transmembrane proteins, therefore. These proteins mainly function as enzymes allowing selective transport of chemical substances including electrons. Within the cell membrane (or the inner mitochondrial membrane in eukaryotic cells) the electron transport chain is embedded by means of which the cell gains energy (Figure 1).

Membranes of the respiratory tract

The lung is an organ designed to bring blood into close contact with the air. This is to allow oxygen to be dissolved in the blood; and carbon dioxide to be discharged into the atmosphere.

Air is pumped down into the lungs through the muscular action of the chest, a combination of activity of the diaphragm and the muscles between the ribs. By the time air reaches the alveoli (Figure 2) it has become warmed and been filtered of particulate materials by hairs.

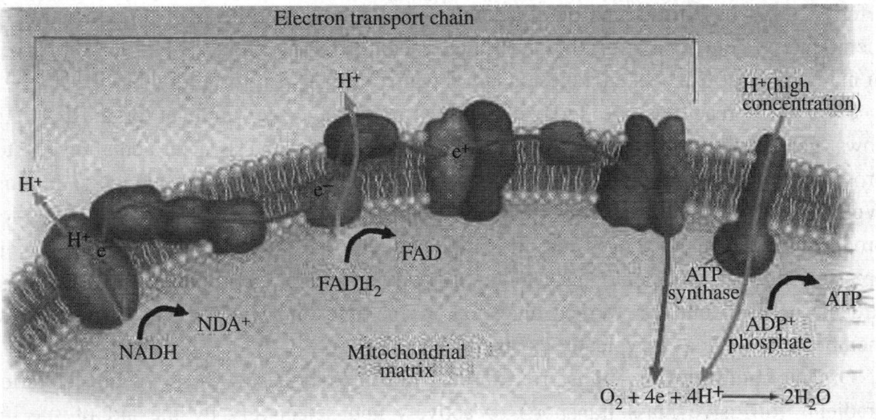

Figure 1 Schematic representation of the electron transport chain embedded in the inner mitochondrial membrane of eukaryotic cells (source: Johnson, 2005)

The gas exchange takes place within the alveoli where air is separated from blood by a membrane, just two cells in thickness. In the lungs of an adult, about three hundred million alveoli provide a surface area of sixty to eighty square metres. The alveoli constitute a membrane system. It separates the gas space filled with air from the blood flowing by.

Oxygen travels through the membrane by force of a concentration gradient, dissolves in the blood and is taken up by the red blood corpuscles. Carbon dioxide in turn travels towards the gas space – by force of a concentration gradient – transforms into its gaseous form, and is blown into the atmosphere as the lungs are emptied.

Membranes of the intestinal tract

The intestinal tract of animals and man is to be understood as an anaerobic reactor in which the organic material contained in the feed is gradually converted through chemical reactions and by the aid of microorganisms into small size molecules. This process is

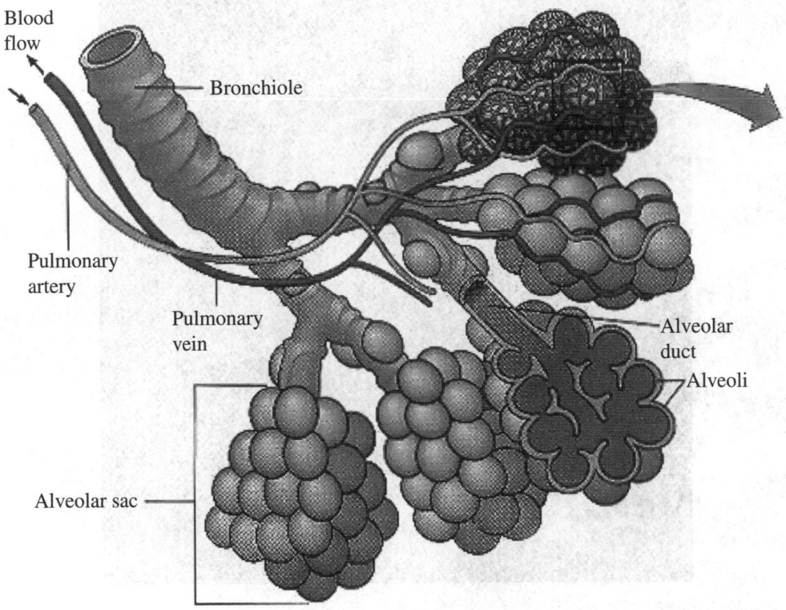

Figure 2 Schematic representation of the gas exchange systems in lungs (source: Johnson, 2005)

referred to as hydrolysis and acidification. The walls of the intestinal tract constitute a membrane system separating the anaerobic reactor from the blood stream, and it serves, in many cases at least, as an adhesion site for microorganisms, forming a biofilm there.

Hydrolytic reactions are most effectively performed by ruminant organisms (e.g. cows, goats and sheep). Again, the rumen is to be understood as a fermentation reactor providing an anaerobic environment, constant temperature and pH, and good mixing. Well masticated substrates are delivered through the esophagus, and fermentation products are either transferred into the blood stream through the wall of the rumen, more precisely through the membrane which separates the rumen environment from the blood stream, or flow out for further digestion in the intestinal tract. Figure 3 shows the membrane module as it typically exists in the rumen.

Organic molecules are transferred through the membranes by force of concentration gradients (diffusive mass transport) or actively and selectively by the aid of specific transmembrane proteins.

Membrane systems in the root zone of aquatic plants

It is worth noticing that roots, in particular roots of aquatic plants form symbiotic associations with soil microbes (Hook and Crawford, 1978). Aquatic plants which typically grow in an anaerobic environment developed a specific form of symbiosis involving ammonia and nitrite oxidizing bacteria (nitrifiers) in particular. Nitrifiers together with other microorganisms grow in the form of a biofilm at the surface of the roots (Figure 4).

Ammonia present in the bulk liquid is transformed into nitrite and nitrate which diffuses into the roots by force of a concentration gradient and is used as a source of nitrogen. In turn, the water plants support nitrifiers and other aerobic microorganisms with oxygen.

Oxygen is required by the root cells anyway. To support the root cells with oxygen, the plants have developed a unique ventilation system. By means of a tubular system within the stems and roots air is "blown" down to the root zone. Excess oxygen not

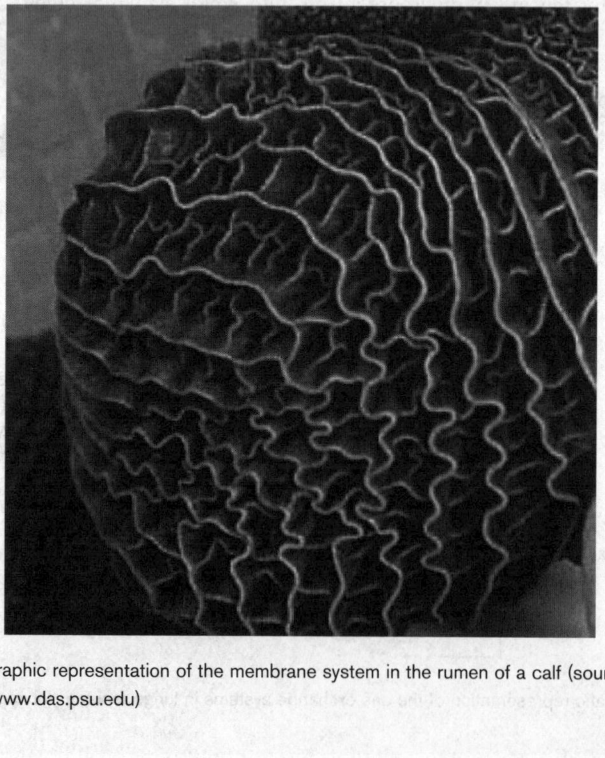

Figure 3 Photographic representation of the membrane system in the rumen of a calf (source: Penn State University, USA, www.das.psu.edu)

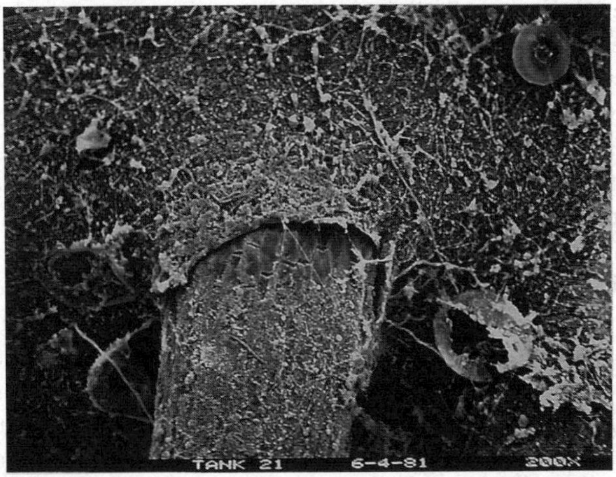

Figure 4 Electron scanning micrograph of a biofilm growing at a root of a water hyacinth plant

utilized by the root cells diffuses through the root structure into the biofilm which adheres to the surface of the roots.

Take home message

- Membranes play a very important role in the metabolism of organisms.
- Mass transport through the membranes is accomplished by development and maintenance of concentration gradients.
- At both sides of the membrane, diffusive mass transfer limitations are minimized by gentle force of either contraction or squeezing or by advection (e.g. flow of blood).
- Selective mass transport is actively controlled by enzymes embedded in the membranes.
- Bio-chemical conversion processes are performed within membranes.
- In many cases membranes are used by microorganisms as an adhesion site (substratum).The biofilms which develop at the surface of the membranes do not inhibit transmembrane flow of material but as representation of symbiosis.
- In natural membrane systems, cross flow is not applied to keep membranes from fouling. Neither pressure nor suction is applied to facilitate transfer of materials through the membranes. Relatively low transfer rates are compensated by provision of large membrane surfaces.

Contemporary membrane technology

Comparison of technical and natural application of membranes reveals some similarities but differences as well. For instance, membranes have been in use for millions of years in organisms whereas man made membrane technology is relatively new.

Currently, application of membrane technology is mainly focused on removal of particulate material from water (micro-, ultra-, nano-filtration) and removal of dissolved material (reverse osmosis). The opposite process, i.e. transfer of material into water is an option which is discussed but rarely applied in the field of water and wastewater treatment. For instance, membrane systems have been investigated for bubble-free transfer of oxygen into water (Wilderer *et al.*, 1985; Lee and Rittmann, 2000), and for bubble-free transfer of ozone into water as well (Picard *et al.*, 2001). In resemblance of root systems, membranes were proposed as a substratum for biofilms (Rothemund *et al.*, 1996; Lee and Rittmann, 2000).

In the uppermost cases, plastic materials (mainly polymers) are used as membrane material. Experiments have also been conducted, however, with sintered materials (ceramics, metals). Membranes applied for desalination (reverse osmosis) or for bubble-free oxygenation do not contain pores, whereas membranes for solid–liquid separation do. In the latter case, we talk about filtration membranes. The membranes are manufactured either as flat sheets or as capillaries. To provide maximum membrane surface per unit volume of the reactor the membranes are arranged in various ways. Pressure or suction is applied to maximize flow of permeate or filtrate, respectively.

During the in-service time the surface of the membranes may get coated with a layer of inorganic or organic material. In this case we are talking about scaling effects. When biofilms grow at the surface we are talking about bio-fouling. In either case, flow through the membrane gets reduced and the efficiency of the membrane module deteriorates. To remove scales and biofilms the membranes are to be backwashed – if possible – or cleaned, mechanically or chemically. The resulting out-of-service time diminishes the efficiency of the membrane system, and additional costs are generated. Keeping membranes from rapid scaling and fouling is therefore of major interest.

Experience shows that fouling can be minimized by exposing the membranes to hydrodynamic shear. Shear can be conveniently generated by letting gas bubbles pass along the membrane surface, perpendicular to the direction of the trans-membrane flow. This option is referred to as "cross flow" operation. Another way of producing a cross flow is shown in Figure 5. Here a stack of ceramic membranes is brought into rotation by which centrifugal forces add to the cross flow effects.

In wastewater treatment plants, membrane modules are mainly used as a substitute for sedimentation tanks. A prominent example of this approach is the so called "Membrane Bio-Reactor" (MBR). Membrane modules are submerged in the activated sludge tank. By sucking the wastewater through the membranes a particle-free, hygienically safe effluent can be produced ready for various re-use purposes such as toilet flushing, watering lawns or crop fields.

Figure 5 Stacks of flat ceramic membranes brought into rotation to control fouling (source: www.kerafol.com)

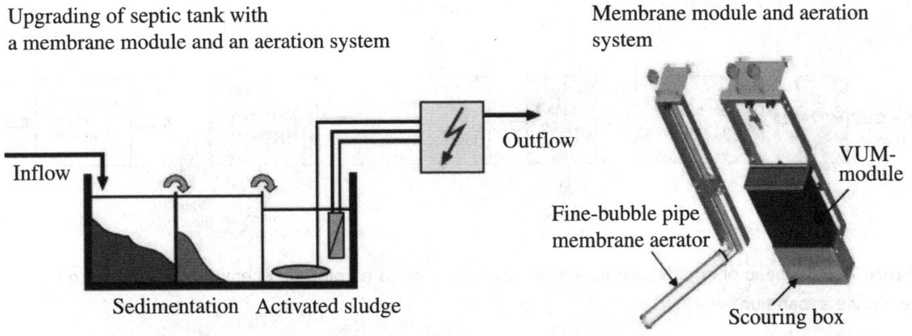

Figure 6 Upgrading of a septic tank with a submerged membrane module (HUBER-VUM-module; VUM = Vacuum Upstream Membrane)

Advanced concepts of membrane application

As mentioned above, particle-free, hygienically safe effluents can be achieved by driving wastewater through membranes. This concept is not only applicable for activated sludge plants. It could also be applied to treat the effluent of septic tanks. Figure 6 shows a configuration as it has been applied at several locations in Germany. Table 1 summarizes the effluent quality obtained.

The quality of the filtrate matches the requirements of the EU bathing water directive. In addition micro-pollutants such as pharmaceutics, hormones and alkyl phenols can be eliminated more efficiently than in conventional wastewater treatment plants by using membrane processes (Meuler *et al.*, 2006).

Membrane modules could also be used as a substitute for primary sedimentation tanks. This concept is visualized in Figure 7. Again, the basic idea is to generate an effluent which is free of particulate matter, free of pathogenic organisms and viruses, in particular. Thus, the permeate can be safely used for toilet flushing, cleaning, watering lawns or for irrigation of agricultural land. Hygienic safety is a must in all these cases. Expensive treatment operations such as nitrification and denitrification are unnecessary, particularly if the permeate is used for irrigation

On first glance, application of membrane separation technology to treat raw wastewater appears impossible because of rapid fouling of the membranes. To counteract fouling processes, removal of readily biodegradable substances is assumed to be necessary. As sketched in Figure 7, this could be achieved by adding precipitants and adsorbents (Teleman *et al.*, 2005) after the wastewater has passed a sequence of mechanical treatment steps.

The results of field tests proved that the addition of precipitants and coagulants increase the separation efficiency by means of a fine screen (Table 2). It appears to be worthwhile to further investigate the concept with membrane separation modules following fine screens.

Table 1 Quality of the water obtained by membrane ultra filtration (source: Meuler *et al.*, 2006)

Parameter	Filtrate*		Treatment efficiency
BOD_5	2	mg/l	99.1%
COD	23.7	mg/l	95.5%
NH_4-N	1.2	mg/l	81.9%
N_{anorg}	24.2	mg/l	
TSS	2.3	mg/l	99.3%
Faecal coliforms	45.8	1/100 ml	99.9%

*mean value

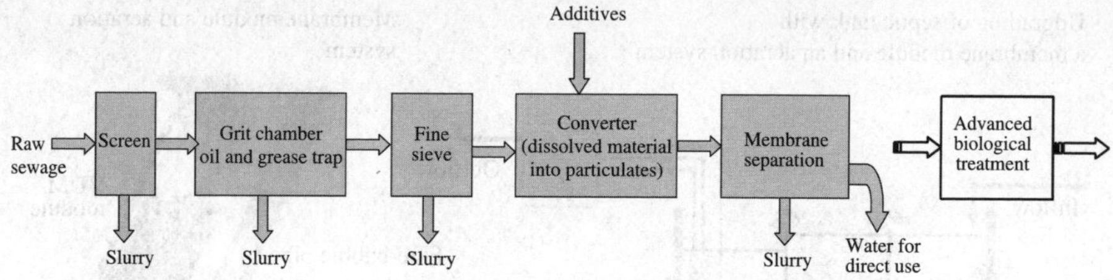

Figure 7 Schematic of an advanced pre-treatment train based on physico-chemical conversion and membrane separation

The same concept could also be applied to treat grey water. In this case, detergents are most probably the main component of organic pollutants. Because detergents are designed to adsorb to surfaces, removal of COD and BOD should be effectively accomplishable by adding powdered activated carbon to the effluent of showers or bath tubs. The loaded activated carbon particles together with other particulate matter (e.g. hair) can be effectively removed by micro- or ultra-filtration. The treated effluent could be directly re-used for showering as has been demonstrated by Leonhard *et al.* (2004).

Various trials have been made to use membrane biofilm systems for the treatment of wastewater. Kolb and Wilderer (1995), for instance, demonstrated the applicability of such a system to aerobically treat process wastewaters containing volatile organic substances (benzene, toluene, xylene and 3-chloro-ethylene) without running the risk of stripping of such problematic substances into the atmosphere.

Mimicking nature: an example

The idea of considering natural membrane systems as a model for technical solutions links directly to the discussion on decentralization of water and sanitation systems and on the idea of replacing the concept of degradation and discharge by recovery and re-use. The technologies briefly described above may be understood as a first step in this direction. We could take a step further, however, in the attempt to take advantage of the experience of Mother Nature. The process schematic of anaerobic treatment of organic matter presented in Figure 8 is just an example of such an approach.

In contrast to the current technical treatment concept, anaerobic processes in nature are not directed to maximizing generation of methane gas. In focus are the generation of building materials for cell synthesis and release of electrons for energy generation.

Anaerobic reactors operated to produce methane gas for energy generation must have a significant size in term of loading to be cost efficient. The amount of gas produced in small reactors is too little to make profitable use of it. In contrast, anaerobic "reactors" in nature are very small in comparison. They can be small and still be efficient because the

Table 2 Comparison of the removal efficiency by means of fine screening of raw municipal wastewater with and without addition of chemicals (source: Fromman *et al.*, 2005)

Treatment	Filterable solids	BOD_5	COD	P_{tot}	Unit
Fine screening	32.5	12–18	24–36	0.04	[g/(E·d)]
	50	20–30	20–30	2.5	[%]
Addition of coagulants + fine screening	62.0	30	60	1.1	[g/(E·d)]
	95	50	50	60	[%]
Addition of precipitants and coagulants + fine screening	62.0	39	78	1.6	[g/(E·d)]
	95	65	65	90	[%]

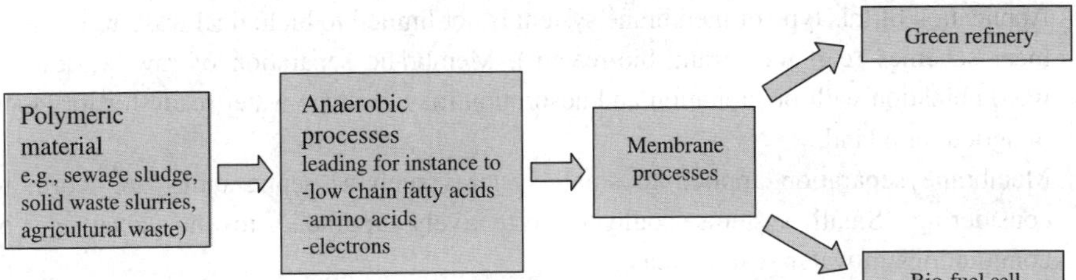

Figure 8 Anaerobic treatment of wastewater involving innovative membrane applications

efficiency is not dictated by amount and value of gas produced but by the value of a variety of materials produced, fatty acids, amino acids and electrons, in particular.

The new generation of anaerobic treatment systems for decentralized application should be based on membrane processes. As in the intestine, membranes should be applied to harvest from the anaerobic environment fatty acids, amino acids and electrons immediately after their release to avoid local accumulation and subsequent inhibitory effects. The membrane system should be designed so that it can be effectively operated at low pressure. Biofilms should be allowed to grow at the surface of the membrane facing the anaerobic reactor but microorganisms should not be driven into the pores of the membrane by force of pressure or suction.

To enhance the flux of anaerobic products, the transmembrane concentration gradient should be kept at a high level. In resemblance of the blood circulation this could be achieved by circulating water along the membrane surface – opposite to the anaerobic zone. The water would serve as a temporary sink. It would be pumped to a secondary "sink" where the picked up materials are harvested from the water for any further use ("green refinery approach"). Electrons could be separately harvested and converted into electrical energy in a bio-fuel cell implemented in the reactor system. Alternatively, it may be possible in future to implement energy generation processes in the membrane as is the case in cell membranes (see Figure 1).

The concept described above may sound futuristic. But this is not the case, actually. First trials have already been made, for instance in the framework of the MELISSA project (www.esa.int/esaCP/ESA9CV0VMOC_life_0.html) conducted by the European Space Agency.

Conclusions

The following list of statements may be treated as incentives for further consideration:

- Membrane systems are very common in nature. Cell metabolism and the metabolism of higher organisms including man are based on membrane processes.
- In organisms, membranes are to be understood as interfaces between the inside and the outside world and as an instrument to transfer substrates, metabolites and gases.
- In most cases, transfer of materials is actively controlled by molecules embedded in the membrane. These substances, mostly enzymes, allow selective mass transfer at high rate.
- In comparison, membrane systems applied in water and wastewater treatment plants are passive and appear rather low in sophistication. Here the mass transfer is performed by a driving force applied from outside (pressure, suction).
- Mostly, membranes are used for solid–liquid separation. Nowadays, the main purpose of technical membrane systems is to gain water free of particulate material and salt. Ultra-filtration allows generation of hygienically safe water for further use, including irrigation.

- Application of this type of membrane system is not limited to biological wastewater treatment schemes (e.g. membrane bio-reactor). Membrane separation of raw wastewater in combination with precipitation and adsorption may produce water ready for irrigation of agricultural land.
- Membrane separation applied to small systems such as septic tanks are worth to considering. Small systems could be effectively upgraded to the advantage of communities in water scarce areas.
- Special attention should be paid to anaerobic processes. The intestine should be treated as a model for innovative solutions, specifically with respect to miniaturization of wastewater treatment plants. If generation of fermentation products for further use is considered as the target – instead of biogas – anaerobic treatment would become applicable even at household level.

In either case, it appears to be of advantage for technology developers to keep a close eye on Mother Nature. There, a great variety of solutions have been tested over millions of years. It is most likely that novel technical solutions will be developed by taking advantage of the "experience" accumulated in the course of evolution.

References

Alberts, B., Johnson, A., Lewis, J., Raff, M., Roberts, K. and Walter, P. (1994). *Molecular Biology of the Cell*, 3rd edition, Garland Publishing, N.Y.

Fromman, C., Bischof, F. and Huber, H. (2005). Advanced MeChem-Technology for Eco-Efficient Wastewater Treatment. *IIWT 2005 – Advances in Industrial Wastewater Treatment*, Chennai, India.

Hook, D.D. and Crawford, R.M.M. (1978). *Plant Life in Anaerobic Environments*. Ann Arbor Sci., ISBN 0-250-40197-5.

Johnson, J.G. (2005). Biology I. www.sirinet.net/jgjohnso/respiration.html.

Kolb, F.R. and Wilderer, P.A. (1995). Activated carbon membrane biofilm reactor for the degradation of volatile organic pollutants. *Wat. Sci. Tech.*, **31**(1), 205–213.

Lee, K.C. and Rittmann, B.E. (2000). A novel hollow-fiber membrane biofilm reactor for autohydrogenotrophic denitrification of drinking water. *Wat. Sci. Tech.*, **41**(4–5), 219–226.

Leonhard, K., Janknecht, P., Schwarzenbeck, N., Dalhoff, R., Tudt, K. and Wilderer, P.A. (2004). Treatment of grey water with adsorption crossflow filtration (ACFF). *Water Res* (submitted).

Meuler, S., Hackner, T., Bischof, F. and Huber, H. (2006). Advantages of membrane technologies for the use in small treatment units. Presented at 7th Specialised Conference on Small Water and Wastewater Systems, Merida, Mexico.

Picard, C., Larbot, A., Sarrazin, J., Janknecht, P. and Wilderer, P.A. (2001). Ceramic membranes for ozonation in wastewater treatment. *Ann. Chim. Sci. Mat.*, **26**(2), 13–22.

Rothemund, C., Amann, R., Klugbauer, S., Manz, W., Bieber, C., Schleifer, K.-H. and Wilderer, P.A. (1996). Microflora of 2,4-dichlorophenoxyacetic acid degradating biofilms on gas permeable membranes. *System. Appl. Microbiol.*, **19**, 608–615.

Teleman, D., Wilderer, P.A., Teodosiu, C. and Koetzle, T. (2005). Enhanced primary treatment of municipal wastewater by Ion exchange and membrane processes. *Env. Eng. Management Journal*, **3**(1), 67–75.

Wilderer, P.A., Bräutigam, J. and Sekoulov, I. (1985). Application of gas permeable membranes for auxiliary oxygenation of sequencing batch reactors. *J. Cons. and Recycling*, **8**, 181–192.

Biological risks to food crops fertilized with Ecosan sludge

B. Jiménez*, A. Austin, E. Cloete***, C. Phasha*** and N. Beltrán******

*Instituto de Ingeniería; Circuito Escolar s/n, Ciudad Universitaria, 04510 México DF, México
(E-mail: *bjimenezc@iingen.unam.mx*)

**CSIR Building and Construction Technology, Pretoria, South Africa

***University of Pretoria, Microbiology and Plant Pathology Department, Pretoria, South Africa

****Institute of Electric Research, Materials and Chemical Processes Department, Morelos, México

Abstract This paper presents the microbial effects of using Ecosan sludge in agriculture. Sludge from KwaZulu Natal in South Africa having a helminth ova content of around 30 HO/gTS, faecal coliforms of 10^6 CFU/gTS, faecal streptococi of 10^6 CFU/gTS and *Salmonella* spp. of 10^5 CFU/gTS were applied to soils to grow carrots and spinach at different rates. Results showed that helminth ova content in crops was always greater in leaves than in stems, with a content varying from 2 to 15 HO/gTS for spinach and sludge application rates of sludge of 0–37.5 ton/ha and from 2–8 HO/gTS in carrots crops for sludge application rates varying from 0 to35 ton/ha. Health risks resulting from crop consumption were calculated using the β-poisson and the single-hit exponential models for *Salmonella* and helminths eggs, respectively.
For *Salmonella*, no risks were found when consuming carrots for all the sludge rates studied while for spinach, risks were high but results were not deemed conclusive due to the technical methods used to measure bacteria. Concerning helminths, it was found that the morbidity rate will increase by 9% for the higher sludge application rates. To increase regional risks by 1% the egg content in crops needs to be less than 0.2 HO/gTS.

Keywords Agriculture; Ecosan; faecal sludge; helminth ova; microbial characterization; risks

Introduction

Ecosan systems are used as an attractive alternative to pit toilets and latrines for sanitation in several countries. According to literature, when applied to land, residues from Ecosan (or Ecosan's sludge) increase agricultural yields (Esrey *et al.*, 1998; Austin and Duncker, 2002), because human excreta contains organic matter, phosphorus and nitrogen compounds that are essential plant nutrients. However, according to Peasey (2002), although dry toilet technology is spreading all over the world and with it the intention to recycle excreta to soils, few have documented the pathogen and parasite die-off or addressed the health problems associated with the revalorization practice. Moreover, few have addressed these problems in developing countries, where health risks due to the use of faeces are higher due to a greater content of pathogens and parasites (Jiménez *et al.*, 2004; Strauss and Blumenthal, 1990). Therefore the objectives of this research were to biologically characterize Ecosan sludge in a developing country, to investigate its biological effects on edible crops and, based on these data, to evaluate the health risks resulting from crop consumption.

Background

Dry sanitation or Ecological sanitation (Ecosan) is considered an appropriate technology for community settings without sewerage and limited water, which is common in arid and semiarid areas of the developing world. It has the advantage of solving many of the

Water Science & Technology **Vol 55 No 7 pp 21–29** © IWA Publishing 2007

doi: 10.2166/wst.2007.122

problems encountered with other sanitation systems (Esrey *et al.*, 1998; Schonning, 2001). One such is that dry toilets store faeces in dry conditions inactivating the biological content to such an extent that sludge (or faecal sludge) can, in principle, be safe hand compost. The use of dry human excreta is not new. It has been documented since the 12th Century in China and until the second half of 1900 in Finland (Olsson, 2001). Even though data concerning the biological inactivation, mainly of parasites, in dry toilets is scarce it has been demonstrated that the dehydrated sludge produced has a lower biological content compared to that from traditional latrine sludge (Schonning, 2001).

Recycling excreta to soil reduces the need for chemical fertilisers; however, pathogens might also be also recycled to humans if improper agricultural practices are used (Hoglund *et al.*, 2000). Concerns about reusing Ecosan sludge include a higher pathogenic content in developing countries compared to sludge from developed countries (Jiménez *et al.*, 2004) and the fact that, being introduced in small-scale systems, it requires more personal involvement from the users (including handling of the waste) constituting a higher human exposure level. Nevertheless, it is considered that where sludge can improve agricultural productivity, independently of its microbial characteristics, it improves the nutritional status of a population lacking food, thus improving public health (Hoglund, 2001; IMWI, 2003). Consequently, there is a need to quantitatively compare health risks associated with the use of Ecosan sludge with the level of health improvement resulting from the increase in nutritional status.

Methodology

In South Africa, Ecosan technology is becoming common (according to Austin and Duncker (2002), it has been installed in its thousands) and there is enough experience in reusing treated sewage sludge in soil (48% of the conventionally treated sewage sludge is applied to soils; Snyman and Van der Walls, 2003). Nevertheless, Ecosan sludge has not been used for agriculture, basically because it has not been fully biologically characterized and because it is believed that the sludge is similar to that coming from septic tanks and, as a consequence, not suitable for agriculture (WRC, 1997). Considering the widespread use of Ecosan in South Africa and the interest in revalorising it due to the poor soils of the region, sludge from this country was used in this study. A 25 kg composite sample of Ecosan sludge that had been mixed with some topsoil (stored at least 12 months according to the South African operational criteria) were collected without further treatment from several toilets from the Durban region in KwaZulu-Natal. For microbial characterization 4 bacteria, 1 fungi and 6 types of helminth ova were measured. Total coliforms were considered due to their presence in faeces, faecal coliforms and faecal streptococcus because they are considered good indicators of faecal pollution by most authors and *Salmonella* spp. because it is part of most sludge regulations around the world and a human pathogen. For fungi, there is no universal indicator and *Aspergillus* spp. was used because it is an opportunistic pathogen belonging to a group of moulds that is found worldwide. Finally, helminth ova were monitored due to their high persistence in the environment and because they are considered quality indicators for most of the reuse practices (WHO, 2000). Helminths are commonly associated with sanitary risks when sludge is used as an agricultural fertilizer (Asaolu and Ofoezie, 2003). To analyze total coliform, faecal coliform, faecal streptoccoci, *Aspergillus* spp. and *Salmonella* spp. a serial dilution technique was used (Islam *et al.*, 2004). The technique used for *Salmonella* was replaced during the second phase of the study with a standard technique owing to the inconsistency of the results (APHA, AWWA and WEF, 1995). TS (total solids), pH, and nitrogen were also determined using standard techniques (APHA, AWWA and WEF, 1995). For helminth ova (HO) detection the Ayres technique modified by Maya *et al.* (2006) was used.

Analyses were performed to characterize (a) the sludge prior to its application, (b) the soils before sowing and after harvesting, (c) the irrigation water during the study and (d) the crops after harvesting. To assess the microbial effects of sludge in agriculture, two kinds of crops were selected: spinach and carrots. These crops were considered because they are usually eaten in South African rural areas, are often consumed raw and for spinach the edible parts grow above the soil, while for the carrots they grow below the ground. Plots of 2×9 m were used. In the experimental farm of Pretoria University each crop was planted in two plots, one was used as a control while the other was divided into three sections. Each plot was treated with a different sludge application rate. To determine the amount of sludge to be added, the following criteria were taken into account: (a) the nitrogen demand by crops (50 kgN/ha for carrots and 100 kgN/ha for spinach); (b) application rates above and below 8 ton/ha, which is the value established as the South African norm and (c) three different helminth ova rates. Following these criteria, sludge was applied to carrots at 0, 7, 12.5 and 35 tons/ha corresponding to 0, 1, 1.7 and 4.8 HO/cm^2, while for spinach 0, 1.3, 19 and 37.5 ton/ha equivalent to 0, 0.18, 2.6, and 5.1 HO/cm^2 were used. The helminth rate was defined as the quantity of total helminth ova applied per square centimetre. Sludge was applied to soil and mixed to a depth of 10 cm. Seeds were planted in the second week of November (summer). The pattern within the blocks was in rows 300 mm apart. The seeds were spaced 50 mm a part within the rows. Spinach was harvested in January (after 7 weeks) and carrots in March (after 12 weeks). In each case, the whole plant was pulled from the soil and cut to collect roots and leaves separately. For sludge and soil analysis, 1 g samples were used. To analyze bacteria and fungi, 5 g of crop samples were taken, while for helminth analysis the sample size was 30 g. To evaluate the health risks, the consumption of 100 g of uncooked spinach or carrot (equivalent to a medium size carrot) was considered. A 2 log reduction during harvesting was assumed and a 1 additional log reduction due to washing for bacteria (WHO, 2006). For helminth ova only a 90% reduction due to washing was considered. The infection probability resulting from the exposure (P) was calculated using a beta Poisson model for Salmonella and a single hit exponential model for helminths (Rose and Gerba, 1991):

β-Poisson Model

$$P = 1 - (1 + ((N/\beta))^{-\alpha} \quad (1)$$

Single-hit exponential model

$$P = 1 - \exp(-rN) \quad (2)$$

where P is the infection risk due to the consumption of pathogens (Salmonella) or parasites (helminths) in crops, N is number of ingested pathogens or parasites; $\alpha = 0.33$; $\beta = 139.9$ and $r = 0.02$.

Results and discussion

Sludge characterization

Ecosan sludge (Table 1) had a TS content of 43 ± 2%, which is higher than that from septage or sludge coming from conventional treatment and dewatering (4% for biological secondary sludge thickened and 28% for sludge exiting filter press using chemicals (Metcalf and Eddy, 2003)). The TS content is not as high as that obtained in sludge heaters (85–92%), a process known for its high disinfection efficiency. While the TS content is high (and hence the humidity is low), the N content (0.2–0.34%) is within the common range for domestic sludge (0.2–0.6%), if the N contribution due to urea is subtracted (90% of the value according to Metcalf and Eddy, 2003). The N content found

Table 1 Ecosan sludge characterization

Parameter	Mean value
N content, %	0.2–0.34
Total coliforms, CFU/g TS	2.2×10^6
Aspergillus spp., CFU/g TS	3.9×10^3
Faecal streptoccoci, CFU/gTS	2.1×10^6
Faecal coliforms, CFU/g TS	1.8×10^6
Salmonella sp., CFU/gTS	2.2×10^5
Total *Ascaris*, ova /gTS	25.3 ± 4.4
Total helminths, ova /gTS	29.8 ± 2.9
Viability	88.8 ± 0.5

in sludge was low compared to other Ecosan sludge, which implied adding higher quantities of sludge in order to fulfill the nutrient demand of crops. This would not have been important if the sludge had no microbial content. However, this was not the case. Concerning helminths, the value found (29.8 ± 2.9, total helminths) seems to indicate that the concentrations are not as high as could be expected for sludge from developing countries (ranging from 67 to 735 ova/gTS according to Jiménez and Wang, 2006) and are even comparable to the one obtained from anaerobic digester sludge in South Africa (2–40 *Ascaris*/gTS, Snyman *et al.*, 2003) a sludge classified as Type B and therefore feasible for agriculture reuse but with some restrictions in order to minimize human exposure.

Helminth ova composition in sludge differs from one region to another reflecting health conditions. In the sludge sampled, 85% (25.3 ± 4.4 ova/gTS) were *Ascaris* eggs but it was noticed that the percentage of *Taenia* eggs was variable and could reach up to 19%, higher than values reported by other authors (Jiménez *et al.*, 2004). *Taenia* is linked with the oncocercosis cysts in meat and Cysticercosis due to *Taenia saginata* (common in Africa) appears to be one of the major pathological threats when sewage sludge is used to fertilize cattle pastures in temperate areas. Therefore they deserve special attention and it might be appropriate for South Africa to establish the maximum content not only of *Ascaris* egss but also all kinds of helminths in sludge, particularly because the analytical method is the same. Eggs' viability in sludge was 88.8 ± 0.5%, a common value for sludge coming from conventional wastewater treatment plants. However, and interestingly, it was observed that *Ascaris* and *Trichuris* egss were larvae at 65%, meaning that for some reason the operating condition in Ecosan promotes the breeding of eggs.

Irrigation water

Water used to irrigate came from a well and was stored in open tanks. The water was not disinfected and very often birds were drinking water from tanks. This is possibly the reason why some microbial pollution was found although at very low concentrations: Total coliforms ranged from 0.1 to 0.3 CFU/100 mL, faecal coliforms from 0.2 to 0.9 CFU/100 mL, faecal streptococci from 0 to 0.1 CFU/100 mL. *Salmonella* spp. and helminth ova were not detected in any of the five samples analyzed.

Soil

Table 2 shows the characterization of soils prior to the study. Soils were slightly alkaline (pH 7.7), and contained microorganisms such as total and faecal coliforms. The first are commonly considered as native in soils, while the second have been reported as native in water in high-temperature countries and therefore could be present in soils. Faecal streptococci and *Salmonella* (using the APHA, AWWA, WEF method) were not found in soils while Helminth ova were present in low concentrations (1.4 ± 0.5 HO/gTS) and

Table 2 Initial soil characteristics

Parameter	Value
pH	7.7 ± 0.21
TS content (moisture)	86 ± 2 (14% ± 2)
Total coliforms	$8.1 \times 10^3 - 2.7 \times 10^5$
Faecal streptococci	0, absent
Faecal coliforms	$2.6 \times 10^3 - 1.1 \times 10^4$
Salmonella spp.	0, absent
Aspergillus spp.	$0 - 7 \times 10^1$
Total helminths, ova /gTS	1.4 ± 0.5
Viability	0–10%

with very low viability (0–10%). According to the records of the farm, the plots had not received any manure application (which can contain helminth ova) for at least 1.5 years. Concerning the genus, almost all the eggs found were *Ascaris* although *Toxocara* was sometimes also found.

Crops

The quantity of sludge applied was equivalent in terms of the actual viable helminth eggs to application rates of 0.9, 1.5 and 4.3 HO/cm^2 for carrots and 0.2, 2.3 and 4.5 HO/cm^2 for spinach. These values should be compared to 0.016 viable HO/m^2 for sludge having 1 viable helminth ova/gTS and applied at a rate of 8 ton/ha as established in South African regulations as well as application rates of 0.4 for carrots or 0.8 HO/cm^2 for spinach obtained using USEPA norms to fulfill the nitrogen demand by crops with sludge containing 0.25 viable helminth ova /gTS. Figure 1 shows the microbial effects of the sludge application only on carrot soils. Total coliforms, faecal streptococci and faecal coliforms (Figure 1a) were present in soils in similar concentrations for all the application rates and only for the highest value can a noticeable increase be seen. Similar results were obtained for spinach soils, although the increase for the highest sludge application rate was less noticeable. In the case of Salmonella, results in spinach soils were erratic indicating that the Islam *et al.* (2004) analytical technique was not appropriate. For carrots using the APHA, AWWA, WEF (1995) technique, *Salmonella* results were negative in all cases. Helminth ova in soils for both carrots and spinach (Figure 1b, only for carrots) show a clear correlation with the rate of sludge applied: the larger the sludge application rate, the greater the content of helminth ova found in soils. In soils, a surprising result was the diminution of the helminth ova viability from the original value of 88.8% to 52 ± 3% for spinach soils and to 39 ± 7% in carrots soils. The greater decrease of viability in carrot soils was likely due to the longer amount of time taken to monitor carrot soils (12 weeks) than that used for spinach soils (7 weeks). The decrease in the viability can be explained,

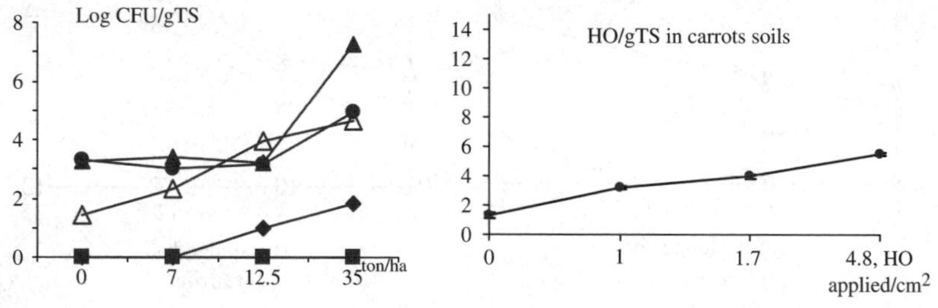

Figure 1 (a) ▲Total coliform; •faecal coliforms; Δfaecal streptococci; ■*Salmonella* spp.; and ◆*Aspergillus* spp., (b) helminth ova content in carrots soil after harvesting

although not conclusively, by the high temperature registered during the summer time in Pretoria (ranging from 27 to 38 °C, during the day) but also to possible prior damage suffered by the eggs during their dehydration in Ecosan. *Ascaris* have been reported to rapidly die at temperatures over 40 °C in different types of media including water, soil, sewage, and crops. The temperature limit could be less if high temperatures are combined with other negative environmental conditions such as high ammonia content or low moisture (Heinonen-Tanski and Van Wijk-Sijbesma, 2004). Another explanation is that most of the eggs were larvae prior to their application, and therefore could not survive in soil.

Figure 2 shows the results of the bacteria numbers in spinach leaves (a) and stems (b). There was not a clear relationship between the quantity of sludge applied and the total microbial number in leaves or stems. For faecal coliforms in stems, the results seem to indicate that bacteria can survive underground but not on the top of the soil, where UV sunlight is available and will kill the organisms. In carrots, total and faecal coliforms as well as faecal streptococci increased as the sludge application rate increased (Figure 2c). *Aspergillus* spp. and *Salmonella* spp. were present in low numbers during all the different treatments. Concerning helminth ova, increasing concentrations were found in both stems and leaves (Figure 3a and b) as the quantity of sludge applied (and hence that of helminths) increased. Contamination is more important in leaves than in stems, seeming to indicate that helminth ova are preferentially attached to plants rather than to soils. Although these results show that crops were polluted even using the smallest application rate, understanding the health significance would require proper epidemiological or toxicological studies that consider the probability of microorganisms, especially helminth ova, actually infecting the host. This would depend on the viability of eggs, the quantity of microorganisms consumed by a person through conventional daily diets in the region and the infective dose. Concerning viability, unfortunately, for spinach crops the data

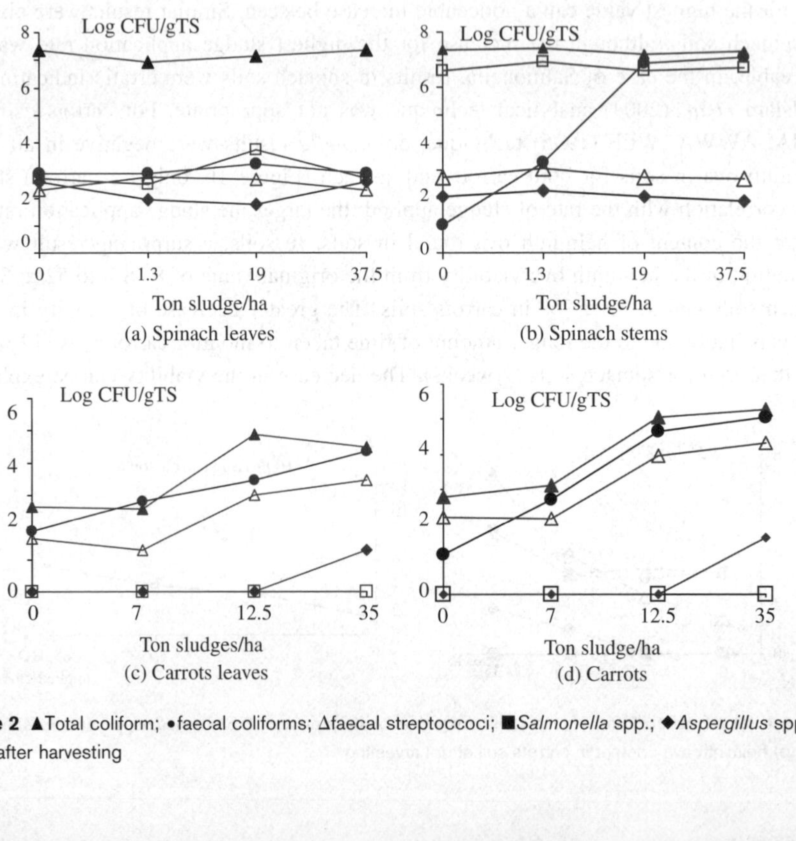

Figure 2 ▲Total coliform; •faecal coliforms; Δfaecal streptococci; ■*Salmonella* spp.; ◆*Aspergillus* spp. in crops after harvesting

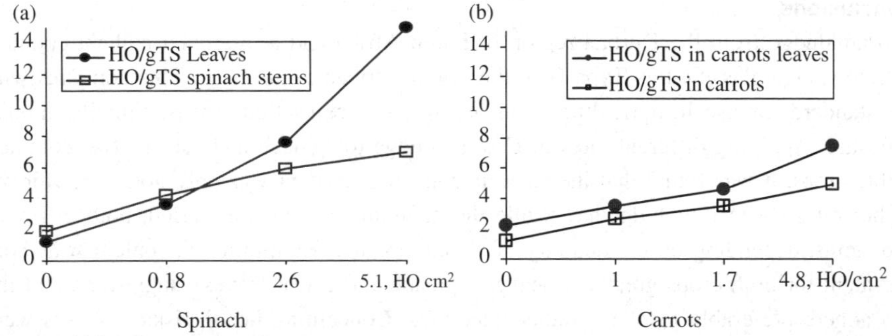

Figure 3 Helminth ova content in crops

was not obtained, but in carrot leaves it was 25 ± 5% while in carrots it was of 20 ± 8%. This indicated that although present, they were mainly in an inactive state reducing the risk to spread the disease through consumption.

Risk evaluation

Biological risk evaluation is a useful tool in predicting possible effects caused by sludge or wastewater reuse and also in defining conditions for its safe revalorization (Haas *et al.*, 1999 and Westrell *et al.*, 2004). The main limitation for its application is data availability, particularly concerning infective doses. For this reason, risk evaluation was only done for *Salmonella* and *Ascaris*. Salmonella risks were calculated for a single consumption of 100 g of spinach or carrots, considering a humidity content of 92% and 88%, respectively. Calculations for spinach were performed although values found were estimated illogically high due to the analytical technique used (Islam technique). Using data obtained, it was estimated that 9% to 60% of spinach consumers will probably get sick. That is to say, risk caused by the produce grown using sludge application rates of 19 to 37.5 ton/ha are unacceptable. In order to have an acceptable risk, *Salmonella* content in spinach need to be less than 3.6 CFU/gTS. When consuming carrots, and because *Salmonella* content was found to be low using the APHA, AWWA, WEF Technique (of around 1 CFU/gTS), the risk was negligible ($P = 2.8 \times 10^{-5}$) for all cases and smaller than considered that acceptable (10^{-4} according to USEPA/USAID, 1992).

Helminthiasis risks due to a single consumption of spinach grown in soils amended with 1.3 to 37.5 ton/sludge.ha (or 0.18 to 5.1 HO/cm^2) were 2×10^{-2} and 9×10^{-2}, while for carrots they were 6×10^{-3} to 1×10^{-2} for sludge application rates of 7 to 35 ton/ha (or 1 to 4.8 HO/cm^2). All these values are greater than the "acceptable risk" proposed by USEPA/USAID, but in communities having helminthiasis morbidity rates from 25 to 90%, as is the case in poor regions from developing countries, maximum calculated risk means 9% additional people infected. To reduce risk caused by Ecosan sludge revalorization to 3% crops we need to have less than 8 HO/gTS and for 1% less than 0.2 HO/gTS. The risk found when using Ecosan sludge should also be compared to the one produced by crops grown on soil without addition but where helminths were found (1.4 HO//gTS with a 5% viability) due to another type of fertilizer applied in previous years that was 1.4×10^{-3}. These results also indicate that unless Ecosan sludge contains very low helminth egg content a risk to disseminate helminthiasis diseases through crop consumption exists. Therefore a decision should be made at a local level to limit sludge revalorization to crops not eaten raw or not growing in soils, or otherwise, to control the risk by consuming only cooked crops.

Conclusions

Ecosan sludge from the Durban region of South Africa had a microbial content comparable to sludge classified as Type B in the South African Legislation. This complies with the standard for use in agriculture considering some restrictions for minimizing human exposure. Applying different rates of Ecosan sludge to spinach and carrots, two common edible crops, it was found that the bacteria and fungi content was only noticeable for the higher rates (>35 ton sludge/ha), while the helminth ova content varied, both in leaves and stems, depending on the quantity of sludge applied. Helminth ova content was more prevalent for both crops greater in leaves (spinach and carrot leaves), suggesting that the ova adhere preferably to plants rather than soils. Concerning health risks, no risks were found from salmonella in carrots grown at all sludge application rates. Risk with spinach resulted high (up to 60%) but due to the analytical methods used to detect this bacteria, the results are not conclusive. Concerning helminthiasis, there is a risk of increasing morbidity rates up to 9% for the higher sludge application rates. To increase the risks by 1% the egg content in crops needs to be less than 0.2 HO/gTS.

References

APHA, AWWA and WPCF (1995). *Standard Methods for the Examination of Water and Wastewater*. 20th edn, APHA/AWWA/WPCF, Washington, D.C.

Austin, A. and Duncker, L. (2002). *Urine-diversion ecological sanitation systems in South Africa*. CSIR, Pretoria.

Asaolu, S.O. and Ofoezie, I.E. (2003). The role of health education and sanitation in the control of helminth infections. *Acta Tropica*, **86**, 283–294.

Esrey, S., Gough, J., Rapaport, D., Sawyer, R., Simpson-Hebert, M., Vargas, J. and Winblad, U. (1998). *Ecological Sanitation*, Sida, Stockholm.

Haas, C.N., Rose, J.B. and Gerba, C.P. (1999). *Quantitative Microbial Risk Assessment*. John Wiley & Sons, Inc, New York.

Heinonen-Tanski, H. and Van Wijk-Sijbesma, C. (2004). Human excreta for plant production. *Bioresource Technology*, **96**, 403–411.

Hoglund, C., Vinneras, B., Stenstrom, T. and Jonsson, H. (2000). Variation of chemical and microbial parameters in collection and storage tanks for source separate human urine. *Environmental Science and Engineering*, **35**, 1463–1475.

Hoglund, C. (2001). *Evaluation of microbial health risks associated with the reuse of source-separated human urine*. Swedish Institute for Infectious Disease Control, Stockholm.

IMWI (2003). *Water Policy Briefing Issue 9 Putting research knowledge into action.*

Islam, M., Morgan, J., Doyle, M.P., Phatak, S.C., Millner, P. and Jiang, X. (2004). Persistence of *Salmonella enterica* Serovar Typhimurium on Lettuce and Parsley and in soils on which they were grown in fields treated with contaminated manure composts or irrigation water. *Foodborne Pathogens and Disease*, **1**(1), 27–35.

Jiménez, B., Barrios, J., Mendez, J. and Diaz, J. (2004). Sustainable management of sludges in developing countries. *Water Science and Technology*, **49**(10), 251–258.

Jiménez, B. and Wang, L. (2006). *Sludge Treatment and Management in Municipal Wastewater Management in Developing Countries*, Ujang, Z. and Henze, M. (eds), IWAP, London, U.K.

Maya, C., Jimenez, B. and Schwartzbrod, J. (2006). Comparison of techniques for the detection of helminth ova in wastewater. *Water Environment Research.*, **78**(2), 118–124.

Metcalf, and Eddy (2003). *Wastewater Engineering Treatment and Reuse*. Fourth edn, McGraw Hill, USA.

Peasey, A. (2002). Health aspects of dry sanitation with waste reuse Task No. 324. *Well Studies in Water and Environmental Health.*

Olsson, G. (2001). The struggle for a cleaner urban environment: water pollution in Malmo. *Ambio*, **30**, 287–291.

Rose, J.B. and Gerba, C.P. (1991). Use of risk assessment for development of microbial standards. *Water Science and Technology*, **24**(2), 29–34.

Schonning, C. (2001). Hygienic aspects on the reuse of source-separated human urine. In *Water and Environmental Microbiology*, Swedish Institute for Infectious Disease Control, Stockholm.

Synman, H. and Van der Walls, L. (2003). Laboratory and field scale evaluation of agricultural use of sewage sludge. WRC Project number k5/1210 Biosolids management at 18 wastewater treatment plants in South Africa – optimization strategies.

Strauss, M. and Blumenthal, U. (1990). *Use of human wastes in agriculture and aquaculture*. IRCWD Report No 08/90, Duebenforf, Switzerland.

US EPA/USAID (1992). *Guidelines for Water Reuse*. United States Environmental Protection Agency, Washington, (Wash Technical Report No. 81, September 1992).

Westrell, T., Schonning, C., Stenstrom, T. and Ashbolt, N. (2004). QMRA and HACCP for management of pathogens in wastewater and sewage sludge treatment and reuse. *Water Science and Technology*, **50**(2), 23–30.

WHO/UNICEF (2000). *Global Water Supply and Sanitation Assessment Report, Joint Monitoring Program for Water Supply and Sanitation*. World Health Organization, Geneva.

WHO (2006). *Guidelines for the Safe Use of Wastewater, Excreta and Grey Water*., Vol. 2, Geneva.

WRC (1997). *Permissible utilization and disposal of sewage sludge*. Water Research Commission TT 85/97.

Performance of very small wastewater treatment plants with pronounced load variations

J. Kegebein*, E. Hoffmann and H.H. Hahn****

*Munters Euroform GmbH, Aachen, Philipsstr. 8, Germany

**Institute for Water Research – Division of Aquatic Environmental Engineering, University Karlsruhe, (Institut für Wasser und Gewässerentwicklung – Bereich Siedlungswasserwirtschaft), Forschungszentrum Umwelt, Adenauerring 20, D-76128 Karlsruhe, Germany (E-mail: *j.kegebein@web.de*; *erhard.hoffmann@iwg.uka.de*; *hahn@iwg.uka.de*)

Abstract The operation of very small biological wastewater systems is strongly influenced by the patterns of wastewater generation. The absence of people in the connected building(s) during holidays or off-season leads to a system underload, other circumstances however may lead to an overload. Experiments have been carried out to display the effects of no-feed conditions on activated sludge biomass and its microbial activity during a 24 hour period after re-feeding. The decrease of the biomass during idle periods can be modelled by a first order equation. The initial specific oxygen uptake rate (OUR) of the remaining biomass decreased with the duration of the preceding no-feed sequence. Four different laboratory-scale treatment plants were operated to demonstrate the system performance when re-started after a break period of 24 and 52 days respectively. The effluent concentration after a break of 52 days had not been sufficient as compared to the discharge requirements given by German laws. In addition a full scale trickling filter was monitored, while the connected building accommodated a maximum of 40 guests interrupted by times with no guest at all. Conclusions are drawn from the experimental results and from literature, giving some hints on how to manage the impacts of varying loads at very small wastewater treatment plants.

Keywords Activated sludge; biomass decay; idle conditions; moving bed biofilm reactor; oxygen uptake rate; rotating bio contactor; trickling filter

Water Science & Technology **Vol 55 No 7 pp 31–38** © IWA Publishing 2007

Introduction

Approximately 10% of the population in Germany discharge their wastewater to decentralised wastewater treatment systems. The percentage goes up 30% in certain federal states particular in Eastern Germany. It has been calculated that about 20% of the total COD emissions from WWTP effluents can be attributed to decentralised systems (Flasche, 2002).

The management of very small wastewater treatment systems has become an important issue in Germany, since simple septic tanks (see Schweizer, 1983) are insufficient for a proper wastewater treatment. A biological upgrade is essential. Treatment systems, which are marketed in Germany usually bear a proof label from the German Institute for Bautechnik (DIBt), which will be assigned after a successful one-year-test. The impact of a 14 day break is nowadays assessed due to a recent refinement of the testing procedure. However, the load variations may dominate and the breaks may be of extended duration in some cases.

From the microbiological point of view, it is important to know about the impact on the total biomass and its activity during idle periods (Urbain *et al.*, 1993; Morgenroth *et al.*, 2000). It determines the treatment capacity of the system in terms of microbiological turn over rates. Beside this, non-biological effects like settling, adsorption on flocs, dilution and hydrolysis play an important role in wastewater treatment. The performance

doi: 10.2166/wst.2007.123

of a complete treatment system cannot be predicted on the base of microbiological turn over rates only. Consequently the system response (Vasel and Jupsin 2003) has been part of the experiments in addition to lab-scale investigations.

Four laboratory-scale treatment plants with a design capacity of 0.1 P.E. each were operated to demonstrate the system performance of a trickling filter (TF), a rotating biological contactor (RBC), an activated sludge- (AS) and a moving bed biofilm system (MBBR) when re-started after a break period of 24 and 52 days respectively. Moreover, the performance of a full scale trickling filter with a design capacity of 20 P.E. was monitored over a period of 85 days, while the connected building (a youth hostel in the Northern Black Forest, Germany) accommodated a maximum of 40 guests interrupted by times with no guest at all.

Materials and methods

Chemical parameters (COD, NO_3-N) were measured with test kits LCK 314 and LCK 339 (Dr. Lange). *Total suspended solids (TSS)* and *volatile suspended solids (VSS)* were measured according to DIN 39409 H2. *Storage of sludge:* Activated sludge was taken from WWTPs in Blankenloch and Berghausen, Germany. It was stored in 10 L Plexiglas cylinders and aerated. Evaporation was compensated by the addition of pure water. Samples were taken for measuring OUR after adding substrate and for SS determination. *Oxygen uptake rates (OUR)* were measured by using an oxygen probe (Greisinger electronic - GMH 3610) and a data logging computer programme (Greisinger EBS9M)). Sludge samples were mixed with substrate to assure an initial VSS and COD content in the range of 1 g/L each. Samples were taken in Karlsruhe-flasks and a sequence of aeration (15 minutes) and oxygen consumption (15 minutes) was applied by using a timer device. Phases of oxygen consumption have been taken for calculation of OUR. Tests were carried out over six weeks using three substrate solutions, i.e. (1) a glucose solution (2) a yellow-water (urine) solution (3) and a brownwater solution (suspended feces). Substrates were taken from a common stock, which had been prepared at the beginning of the experiments and was frozen in icecube trays until use.

Laboratory scale WWTPs were designed for 0.1 P.E. each. Design data are given in Table 1. Wastewater was pumped from a sanitary sewer in a residential area in Eggenstein-Leopodshafen (near Karlsruhe, Germany) at a rate of 10–20 L/d each. The influent COD was 1165 mg/L on average. Experimental breaks were applied when the WWTPs showed stable conditions. During break-times aeration was applied one minute per 15 minutes, as was the trickling filter pump and the disk motor of the RBC in order to keep the biofilms alive. Re-feeding started after 24 and 52 days without inflow.

The *full scale trickling filter* consisted of a three-chamber settling basin (8 m^3), a trickling filter with stone package (5.5 m^3) and a secondary settling tank (4.2 m^3). Prior to the investigation an internal recycle from the secondary settling tank to the head of the trickling filter had been implemented. Beside this, an external recycle from the secondary settling to the first chamber of the settling tank was taken in operation (Figure 1). Samples were taken from the final effluent and from the effluent of the primary settling tank. Samples were frozen until being taken to analysis.

Results

Biomass decay

Changes of the organic content of the stored sludge are displayed in Figure 2 along with an exponential regression function. The exponential constants of -0.0197 and $-0.0258\,d^{-1}$ represent the decay rate. The decay rate constants are within the scope of values from literature. According to Metcalf and Eddy (1991) the range is from

Table 1 Design data of the laboratory-scale WWTPs

	AS	MBBR	RBC	TF
Primary settling (L)	16	16	16	16
Reactor / trough (L)	32	32	32	40L TF; 32L trough
Secondary settling (L)	16	16	16	16
Biomass or area/volume for attached growth	64 gTSS	3,2 m^2 Biolox 10* (5 L)	3 m^2 12 disks	40 L RFK 50 L*
	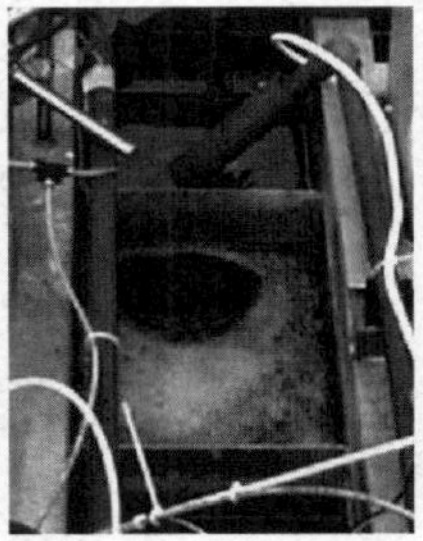			

*Carrier media manufactured by Rauschert Verfahrenstechnik GmbH, Germany; AS-activated sludge, MBBR-moving bed biofilm reactor, RBC-Rotating bio contactor, TF-Trickling filter

Figure 1 (a) Scheme of the trickling filter; (b) Youth hostel Raumünzach (Black Forest, Germany); (c) Distribution of internal water recycle

$-0.025\,d^{-1}$ to $-0.075\,d^{-1}$. Results from Coello Oviedo *et al.* (2002) indicate a constant of $-0.06\ d^{-1}$. It should be noted, that the sludge has been stored under aerobic conditions, which are supposed to have a significantly higher decay rate (due to the activity of protozoa) as compared to anaerobically stored sludge (Batla and Gaudy, 1965; Lopez and Morgenroth, 2002).

OUR after re-feeding

Specific cumulated oxygen uptake rates are displayed in Figure 3. OUR was maximum when yellow water has been added as substrate. This can be attributed to the activity of nitrifying bacteria. Carbon substrates (glucose, brown water) showed a lower uptake. Although the biomass without preceding storage always exhibited the highest OUR, a lack of consistency has been observed, since the sludge with the most extended storage duration (six weeks) did not always represent the lowest activity.

However, initial oxygen uptake reflected the duration of the preceding storage time. Figure 3a shows the relative uptake within the first two hours after substrate was added. The degression with storage time became indistinct within 24 hours (Figure 4b).

It can be concluded that the sludge almost regained its specific activity within 24 hours. This finding is of importance for the assessment of idle conditions: it can be stated that the major task for operating very small WWTPs during breaks is keeping the mass of biomass so that its activity will recover within a short term period.

Performance of the laboratory-scale systems

Table 2 shows the average performance data during steady state conditions. The influent taken from a sanitary sewer exhibited high COD concentrations leading to high loading

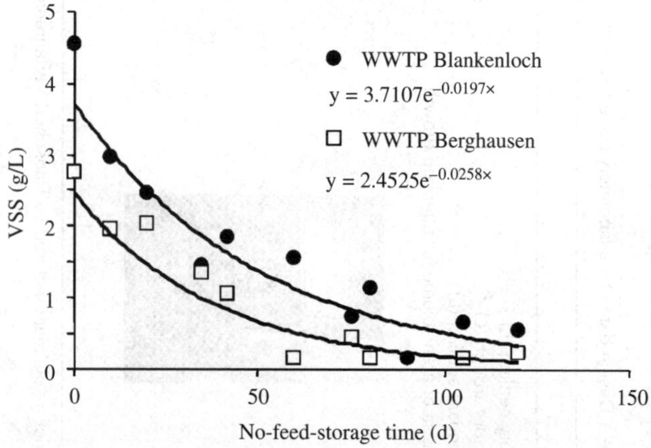

Figure 2 Decrease of VSS of two different stored sludges

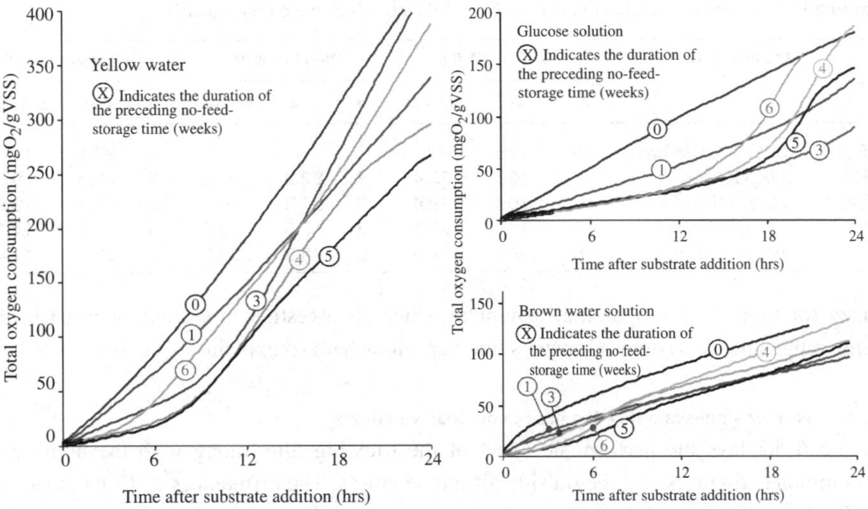

Figure 3 Total oxygen consumption with respect to the preceding no-feed storage time, substrate and time after substrate addition

rates for the treatment systems, thus nitrification occurred to a limited extent only. However, COD-elimination was reasonable and experimental breaks were applied.

The effluent COD dropped below 100 mg/L during break conditions, except for the activated sludge system in the second experimental run. Re-feeding the systems caused an increase of the effluent COD with maximum values in the range of 160 mg/L after the 24 days-break and values up to 260 mg/L after the 52 days-break (Figure 5).

The feed pump was accidentally set to 20L/WWTP·d during the second re-start, followed by a pump clogging after three days, which disabled further feeding. Design feeding of 10 L/WWTP·d was reconstituted at the sixth day after re-start. Even though the effluent quality was worse as compared to the steady state conditions, COD elimination was in the range of 80%. The COD requirements for wastewater at the point of discharge of 150 mg/L as defined by German Water Regulations may be exceeded after a 24 days break, whereas a break of 52 day ultimately leads to an unacceptable effluent quality.

Although it was intended by the experiments to work out some kind ranking of appropriate systems for varying load conditions it has to be stated, that the results seem too

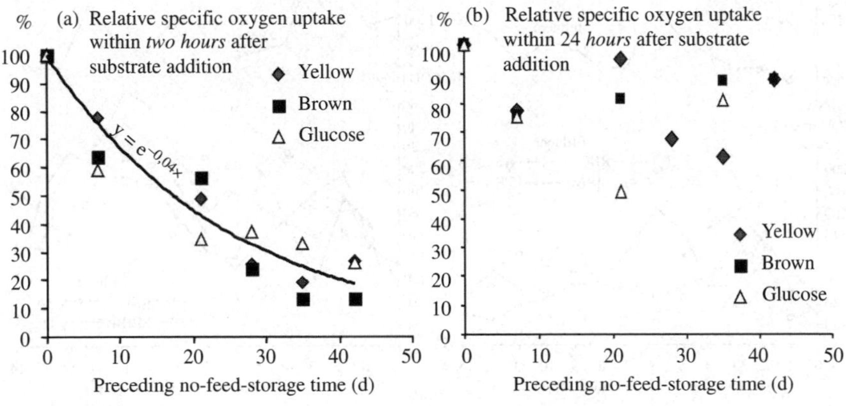

Figure 4 Relation of the oxygen uptake of sludge from the storage container (0 weeks = 100%) within two and 24 hours respectively

Table 2 Performance of the laboratory scale WWTP during steady state conditions

	Loading rates	COD (mgO_2/L)			Nitrate (mgN/L)			Total Nitrogen (mgN/L)		
		n	Ave.	Std. Dev.	*n*	Ave.	Std. Dev.	*n*	Ave.	Std. Dev.
Influent	11.7 gCOD/d each	16	1165		–	–	–	8	98.7	11.2
RBC	2.0 $gCOD/m^2d$	16	133	32.4	8	33.2	16.3	4	78.3	9.3
MBBR	1.9 $gCOD/m^2d$	16	109	29.4	8	25.3	5.6	4	63.2	18.9
AS	0.09 gCOD/gSSd	16	122	22.1	8	15.6	5.9	4	54.2	14.1
TF	291 $gCOD/m^3d$	8	98	29.6	8	45.3	22.6	4	85.3	11.4

fuzzy for supporting any similar statement, since all investigated systems worked in the same range and all systems showed some variations hard to explain.

Performance of a full-scale trickling filter under load variations

Figure 6 displays the performance data of the trickling filter along with the number of accommodated guests and the outside air temperatures. The differences of COD measured at the final effluent and the effluent of the primary settling tank appeared very low. This finding is due to the recycle line from the secondary to the primary settling tank and the subsequent dilution.

Even though the recycle line brings nitrate into the primary settling tanks, it could hardly be detected there. This indicates that the primary settling tank serves as a denitrification reactor. Nitrate could only be detected in the settling tank when the building was closed and the carbon feed ceased.

These observations demonstrate, that the settling tank cannot be considered as a sole physical treatment unit, but also as an upstream denitrification plug flow reactor unit. Although there are no limitations (for very small WWTPs in Germany) on discharging nitrogen, it should be noted that an enhanced denitrification process goes hand in hand with COD removal and saves energy for aeration.

Figure 6 shows that the interim breaks did not affect the overall performance, since COD effluent concentrations were well below 150 mg/L. However, the maximum break duration has been 14 days only (end of November). Therefore the result is in accordance with the laboratory-scale experimental data. It should be noted, that the trickling filter is designed for a maximum capacity of 20 P.E., whereas it served 40 P.E. sometimes.

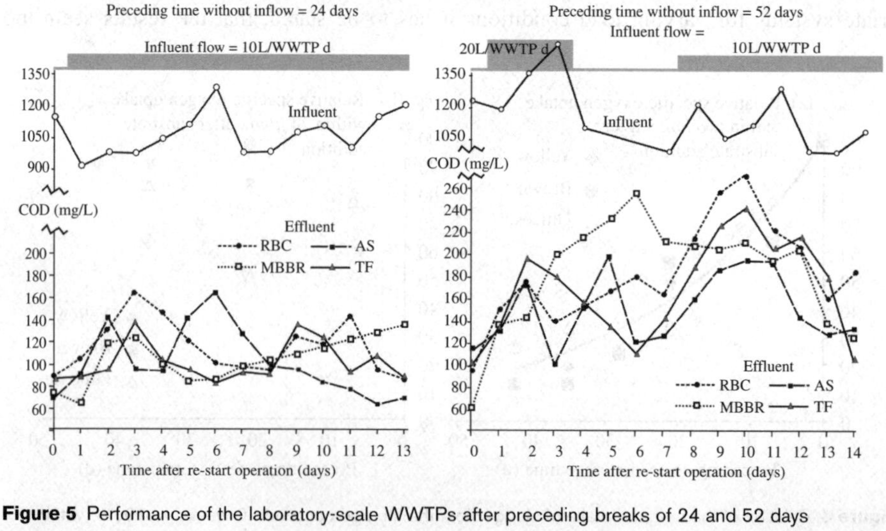

Figure 5 Performance of the laboratory-scale WWTPs after preceding breaks of 24 and 52 days respectively

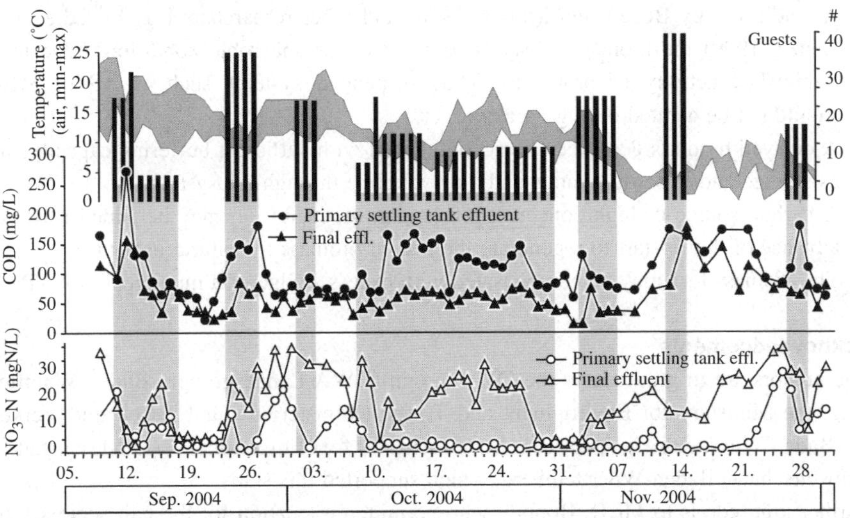

Figure 6 COD, nitrate, temperature and accommodation of a youth hostel trickling filter from early September to early December 2004

Summary and conclusions

It can be stated, that the loss of biomass significantly affects the treatment capacity of a biological system whereas the specific activity – measured as OUR – recovers within a reasonable time. However, biomass decay does not start immediately after the feed ceases. The performance data of the youth hostel trickling filter demonstrate that residual substrates provide further feed for the bacteria, particularly when water is recycled through the system. For practical reasons it is interesting to know how long a system can stay idle without affecting the treatment efficiency. The answer will not be uniform, since it depends on the specific boundary conditions, such as process scheme, temperature, regular loading, buffering capacity and more. Moreover, the definition of a sound treatment in terms of effluent limits is not uniform throughout the countries of their application.

Anyway, the following benchmark data can be summarised from the experiments and from literature data as an approach.

- The Alpenverein states on its internet homepage a startup time for a trickling filter of 12 days after an extended off-season (Alpenverein, 2000)
- Begert and Müller (1976) investigated the impact of a 70 days break on an activated sludge laboratory scale system. They concluded a re-start-time of 7 days, which could be shortened to three days, when primary sludge was filled into the reactor at the beginning of the break.
- The displayed laboratory-scale experimental results can be interpreted in a way, such that a 24 days break highlights the maximum break time without exceeding the COD limit of 150 mg/L during re-start. A break time of 52 days required a start-up in the range of 1–2 weeks.
- No impact must be expected when the break time is in the range of 14 days. This conclusion is supported by the performance of the youth hostel trickling filter and by investigations from Barjenbruch and Al Jiroudi (2004) in the context of the testing procedure for very small WWTPs for the assignment of a proof label and by early experiments from Hörler and Pracek (1964).
- Measures for managing breaks may either aim at keeping the total biomass or at regenerating the loss.

- As indicated by Begert and Müller (1976) and other researchers (i.e. Loosdrecht and Henze, 1999) the biomass decay is lowered under anaerobic conditions (due to the diminished activity of protozoa). Thus suspended systems such as AS or MBBR should not be aerated during breaks.
- A delayed biomass decay can be accomplished with sufficient buffering capacity, such as storage and/or settling tanks and a recycle line through the system.
- A carbon source of high concentration (sugar, methanol …) may be added a week in advance of the re-start to regenerate the loss of biomass by natural growth.
- The biomass loss may be compensated with excess sludge of a municipal WWTP.

Acknowledgements

The authors are in gratitude to the DWA group (DWA Landesgruppe) Baden-Wuerttemberg, the Ministery for Environment and Traffic (Ministerium für Umwelt und Verkehr) of Baden-Wuerttemberg and the State-Institute for Environment (Landesanstalt für Umweltschutz) Baden-Wuerttemberg, which supported this study.
Further gratitude is to Mr D. Boppre, who permitted a location for the half scale setup in the vicinity of a sanitary sewer and to Mr D. Pfaff, the manager of the youth hostel, who participated the project with passion.

References

Alpenverein, (2000). Inbetriebnahme, www.alpenverein.de/huw/life.

Barjenbruch, M. and Al Jiroudi, D. (2004). Vergleich von Kleinkläranlagen – Ergebnisse vom Demonstrationsfeld in Dorf Mecklenburg, Universität Rostock – Institut für Umweltingenieurwesen, 5. Rostocker Abwassertagung, 133–148 (im German).

Batla, M.N. and Gaudy, A.F. (1965). Role of protozoa in the diphasic excertion of BOD. *J. San. Eng. Div. ASCE*, **91**(SA3), 63–69.

Begert, A. and Müller, W. (1976). Der Einfluss abnormaler Betriebsverhältnisse auf die Wirksamkeit von Kläranlagen. *ÖAR*, 71 ff (in German).

Coello Oviedo, M.D., Sales Marquez, D. and Quiroga Alonso, J.M. (2002). Evolution of the activated sludge system under starvation conditions, *Water Intelligence Online.* (unique ID: 2002 12009).

Flasche, K. (2002). Einsatzmöglichkeiten und Leistungsfähigkeit von Kleinkläranlagen, Disseratation ISAH, Universität Hannover (in German).

Hörler, A. and Pracek, V. (1964). Ergebnisse von Untersuchungen der EAWAG mit dem Kombibecken. *Schweiz. Zeitschrift für Hydrologie*, **XXVI**, 427 ff (in German).

Loosdrecht, M.C.M. and van Henze, M. (1999). Maintenance, endogenous respiration, lysis, decay and predation. *Wat. Sci. Tech.*, **39**(1), 107–117.

Lopez, C. and Morgenroth, E. (2002). Maintenance and decay in enhanced biological phosphorus removal in activated sludge during long-term starvation, WEFTEC.

Metcalf and Eddy (1991). *Wastewater Engineering*, Tata McGraw-Hill, New Delhi.

Morgenroth, E., Obermayer, A., Arnold, E., Brühl, A., Wagner, M. and Wilderer, P.A. (2000). Effect of long-term idle periods on the performance of sequencing batch reactors. *Wat. Sci. Tech*, **41**(1), 105–113.

Schweizer, V. (1983). Leistungsfähigkeit von Mehrkammergruben, *Korr. Abwasser*, 7/83, 473–480 (in German).

Urbain, V., Pys, E., Block, J.C. and Menam, J. (1993). Composition and activity of activated sludge under starvation conditions. *Environmental Technology*, **14**, 731–740.

Vasel, J.L. and Jupsin, H. (2003). A system for on-site performance evaluation of small wastewater treatment plants according to the European procedure. *Wat. Sci. Tech.*, **48**(11), 401–409.

Treatment of lower load graywater by using a controlled soil natural treatment system

M.A. López-Zavala

Centro de Estudios del Agua, Instituto Tecnológico y de Estudios Superiores Monterrey, Av. Eugenio Garza Sada Sur No. 2501, Col. Tecnológico, 64849 Monterrey, N.L. México (E-mail: *miganloza@itesm.mx*)

Abstract Experiences of treating graywater by soil natural treatment systems have not been widely reported. In general terms, graywater has a lower concentrations of organic matter; nutrients (e.g. nitrogen), and microorganisms than combined wastewater; therefore, the graywater treatment capacity of soil should be evaluated. In this study, the performance of a natural soil treatment system was evaluated when shower graywater from a house located at Sapporo City, Japan, was treated by a layer of 30 cm of soil from the Hokkaido University Campus. Results showed very high removal efficiencies of chemical oxygen demand (98%) and Kjeldahl nitrogen (95%) at high infiltration rates (214 cm/d), which denotes that natural soil treatment systems could be a good alternative to treat lower load graywater.

Keywords Graywater; onsite wastewater treatment; soil treatment

Water Science & Technology **Vol 55 No 7 pp 39–45** © IWA Publishing 2007

Introduction

The Onsite Wastewater Differentiable Treatment System (OWDTS) introduced by Lopez Zavala *et al.* (2002), Figure 1, conceives the fractioning of household wastewater into three categories, reduced-volume black water, higher-load graywater (HLGW) and lower-load graywater (LLGW). The treatment of HLGW is thought to be conducted by a membrane bioreactor, characterized by high performance and small footprint, whose effluent will be disposed into a controlled soil natural treatment system (CSNTS). The LLGW will be directly disposed into a CSNTS, where degradation of pollutants will be naturally performed by physicochemical and biological processes occurring in the soil–water ecosystem. Here, it is called a controlled system because the disposal or reuse of the treated effluent is conducted under controlled conditions.

Experiences of treating graywater by soil natural treatment systems have not been widely reported. In general terms, graywater has lower concentrations of organic matter, nutrients (e.g. nitrogen) and microorganisms than combined wastewater. However, the concentration of phosphorus, heavy metals and xenobiotic organic pollutants are around the same levels (Jenkins, 1998; Santala *et al.*, 1998; Almeida *et al.*, 1999; Nolde, 1999; Gunther, 2000; Ledin *et al.*, 2001; Eriksson *et al.*, 2002; Garland *et al.*, 2004). Low concentration of nitrogen may limit the biological processes in the soil system; therefore, evaluation of biodegradation efficiency of LLGW pollutants in a controlled SNTS (lab-scale) was the aim of this study. Because aerobic microbial activity in the soil system mainly occurs in the first 30 cm from the soil surface, three soil columns of 30 cm each were used during the experiment. The soil surface was not covered with vegetation because the microbial activity only was the target.

Materials and methods experimental device

The experimental device used in this study consisted of three soil columns through which graywater was infiltrated for treatment. The soil containers were made of Lucite

doi: 10.2166/wst.2007.124

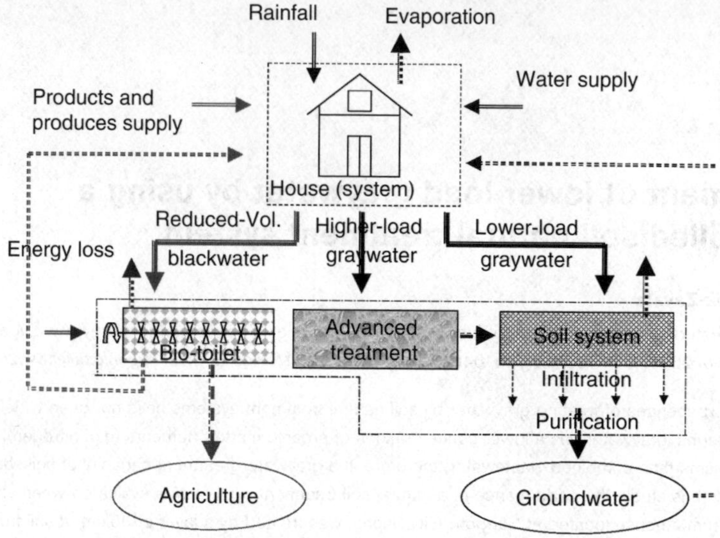

Figure 1 The onsite wastewater differentiable treatment system (Lopez Zavala *et al.*, 2002)

(Figure 2). The inner diameter of the soil containers was 9.4 cm; the length of all containers was approximately 50 cm and the bottom ended in a conical shape. Graywater was supplied into the soil columns by using a high precision pulse-pump with a maximum discharge capacity of approximately 50 ml/min and ten flow rate levels. Graywater was pumped from plastic containers (pet bottles) with 2.0 L capacity.

Treated graywater (water drained from soil columns) was also collected in 2.0 L capacity pet bottles. All graywater used was stored in a refrigerator at a temperature of about 3 to 4°C.

Determination of soil particles size distribution

Soil texture and structure are important physical properties due to their influence on soil hydraulic characteristics, such as infiltration rates and permeability. In this study, particles size distribution of soil from Hokkaido University Campus was determined by sieving using nine different sizes of mesh. Thus, nine different fractions were set, from <0.25 mm to >2.36 mm.

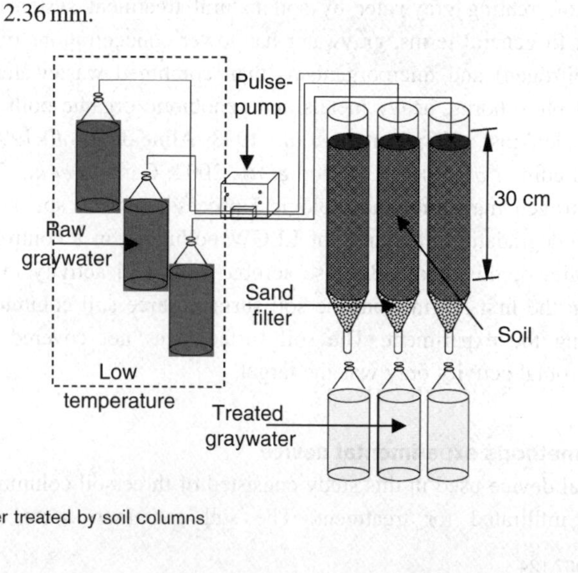

Figure 2 Graywater treated by soil columns

Graywater characterization

Graywater from a shower of a house located in the city of Sapporo, Japan was used in this study. Several graywater samples were collected when taking the shower and stored in a refrigerator at about 3 to 4°C. No more than 1 day old 100 ml samples were filtrated by using 0.45 μm filter paper. Chemical oxygen demand (COD) analyses were conducted for raw graywater and filtered graywater samples. Thus, soluble and particulate organic matter contained in graywater was determined for several samples. Additionally, parameters such us pH, electrical conductivity (EC), Kjeldahl nitrogen (K-N), ammonia nitrogen (NH_4-N), fluorine (F), chlorine (Cl), nitrites (NO_2), nitrates (NO_3), phosphates (PO_4), sulfates (SO_4), sodium (Na), potassium (K), magnesium (Mg), and calcium (Ca) were determined in each sample. All determinations were conducted by using standard methods for water and wastewater analysis (APHA *et al.*, 1989).

Graywater treatment tests

After preparing the soil columns, distilled water was added into each column to cause the lixiviation of dissolved organic matter already contained by the soil. Thus, several "washing events" of soil were conducted until reaching a quite low turbidity in water drained from the soil columns.

Graywater was treated into two soil columns as follows. Two litres of raw graywater were pumped from the 2 L plastic containers using a high precision pulse-pump and then supplied to the soil column. Additionally, two litres of distilled water were also supplied into a third soil column in order to determine the contribution of the organic matter already contained by the soil on the graywater treatment process (blank). Water effluents drained from the soil columns were collected into 2 L plastic containers. Eight different infiltration rates 13.1, 24.3, 49.8, 74.7, 121.3, 155.9, and 213.7 cm/d were evaluated.

Soil temperature under which the treatment process occurred and the volume of water effluents collected from each soil column were measured. A 100 ml sample from each collected effluent was filtrated by using 0.45 μm filter paper. Chemical oxygen demand (COD) analyses were conducted for raw effluent and filtered effluent samples. Additionally, parameters such us pH, electrical conductivity (EC), Kjeldahl nitrogen (K-N), ammonia nitrogen (NH_4-N), fluorine (F), chlorine (Cl), nitrites (NO_2), nitrates (NO_3), phosphates (PO_4), sulfates (SO_4), sodium (Na), potassium (K), magnesium (Mg), and calcium (Ca) were determined in each sample. All determinations were conducted by using standard methods for water and wastewater analysis.

Results

Graywater water characterization

From the chemical analyses of 8 samples, it was found that the shower graywater has an average organic load of 374.4 mg COD/L. From this, approximately 65% is particulate and 35% soluble (Figure 3). Nitrogen load resulted to be about 11.6 mg T-N/L from which 88% was O-N, 6% NH_4 and 6% NO_3 (Figure 4). Average values of the rest of the parameters are summarized in Table 1. Comparing these data with that reported for the

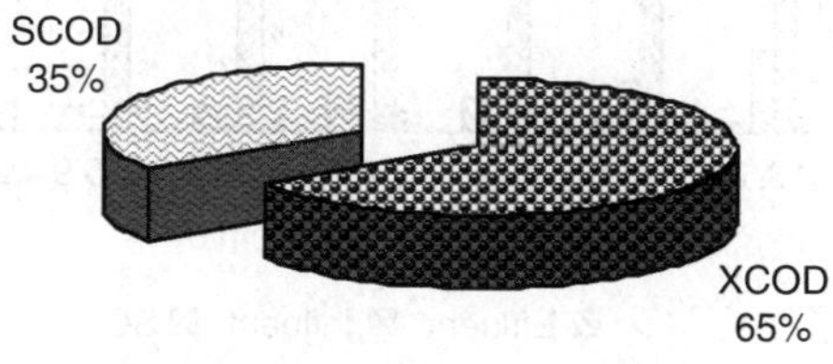

Figure 3 Organic composition of graywater from the shower

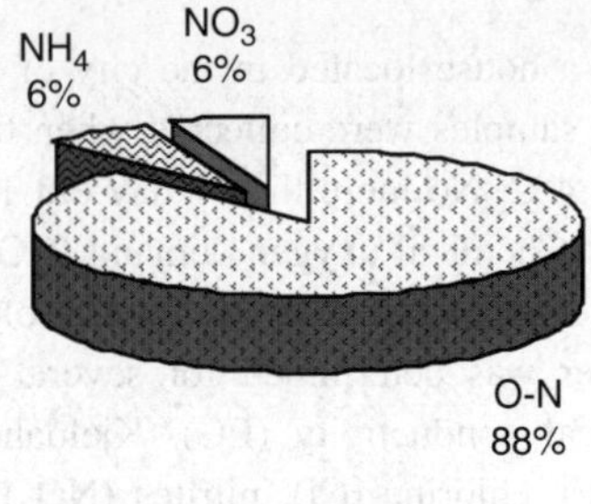

Figure 4 Nitrogen composition of graywater from the shower

Table 1 Average characteristics of shower graywater from a house located in Sapporo City, Japan

Anions	Mg/L	Cations	mg/L	Other parameters	
F	0.79	Na	21.38	COD (mg/L)	374.43
Cl	27.89	NH_4	0.74	XCOD (mg/L)	245.16
NO_2	0.00	K	3.53	SCOD (mg/L)	129.27
NO_3	0.69	Mg	0.00	T-N (mg/L)	11.59
PO_4	0.00	Ca	0.00	O-N (mg/L)	10.16
SO_4	17.00			EC (mS/cm)	237.46
				pH	6.52

primary effluent from the municipal wastewater treatment plant (Henze *et al.*, 1995), we found that graywater from shower contains 5% more soluble organic matter and more O-N than primary effluent from municipal wastewater treatment plants.

Organic load removal

The effect of infiltration rates on the organic load of the effluent from soil columns can be seen in Figure 5. At low infiltration rates it is interesting to note that COD of the distilled water effluent is greater than the shower treated effluent. This means that presence of organic matter in graywater contributed to better performance of removal processes (biological, physical, or both) that occurred into the soil. At high infiltration rates, the COD of distilled water effluent and graywater treated effluent are of similar magnitude. This means that COD contained in the graywater treated effluent is given by the natural soil contribution. In other words, naturally, the ground loses organic matter through its profile during the water infiltration process. These losses nevertheless do not affect the removal rates of organic matter contained in graywater, further more such removal rates

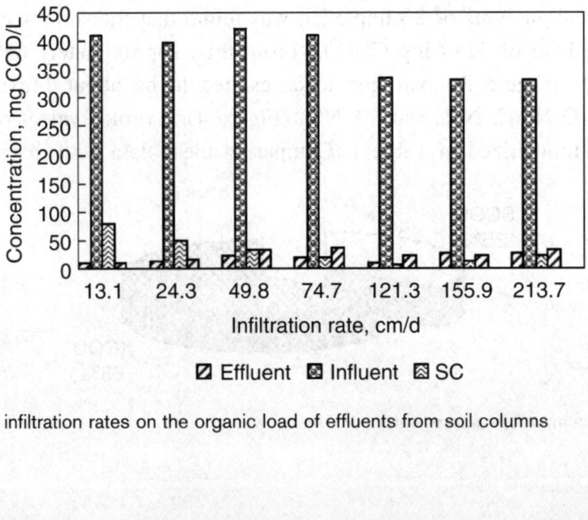

Figure 5 Effect of infiltration rates on the organic load of effluents from soil columns

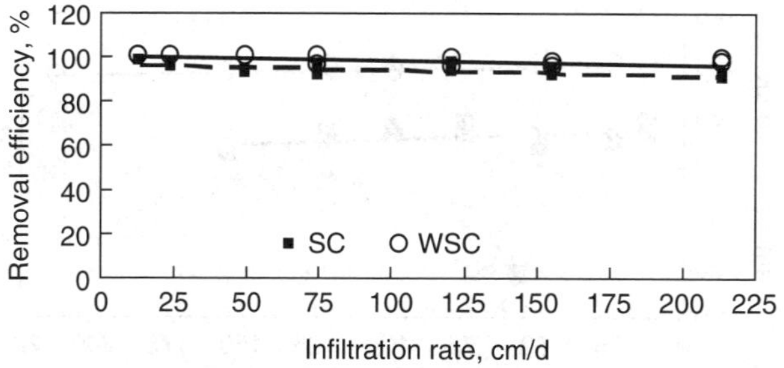

Figure 6 Removal efficiencies of COD by the soil treatment system. SC: soil contribution, WSC: without soil contribution

seem to be remarkably unaffected by the magnitude of infiltration rates. However, more long term research is needed to have more conclusive results in this respect.

Figure 6 shows the organic load removal, in terms of removal efficiency of COD, observed during the performance of graywater treatment tests for different infiltration rates. In this figure two results are plotted. The data denoted by rectangles correspond to total removal efficiencies of COD that include the contributions of the soil to the effluent organic load because soil already contained soluble organic matter which is drained during tests. The removal efficiencies denoted by circles were estimated after subtracting the contributions of soil to the organic load of the effluent drained from the soil columns. The contributions of soil were estimated from the tests conducted exclusively with distilled water. In both plots, a steady decreasing tendency of removal efficiency was observed as infiltration rates increased. Even though, the removal efficiencies are quite high at the highest infiltration rate, 90% when soil contribution is considered and 98% in the opposite case.

Nitrogen removal

Regarding the nitrogen removal, it was observed that concentrations of Kjeldahl nitrogen (K-N) in the treated effluents were of the order of 3 mg/L, irrespective of the infiltration rates (Figure 7). Furthermore, such K-N loads in the treated graywater effluents were similar to those K-N concentrations measured in effluents produced when only distilled water was supplied. This means K-N loads in treated graywater effluents are given by soil contributions. In Figure 8, removal efficiencies of K-N are plotted. Similarly to Figure 6, rectangles denote removal efficiencies when soil contributions are considered and circles represent results without soil contributions. Both plots presented steady decreasing trends as infiltration rates increased; however, the difference in removal

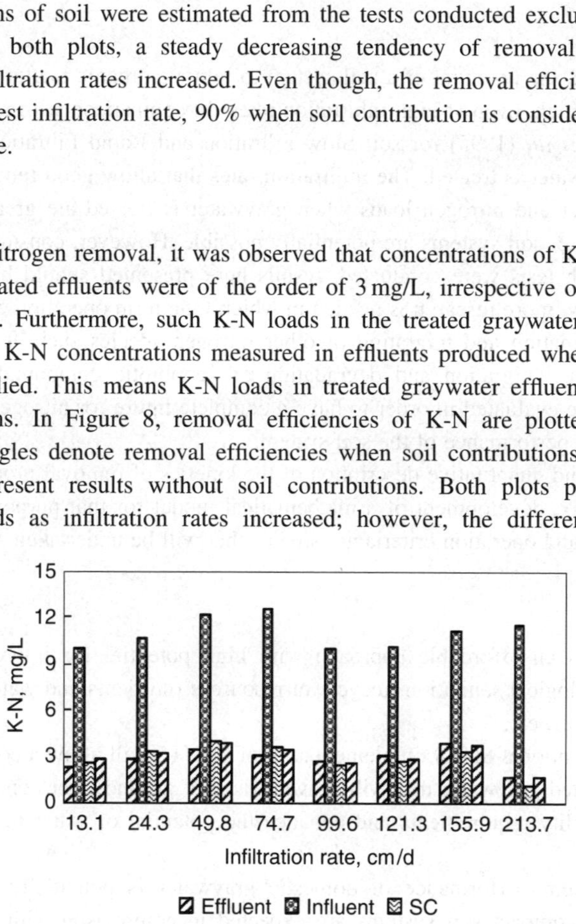

Figure 7 Effect of infiltration rates on the nitrogen load of effluents from soil columns

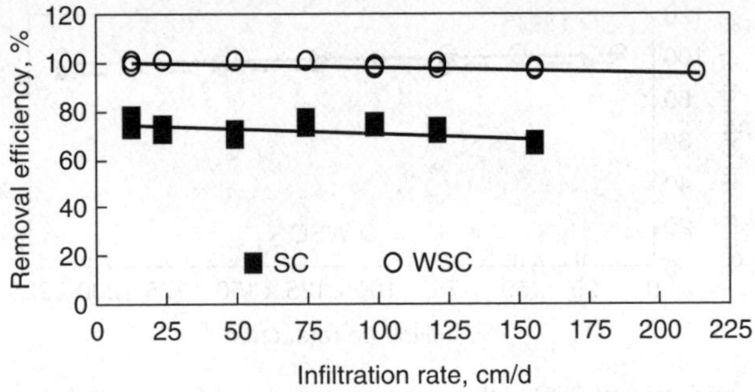

Figure 8 Removal efficiencies of K-N by the soil treatment system. SC: soil contribution, WSC: without soil contribution

efficiencies at low and high infiltration rates is on the order of 5% or even smaller. In this figure, the soil contributions on removal efficiencies of K-N are very clear in comparison with COD removal efficiencies, where contributions of soil reduced approximately 4% the removal efficiency; whereas in the case of nitrogen, the soil contributions decreased 25% the removal efficiencies.

Regarding no soil contributions, the removal efficiency of K-N at the highest infiltration rate evaluated was about 95%, which is quite a good result.

Discussion

As shown above, the potential of a soil system to remove organic matter and nitrogen (as a K-N) is quite high, even at high infiltration rates. In comparison with recommendations made by Reed *et al.* (1995) for soil Slow Filtration and Rapid Filtration systems when domestic wastewater is treated. The infiltration rates that allow good removal efficiencies of organic matter and nitrogen loads when graywater is treated are greater. This means that more compact soil systems are potentially possible. However, considering that in this study only batch tests were conducted, results here presented should be managed with care. Necessarily, more research is needed in which long term operation can be evaluated. In addition, formation and migration of other nitrogen species such as nitrates, nitrites and nitrogen gas, migration and degradation of xenobiotic compounds and traces of organics must be evaluated in order to have a complete figure for nitrogen transformation and degradation performance of the soil system.

Qualitative and quantitative description of the kinetics of removal processes that occur in the soil matrix, development of a mathematical model for that purpose and establishment of design and operation criteria are studies that will be undertaken shortly.

Conclusions

The OWDTS is an affordable approach with high potential for improving traditional OWTS, dry ecological sanitation, recycle of resources (nutrients and water), conservation of water resources, etc.

Unlike conventional SNTS, implementation of CSNTS will allow a controlled disposal or reuse of treated graywater that will prevent risks of groundwater contamination, guarantee water quality requirements and enhance the potential of water reuse at household level.

Good treatment performance of domestic graywater is potentially achieved in the CSNTS. More compact soil systems are expected in comparison with SNTS that treat domestic wastewater.

References

Almeida, M.C., Butler, D. and Friedler, E. (1999). At-source domestic wastewater quality. *Urban Water*, **1**, 49–55.

APHA, AWWA, WPCF (1989). *Standard Methods for the Examination of Water and Wastewater*, 17th edition, USA.

Eriksson, E., Auffarth, K., Henze, M. and Ledin, A. (2002). Characteristics of gray wastewater. *Urban Water*, **4**, 85–104.

Garland, J.L., Levine, L.H., Yorio, N.C. and Hummerick, M.E. (2004). Response of graywater recycling systems based on hydroponic plant growth to three classes of surfactants. *Water Research*, **38**, 1952–1962.

Gunther, F. (2000). Wastewater treatment by greywater separation: Outline for a biologically based greywater purification plant in Sweden. *Ecological Engineering*, **15**, 139–146.

Henze, M., Gujer, W., Mino, T., Matsuo, T., Wentzel, M.C. and Marais, G.V.R. (1995). *Activated Sludge Model No. 2*, IAWQ Scientific and Technical Reports, No. 3, London.

Jenkins, D. (1998). The effect of reformulation of household powder laundry detergents on their contribution to heavy metals levels in wastewater. *Water Environment Research*, **70**(5), 893–980.

Ledin, A., Eriksson, E. and Henze, M. (2001). Aspects of groundwater recharge using grey wastewater. In *Decentralized Sanitation and Reuse. Concepts, Systems and Implementation*, Lens, P., Zeeman, G. and Lettinga, G. (eds), IWA Publishing.

Lopez Zavala, M.A., Funamizu, N. and Takakuwa, T. (2002). Onsite wastewater differentiable treatment system: modeling approach. *Water Science and Technology*, **46**(6-7), 317–324.

Nolde, E. (1999). Greywater reuse systems for toilet flushing in multistory buildings – over ten years experience in Berlin. *Urban Water*, **1**, 275–284.

Reed Sherwood, C., Crites Ronald, W., and Middlebrooks, E. Joe (1995). *Natural Systems for Waste Management and Treatment*, Second edition, McGraw-Hill, Inc.

Santala E., Uotila J., Zaitsev, G., Alasiurua, R., Tikka, R., and Tengvall, J. (1998). Microbiological greywater treatment and recycling in an apartment building In AWT98 – *Advanced Wastewater Treatment, Recycling and Reuse*: Milan, 14–16 September, 1998 (319–324).

Evaluation of an optimal fill strategy to biodegrade inhibitory wastewater using an industrial prototype discontinuous reactor

G. Buitrón*, I. Moreno-Andrade*, J.A. Linares-García*, J. Pérez*, M.J. Betancur** and J.A. Moreno**

*Environmental Bioprocesses, Universidad Nacional Autonoma de Mexico (UNAM), Circuito Escolar, Edif. 5, Ciudad Universitaria, 04510 Mexico D.F., Mexico (E-mail: *gbm@pumas.iingen.unam.mx*)

**Automation Departments, Institute of Engineering, Universidad Nacional Autonoma de Mexico (UNAM), Circuito Escolar, Edif. 5, Ciudad Universitaria, 04510 Mexico D.F., Mexico

Abstract This work presents the results and discussions of the application of an optimally controlled influent flow rate strategy to biodegrade, in a discontinuous reactor, a synthetic wastewater constituted by 4-chlorophenol. An aerobic automated discontinuous reactor system of 1.3 m^3, with a useful volume of 0.75 m^3 and an exchange volume of 60% was used. As part of the control strategy influent is fed into the reactor in such a way as to obtain the maximal degradation rate avoiding inhibition of microorganisms. Such an optimal strategy was able to manage increments of 4-chlorophenol concentrations in the influent between 250 and 1000 mg/L. it was shown that the optimally controlled influent flow rate strategy brings savings in reaction time and flexibility in treating high concentrations of an influent with toxic characteristics.

Keywords 4-chlorophenol; inhibitory wastewaters; optimal control; SBR

Water Science & Technology **Vol 55 No 7 pp 47–54** © IWA Publishing 2007

Introduction

The majority of industrial wastewaters can be degraded using microorganisms. Nevertheless there exist more than 70,000 compound synthetic organic chemicals (xenobiotic compounds) and many of them are very difficult to degrade using conventional biological processes. Generally, when some of these compounds are present in industrial wastewater, problems in the biological treatment occurs because of the low biodegradability and inhibition of the biomass. The possibility that a toxic compound will be degraded depends mainly on the presence of a specialized microorganism and the acclimation period (Buitrón and Moreno, 2004).

Phenols are very common pollutants in industrial wastewaters from industries such as pharmaceutics, chemistry, petrochemical, pulp and paper industries, wood preservative production, pesticides, herbicides, fungicides, etc. For example, chlorophenols can migrate easily from a liquid environment to another due to its relatively high solubility. When the wastewaters that contain chlorophenolic compounds are discharged into rivers they can produce metabolic changes in the populations that live in this environment. It has been reported that these compounds have carcinogenic and teratogenic properties. In the human case, brominated and chlorinated phenols were found in plasma from Swedish blood donors (Hovander *et al.*, 2002). Also, chlorophenols have possible endocrine-disrupting effects, exerted by interference with the transport of thyroid hormones (van den Berg, 1990). Due to the high quantities of chlorophenols produced in the industry and the adverse effects in living organisms, these compounds have been classified as high-priority polluting agents by the United States Environmental Protection Agency, US EPA (Wentz, 1989). In consequence, it is important to produce an efficient treatment for the removal of phenolic compounds present in industrial wastewater.

doi: 10.2166/wst.2007.126

Industrial wastewater is usually treated by the classical continuous activated sludge processes (Eckenfelder and Musterman, 1995). Nevertheless, because the industrial waters are very variable in the flow and concentration (shock loads) of toxic compounds, their design requires large volumes of the reactor tank in order to provide a long enough retention time to achieve the biodegradation, and sometimes removal efficiencies are not satisfactory. In order to avoid these problems and to increase the biotreatment efficiencies of industrial wastewater, alternatives technologies such as the Sequencing Batch Reactor (SBR) have been proposed. The SBRs are flexible in the control operation and can be automated.

The SBR operates under five phases: fill, react, settle, draw, and idle (Wilderer *et al.*, 2001). In the usual operation mode, the duration of these phases is typically determined by an expert operator based on his experience and exhaustive testing in the laboratory with a pilot plant. In particular, the reaction phase is sufficiently long to allow the toxic substances to be effectively biodegraded. The settle and draw phases are fixed in duration by the characteristics and constraints of the activated sludge and the reactor itself.

A SBR operated in the usual form presents several constraints when applied to toxic wastewater degradation: inhibition of the microorganisms, problems with shock loads of toxic compounds, deacclimation and problems of microorganism starvation, and low efficiencies regarding the removal of toxic compounds (Buitrón *et al.*, 2005). In order to overcome the problems discussed above and to exploit the maximum capabilities of the discontinuous reactors, Betancur *et al.* (2004) described the mathematical development of a new optimal controlled influent flow rate strategy to robustly maximize the reaction rate of a discontinuous process treating inhibitory compounds. The optimal control strategy was successfully tested in a lab-scale pilot plant for the degradation of different concentrations of 4-chlorophenol (up to 625 mg/L) (Moreno-Andrade *et al.*, 2006). The optimal control strategy showed a reduction in the degradation time (around 52%) and in the supplied air volume when the optimal strategy was compared with the usual mode.

This work evaluates and discusses the application of an optimal control strategy to biodegrade wastewater containing 4-chlorophenol in an industrial prototype sequencing batch reactor.

Methodology

An aerobic automated SBR system of 1.3 m^3, with a useful volume of 750 L and an exchange volume of 40% was used (Figure 1). The airflow rate was 100 L min^{-1} and operated at ambient temperature (around $19 \pm 2\,°C$). The reactor was inoculated with microorganisms from a municipal activated sludge treatment plant (2000 mgVSS/L).

Figure 1 Industrial-prototype SBR used in the degradation of wastewater containing 4-chlorophenol

A synthetic wastewater containing 4-chlorophenol (4CP), (Aldrich, purity $\geq$99.0%) was used as a sole source of carbon and energy. Nutrients (nitrogen and phosphorus) and oligo-elements were added as recommended by AFNOR (1985). Phenol concentration was measured taking samples and processing them offline using a modified colorimetric technique of 4-aminoantipyrine (*Standard Methods*, 1992). Total and volatile suspended solids (TSS and VSS), as well as Chemical Oxygen Demand (COD), were determined according also to *Standard Methods* (1992). The SBR was operated under the following strategy: pre-aeration time (15 min), fill and reaction time (variable depending on the necessary time to reach a total degradation of the mixture) and settling time (30 min), draw (6 min).

The reactor was operated for 57 cycles (65 days). The first 15 cycles (24 days) were used for the biomass acclimation to 4CP. The last cycles (cycles 16 to 57) were operated with the optimal strategy. Under standard operation the reactor degraded an influent synthetic wastewater containing 250 mg/L of 4-chlorophenol. In the last degradation cycle (57) a punctual increase of 1000 mg 4CF/L in the influent concentration was tested in order to evaluate the response of the optimal strategy when a concentration peak is present. This shock load represents an increase of 400% in the influent concentration compared with the usual concentration applied during the operation of the reactor.

The biomass was acclimated using a variable cycle strategy, i.e. the reaction phase duration was variable and stopped when the removal of 4CP was equal to or greater than 95%. After the acclimation period, the reactor was operated using the optimal strategy. The optimal strategy controls the fill and reaction phases by using a variable named oxygen mass uptake rate (OMUR) directly related to the reaction rate. Such a variable can be estimated in real time by using the DO concentration and the volume of the reactor, as described in detail by Moreno-Andrade *et al.* (2006). Considering that an inhibitory wastewater follows the Haldane behavior, there exists a certain substrate concentration (S^*) where the maximal degradation rate is obtained. When the influent concentration is higher than S^*, the inhibition starts. The optimal strategy controls the influent flow rate in such a manner that the reaction rate is maintained close to its maximal value. Thus the reactor operates with a controlled periodic influent as a fed-batch. In theory it is possible to biodegrade any toxic concentration since the strategy feeds only the necessary amount of wastewater to maintain the maximal degradation rate and to avoid inhibition.

Results and discussion

Acclimation

The reactor was acclimated to degrade 250 mg/L of 4CP using the usual operation mode (i.e. no control strategy was used for the biomass acclimation). The acclimation of the microorganisms was reached in 15 cycles (24 days). During the acclimation, the 4CP was degraded with efficiencies higher than 99.9% as 4CP and 97% as DOC and COD. During the acclimation, the reaction phase duration was reduced from 90 h to 2 h, after 19 operation days (from cycle 1 to cycle 15). The duration of the reaction phase was maintained in 2 h. Once the biomass was completely acclimated to the 4CP the degradation time is constant as described by Moreno-Andrade and Buitrón (2004).

As the acclimation was carried out, the sludge volumetric index (SVI) decreased (Figure 2). It is clear that filamentous bacteria found in the inoculum used were eliminated by the toxic compound. In general, the microorganisms degrading the 4CP presented excellent settling proprieties. Even for a problematic inoculum, SVI decreased from 364 to 132 mL/g after 15 cycles. Also, during the biomass acclimation TSS in the effluent were reduced from 55 to 23 mg/L, confirming the good settling proprieties of the biomass.

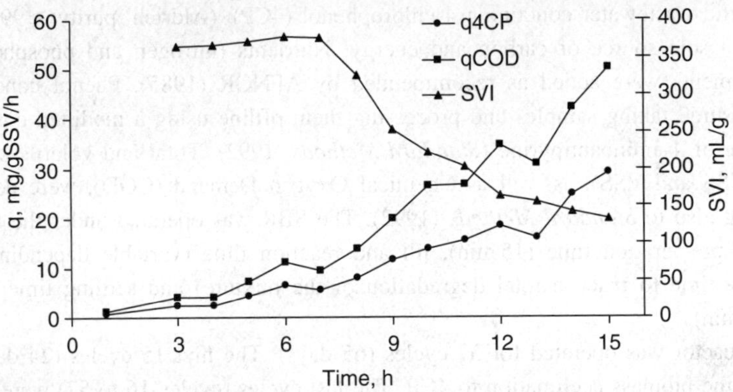

Figure 2 Specific degradation rate (q), and sludge volumetric index (SVI) variation during the acclimation of the microorganisms to degrade 250 mg/L of 4CP

Figure 3 presents the kinetic curves for the operation of the reactor once the biomass was acclimated. It is possible to observe that, in the usual operation mode, the fill of the reactor is very fast (5 min) and the 4CP concentration increases up to 95 mg/L in the reactor. The 250 mg/L are diluted by the remaining water present in the reactor. The behavior of the dissolved oxygen (DO), OMUR and substrate concentration (measured as 4CP) is observed in Figure 3. In this case the fill of the reactor is very fast and the DO decreases to 5 mg/L in 5 min. The OMUR increases suddenly; however, when toxic substrate concentration is higher than S^*, the inhibition starts, and the OMUR decreases again.

The reaction phase takes around 2.5 h. During this period the DO concentration decreases gradually and the OMUR increases indicating the improvement of the microbial activity because the substrate concentration is arriving at S^*. In the usual operation mode the maximal microorganism's activity is near to the maximal few minutes. Once the substrate is very low, the DO concentration increases and the OMUR decreases because there is no more substrate in the water, and then the microorganism activity is reduced. At the end of the reaction phase only the endogenous respiration prevails and the OMUR reaches its minimal value. It is clear to observe that with this operation mode the degradation activity and velocity are only at the maximal value at the end of the reaction and only for a few minutes, because the toxic substrate concentration is superior to S^* in the major part of the reaction phase and there is inhibition of some microorganisms.

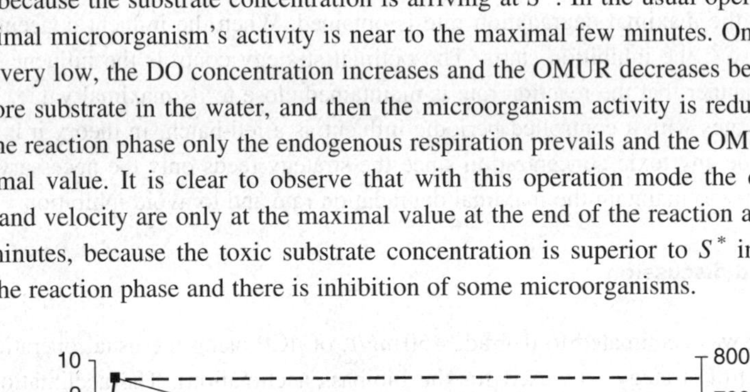

Figure 3 Kinetic profiles of 4CP, OMUR and DO for the degradation of 250 mg 4CP/L using the usual operation mode

Evaluation of the optimal control strategy

A good performance of the reactor operated with the optimal control strategy was obtained since the degradation of the 4CP was efficiently completed. The average removal efficiencies were up to 98% as chemical oxygen demand and >99.9% as 4CP.

Figure 4 presents the behavior of the substrate concentration, measured as 4CP during a cycle operated under the optimal control strategy (250 mg 4CP/L). It is possible to distinguish how the control operates following the substrate, the DO and the OMUR curve. Once the influent begins to be fed to the reactor, the reaction starts. The DO decreases (first 0.2 h, Figure 4) as the metabolic activity of the biomass increases to degrade the 4CP. Simultaneously, the estimator calculates OMUR and follows its value. When a maximal point of OMUR is detected (before a minimum in the DO concentration), the feeding pump is turned off. The microorganisms degrade the substrate. At this point, OMUR decreases because the concentration of substrate is lower than S^* (it has passed from 13 to 0 mg 4CP/L, at 0.5 h in Figure 4).

Thus, the feeding pump is switched on again, and a new charge of substrate is fed to the reactor. This procedure is repeated until the maximal volume has been reached and then, the degradation proceeds in a batch manner, ending the reaction when the DO concentration increases to a constant value and OMUR is decreased and stabilized at a low value. It is possible to observe in Figure 4 how the concentration of 4CP in the reactor is maintained around 12 mg/L, which corresponds to a concentration close to the maximal OMUR value and S^*. The operation mode can be explained as if in the reactor several mini-batches took place during a cycle. For the example shown, the inhibition never occurred due to the fact that 4CP concentration in the bulk of the liquid in the reactor was around S^* and was never higher than 14 mg/L during the whole reaction.

The degradation of 250 mg 4CP/L with the optimal control strategy was obtained in 2 h, this represents a 15% less than the reactor using the usual operation mode (Figure 3). The reduction in the time is because the control maintains the microorganisms at this maximal activity and thus the maximal degradation velocity.

The bioreactor was operated during 41 cycles using the optimal control strategy. Figure 5 presents the evolution of the specific degradation rate, the total suspended solids in effluent and sludge volumetric index during the long term evaluation of the optimal control strategy. The operation of the reactor was stable, with efficiencies higher than 99.9% for the 4CP degradation, and 91 ± 3% as COD. The effluent presented 38 ± 15 mg COD/L and 30 ± 12.5 mg TSS/L. On average, the specific degradation rate

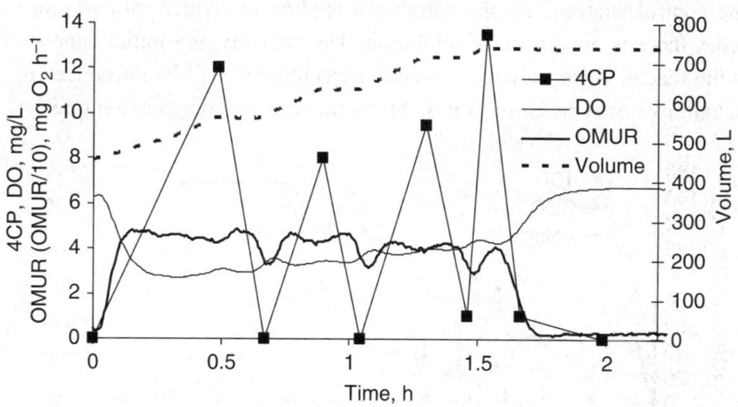

Figure 4 Evolution of the substrate concentration, measured as 4CP and DOC during a cycle using the optimal control strategy. Note that the changes in the γ value agree with the changes in the DO concentration, since the DO is used to estimate OMUR

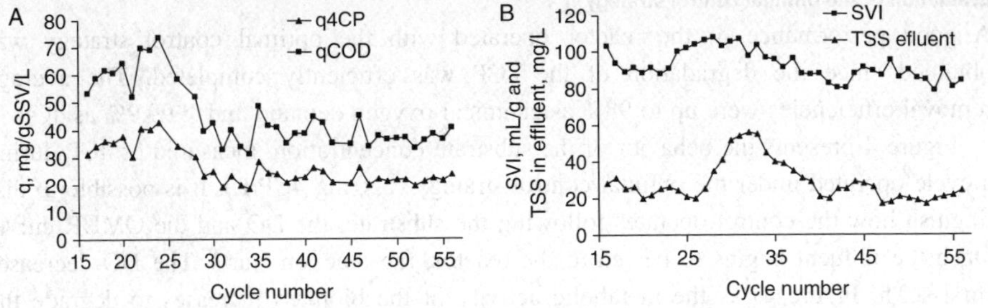

Figure 5 (A) Specific degradation rate, (B) Total suspended solids in effluent and sludge volumetric index variation during the operation by the optimal control strategy

was 45 ± 12 mgCOD/mgVSS/h for the COD and 26 ± 7 mg4CP/mgVSS/h for the 4CP during the operation with the optimal control strategy. In general the reactor shows that the practical implementation of the optimal fill control strategy was possible with stable and reproducible results in an industrial prototype discontinuous reactor.

Degradation of a shock load of 1000 mg 4CP/L

As discussed in the introduction, the main problem with toxic wastewaters is the shock loads that could inhibit the biomass causing reactor failure. For this reason in the last degradation cycle, the influent concentration was increased to 1000 mg 4CP/L to evaluate the response of the optimal strategy. The concentration peak represents an increase of 400% in the 4CP concentration, compared with the standard condition of the reactor. The behavior of the degradation peak is present in Figure 6. The performance of the control was similar to the degradation of 350 mg/L, but in this case more mini-batches were employed for the total degradation of the influent volume and the total degradation time also increased. The degradation time needed for the total degradation of the substrate was 8 h.

An important point to be considered is that concentrations higher than 350 mg/L produce inhibition problems and it is difficult to degrade with the usual biological processes (Buitrón *et al.*, 2003). For example, when wastewaters containing high concentrations of phenolic compounds are discharged in a conventional treatment plant, it is necessary to dilute the influent in order to avoid inhibition. Thus, several cycles will be required, besides the water used for dilution. On the contrary, with the optimal control strategy, there is no need for dilution, since in the reactor the necessary amount of wastewater is fed by the control strategy. As the automated feeding is divided into several controlled mini-batches there is no chance of inhibition. Theoretically any initial concentration can be fed to the reactor since only the adequate concentration will be introduced to maintain the degradation velocity as close as possible to the maximal degradation rate.

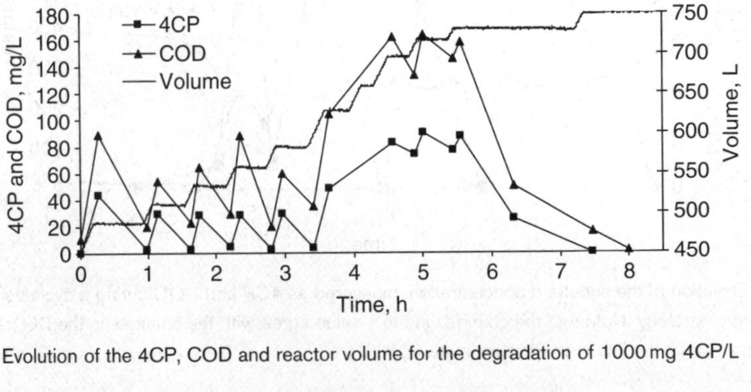

Figure 6 Evolution of the 4CP, COD and reactor volume for the degradation of 1000 mg 4CP/L

Another advantage of the control strategy is that it is not necessary to know the initial toxic wastewater concentration because the control is based on the oxygen consumption of the microorganisms and thus, directly related to the microbial activity. This is significant because in the usual operation mode it would be necessary to know the initial concentration when a shock load is arriving in order to make the necessary dilution to avoid inhibition and to proceed with the satisfactory degradation of the wastewater.

Conclusions

The optimal fill strategy in the biodegradation of inhibitory wastewater in an industrial-prototype was tested. The optimal control strategy showed that it is robust and stable in long operation periods (more than 2 months). The degradation efficiencies of the 4-chlorophenol were as higher as 99.9%, and 98% as chemical oxygen demand, the parameters evaluated demonstrate the excellent behavior of the process. The optimal control strategy was able to manage increments of toxic concentrations in the influent up to 1000 mg/L of 4-chlorophenol without inhibition problems.

Theoretically, any initial concentration can be fed to the reactor since only the adequate concentration will be introduced to maintain the degradation velocity as close as possible to the maximal degradation rate. Another advantage of the control strategy is that it is not necessary to know the initial toxic wastewater concentration because the control is based on the oxygen consumption of the microorganisms and thus, directly related to the microbial activity.

Acknowledgements

This paper includes results of the EOLI project that is supported by the INCO program of the European Community, Contract number ICA4-CT-2002-10012. The scientific responsibility rests with the authors. J.A. Linares-García and Iván Moreno-Andrade thanks CONACYT.

References

AFNOR (1985). Evaluation en mileu aqueux de la biodegradabilité aérobie “ultime” des produits organiques solubles. *Normalisation française, NFT*, 90–312.

Betancur, M.J., Moreno, J. and Buitrón, G. (2004). Event-driven control for treating toxicants in aerobic sequencing batch bioreactors. *Proceedings of the 9th International Symposium on Computer Applications in Biotechnology (CAB9), CDROM file 1074*, 28-31 March 2004. Nancy, France.

Buitrón, G., Schoeb, M-E., Moreno-Andrade, I. and Moreno, J. (2005). Comparison of two Control Strategies for an Automated Sequencing Batch Reactor under Extreme Peaks of 4-Chlorophenol. *Water Research*, **39**(6), 1015–1024.

Buitrón, G. and Moreno, J. (2004). Modeling of the acclimation/deacclimation process of a mixed culture degrading 4-chlorophenol. *Water Science and Technology*, **49**(1), 79–86.

Buitrón, G., Schoeb, M.-E. and Moreno, J. (2003). Automated sequencing batch bioreactor under extreme peaks of 4-chlorophenol. *Water Science and Technology*, **47**(10), 175–181.

Eckenfelder, W.W., Jr. and Musterman, J.L. (1995). *Activated Sludge Treatment of Industrial Wastewater*, Technomic Publishing Company, Lancaster Pennsylvania, pp. 281.

Hovander, L., Malmberg, T., Athanasiadou, M., Athanassiadis, I., Rahm, S., Bergman, A. and Klasson, E. (2002). Identification of hydroxylated PCB metabolites and other phenolic 65 halogenated pollutants in human blood plasma. *Archives of Environmental Contamination and Toxicology*, **42**, 105–117.

Moreno-Andrade, I. and Buitrón, G. (2004). Variation of the microbial activity during the acclimation phase of a SBR system degrading 4-chlorophenol. *Water Science and Technology*, **50**(10), 251–258.

Moreno-Andrade, I., Buitrón, G., Betancur, M.J. and Moreno, J.A. (2006). Optimal degradation of inhibitory wastewaters in a fed-batch bioreactor. *Journal of Chemical Technology and Biotechnology*, **81**, 713–720.

Standard Methods for the Examination of Water and Wastewater (1992). 18th edn, APHA/AWWA/WEF, Washington DC, USA.

van den Berg, K.J. (1990). Interaction of chlorinated phenols with thyroxine binding sites of human transthyretin, albumin and thyroid binding globulin. *Chemico-Biological Interactions*, **76**, 63–75.

Wentz, C.A. (1989). *Hazardous Waste Management*, McGraw-Hill Book Co, New York, USA.

Wilderer, P.A., Irvine, R.L. and Goronszy, M.C. (2001). *Sequencing Batch Reactor Technology. (Scientific and Technical Report No 10)*. IWA Publishing, London, 76pp.

Characterization of the activated sludge and of the operating conditions of six SBR treating dairy effluent and operated at industrial scale

Joseph V. Thanikal*,, M. Torrijos*, Ph. Sousbie* and J.Ph. Delgenes***

*INRA, LBE, Avenue des Etangs, 11100, Narbonne, France (E-mail: *torrijos@ensam.inra.fr*)

**Kumaraguru College of Technology, Coimbatore – 641 006, India (E-mail: *jthanikal@gmail.com*)

Abstract The sludge from six SBRs treating dairy effluent and located at same geographical location, in North East of France, were collected to study their characteristic behavior. The six plants were designed and constructed by the same manufacturer and are working under quite similar operating conditions. The objective of the study was to observe if any similarity existed in the characteristics of the sludge collected from the SBRs. The sludge was characterized for morphological properties (filament index, floc size), settling, compressibility, suspended solids (SS) concentration. The sludge from each plant was different from the others in most of the characteristics. One sludge out of six (sludge G) was completely different from the others with a very degraded structure and low discrete settling and compression. This reactor was not working fully satisfactorily with a too high COD at outlet, probably because this SBR was undergoing repetitive overloading linked to a very bad recovery of the whey by the cheese maker. The five other SBRs were working fully satisfactorily but the characteristics of the five sludges were quite different from one sludge to another. The size of the flocs seemed to be the only parameter measured which could be correlated to the settling characteristics of the sludge. The sludge characteristics and the parameter correlations were also compared with that of municipal activated sludge and were found to be very different.

Keywords Activated sludge; dairy effluent; floc size; flocculation; Sequencing Batch Reactor; sludge volume index

Water Science & Technology **Vol 55 No 7 pp 55–63** © IWA Publishing 2007

Introduction

From the mid-90s, the SBR process has been applied in France to the treatment of cheese making plants, particularly in the Jura Mountains in north-east of France, where there are some 200 small cheese-making plants. Each plant employs an average of 2 people and processes between 400,000 to 18,000,000 litres of milk a year, the average being 2,500,000 litres/year. Pollution from these cheese-making units is generally at the level of a 200–800 population-equivalent (P.E.) in hamlets with often as few as 100 to 150 inhabitants. For wastewater treatment, the cheese makers must choose between either an independent solution or being connected to the public sewage system for a collective treatment solution, when it exists.

The first SBR plant for an individual treatment was installed in 1996 at the cheese making unit of Le Fied in the Jura *department* (administrative entity). This plant was monitored over one year, to measure the actual technical performance, to assess the maintenance requirements and to determine the operating costs (Torrijos *et al.*, 2001). Nowadays, there are more than 30 SBR plants in operation in the Jura mountains. These SBRs are operated by the cheese makers working in the cheese plant. An independent control office carries out sporadic checks (1 to 3 per year) to verify the behavior of the wastewater treatment plants.

In this work, the activated sludge from six SBRs treating dairy effluents and in operation in the Jura mountains, were characterized (morphological properties, settling,

doi: 10.2166/wst.2007.127

compressibility, suspended solids (SS) concentration,…). These SBRs have been operated for 3 to 9 years at industrial conditions, which means that the effluent produced each day by the cheese plants is treated without any knowledge of its concentration or volume and that the SBRs are operated by the cheese maker without any extra help. The aim of this study was to evaluate the level of similarity between sludge in different SBR reactors working in close conditions, that is to-say: (i) the SBRs treat an effluent coming from the production of the same type of cheese (Comté); (ii) they are located in the same geographic area; (iii) they were built by the same manufacturer and their design was based on the same criteria.

Materials and methods

SBR plants

The six cheese making plants from where the sludge samples were collected are located in the same area (Jura mountains) and the distance between the two most separated plants is 80 km. The SBRs were identified as: A, Ar, F, G, N and O according to the first letter of the village where they are located. The cheese making plants are very small and run by 1 or 2 cheese makers. Between 1,200,000 and 5,000,000 litres of milk are processed per year (Table 1). Production is mainly the variety of cheese, Comté, a pressed and heated cheese. The pollution rejected by the cheese plants represents between 190 and 550 people-equivalent (PE) according to the size of the production units.

The six SBR plants were constructed between 1996 and 2002, by the same manufacturer, with the same principles for design based on a low loading rate, that is to say, a volumetric loading rate of 0.35 kg BOD_5/m^3.d and a mass loading rate of 0.1 kg BOD_5/kg of VSS.d. The design hydraulic retention time (HRT) is between 4 and 6 days. The SBRs are operated by the cheese makers and work with one treatment cycle a day. The effluent generated during one working day is held temporarily in a first buffer tank. At the start of a SBR cycle, the aeration tank is filled with the effluent to be treated and then is continuously aerated for 20 h, followed by settling which lasts 4 h. During the last hour of a cycle, the volume initially introduced is run off and, when necessary, a portion of the sludge removed from the bottom of the aerated tank.

Sludge sample collection

The sludge samples were collected inside the SBR tanks at the end of the aeration phase, just before the start of the settling period. Transportation time between the SBR plants and the laboratory was around 7 h. Sludge was maintained at ambient temperature during transportation and refrigerated to 4 °C as soon as it reached the laboratory. Tests were completed within 48 h.

Table 1 Daily volume of milk processed by the dairies, rated characteristics of the effluents and volume of the SBRs

	Reference of the SBR plants					
	A	**Ar**	**F**	**G**	**N**	**O**
Date of start	Nov 2001	March 2002	June 1996	March 2002	March 2002	Oct 1999
Volume of milk processed/day (L/day)	1,200,000	5,000,000	2,700,000	1,400,000	4,500,000	3,000,000
Rated volume of effluent (m^3/d)	5.5	21	14	9	27	11
Rated quantity of BOD_5 (kg/d)	12	40	23	14	35	21
Volume of the SBR (m^3)	35	115	65	40	100	60

Sludge volume index and zone settling velocity (SVI & ZSV)

The tests were carried out in a transparent cylinder of 1 L and 10 cm in diameter as per *Standard Methods* (APHA, 1995). The sludge was diluted when the settled volume after 30 minutes exceeded 25% of the volume of the cylinder. The settled sludge volume was measured every 5 minutes up to 30 minutes. The SVI was expressed as settled sludge volume in mL/g of MLSS. The slope of the initial straight line portion of the curve plotted between time and settled sludge volume gave the ZSV. To simulate the behavior of the sludge in the SBR, a settling test was also carried out with non-diluted sludge and the height of the settled sludge was measured after 30 minutes, 1, 2 and 4 h.

Floc size distribution

Floc size distribution was measured by using a Beckman Coulter LS200 granulometer. The instrument uses the Fraunhofer method of diffraction of light for measurement of size of floc in the range of 0.2–2,000 μm. The samples were added to the vessel containing water, which is provided with sensors, detecting the percentage concentration of sample. The addition of sample was done until the concentration was between 9 to 11%. The light scattered was detected by a detector that converts the signal to a size distribution based on volume.

Filament index

The sludge samples were examined for filaments by phase contrast microscopy using 10x objective magnification, with an Olympus BX60 microscope. The images were captured using a Nikon digital camera DXM1200F and ACT-1 software, attached to the microscope. The filamentous organisms were quantified as filament index using the method described by Jenkins *et al.* (1993). The filament index was rated on a scale of 0–6. The filaments were described (branching, filament shape, location, cell shape) using the criteria described by Jenkins *et al.* (1993). The observations were made for presence of true and false branching, straight, smoothly curved, bent or irregularly shaped chains of cells. Also, whether the filaments are extending from the floc, lying mostly within the floc or present in the liquid between flocs.

Analysis

When necessary, samples were centrifuged at 6000 g for 10 minutes before analysis to remove suspended solids. COD was measured according to the Dichromate reactor digestion method test (Hach 0–1500 mg/l vials). Other parameters were measured following *Standard Methods* (APHA, 1995).

Results

Influent and effluent characteristics for the 6 SBR plants

Characteristics of the raw wastewater. The effluent treated in the SBRs is made up of the water used for cleaning the tanks, the pipes, the cheese moulds and the floor of the cheese dairy. The sweet whey obtained during the forming, transfer, and palletisation during the first pressing stages and the acidic whey obtained at the end of pressing, are carefully recovered and kept apart from the rinsing water. The data presented for the raw wastewater (Table 2) are average values coming from the analysis achieved by an independent control office which carries out periodic checks (1 to 3 per year) to verify the behavior of the wastewater treatment plants. The raw wastewater from 3 cheese dairies (F, N and O) had a low organic content with COD between 1.45 and 1.8 g/L. On the other hand, the organic content of the effluent from A, Ar and G were much higher with COD between 4 and 5 g/L.

Table 2 COD at inlet and outlet of the 6 SBRs

	Reference of the SBR plants					
	A	**Ar**	**F**	**G**	**N**	**O**
Raw wastewater COD[(1)]	4.3 g/L (± 1.78)	5 g/L (± 1.35)	1.6 g/L (± 0.54)	4.2 g/L (± 2.05)	1.8 g/L (± 1.53)	1.5 g/L (± 0.72)
Treated wastewater COD[(2)]	48 mg/L	133 mg/L	18 mg/L	312 mg/L	70 mg/L	26 mg/L

[(1)]Results from the independent control office
[(2)]Results from the isolated outlet sampling carried out the day of sludge sampling

Treated wastewater characteristics. When collecting sludge samples, samples were also taken at the outlet of the SBRs at the end of the cycle in progress (Table 2). Five samples out of six had a COD at the outlet lower than 140 mg/L and among these 5, 4 sludge (A, F, N and O) had a very low COD with values less than 70 mg/L. Sample G had a high soluble COD (312 mg/L) exceeding the limit for discharge and indicating an insufficient purification efficiency.

Sludge characteristics

Suspended solids in the SBRs. Three SBRs (Ar, F and G) had close SS concentrations (Table 3) with values between 4.8 and 6.2 g/L. Sample N had a very high SS concentration with 10.4 g/L and two SBRs (A and O) had quite a low SS concentration with 2.06 and 2.85 g/L respectively.

Settling characteristics (SVI and ZSV). Zone settling velocity (ZSV) characterizes the settleability of the sludge and sludge volume index (SVI) characterizes its compactability. SVI is frequently used to monitor settling characteristics in routine operation of activated sludge processes (APHA, 1995). These two parameters were measured for the six SBR samples to characterize the settling characteristics of the sludge. Four samples A, F, N and O had quite low SVIs with values between 40 and 75 mL/g and two sludges (Ar and G) had much higher SVIs with 153 and 140 mL/g respectively. Sludge G was the only one with a very low ZSV (1.19 m/h) when the five other sludges had quite close ZSV with values between 3 and 4 m/h which corresponded to the highest values found in domestic wastewater treatment (Daigger and Roper, 1985; Jin *et al.*, 2003).

In addition to the standard measurement of the SVI after half an hour with dilution of the sludge if necessary, the height of the un-diluted sludge in a 1 L cylinder after 4 h of settling (called hereafter: non-diluted SVI) was measured to fit to the real conditions of operation of the SBRs, as the settling period of a treatment cycle is generally 4 hours. Sludge G was the only one with a high non-diluted SVI (89 mL/g). This parameter was between 39 and 62 mL/g for the 5 other sludges.

Table 3 Summary of physical properties of the activated sludge and morphological properties of the flocs

	Unit	Reference of the SBR plants					
Parameter		**A**	**Ar**	**F**	**G**	**N**	**O**
SS concentration	g/L	2.06	6.18	4.8	5.72	10.35	2.85
SVI	mL/g	63	153	54	140	72	42
ZSV	m/h	3.6	3.1	3.9	1.2	3.7	4
Height of sludge after 4 h of settling	mL/L	110	260	220	510	645	110
Non diluted SVI	mL/g	53	42	46	89	62	39
Flocsize	µm	210	86	65–300	54	107	175
Filament index	–	4–5	3	4	3	2	3

Microscopic examination and filament index. The sludge samples were examined microscopically at 10× magnification, to study the floc structure and the presence of filaments and protozoa. Filamentous organisms play an important role in floc structure as they provide networks or backbones for the flocs. The networks direct flocs to grow into shapes other than spherical and allow the flocs to grow larger (Sezgin, 1982).

The floc structure and shape were different for all the sludge samples. Filaments were present in all samples but they were present in abundance in samples A and F. Sludge A was made-up of dense flocs to which the filaments were attached. They were present in abundance (FI of 4–5). The filaments observed were straight, smoothly curved, bent and extending from the surface of the floc. Floc sizes and shapes of sample F were quite irregular. Mycelial filaments with true branching extending from the floc surface were present in plenty with a FI of 4. Filaments were bridging the small flocs together making a larger size of floc. A cloudy layer was observed around the flocs. The sludge from O showed the presence of smaller flocs, comparatively to the previous sludge, but they were dense in nature with presence of filaments and protozoa. FI was 3 and the filaments attached to the floc were bent, smoothly curved, small, extending from the floc surface. The sludge Ar and N were quite similar with flocs irregular in size and the presence of mycelial filaments with branches. Small straight filaments were also observed free in the liquid between the flocs. On the whole, the sludge were rated with a FI of 3 for Ar and 2 for N. The sample G was completely different from all the other samples with highly disrupted and very irregular flocs. Filament content was low and the sludge was rated for FI as 2.

Floc size distribution. The floc size, determined using a Beckman Coulter granulometer, showed different particle size distributions for the various sludges (Figures 1 and 2). The diameter of the flocs covered quite a wide range indicating very different floc sizes for a given sludge, particularly for sludge A, F, G and N. Sludge Ar and O had a particle size distribution close to the distribution generally observed with municipal sludge with a clearly monomodal distribution (Jin *et al.*, 2003; Govoreanu, 2004). Sludge F showed a clear bimodal distribution with very large flocs of 300 μm mean diameter and small flocs of 65 μm mean diameter. Sludge G was quite different from the other sludge with a very high proportion of small units and a mean diameter of 54 μm, confirming that this sludge is very disrupted.

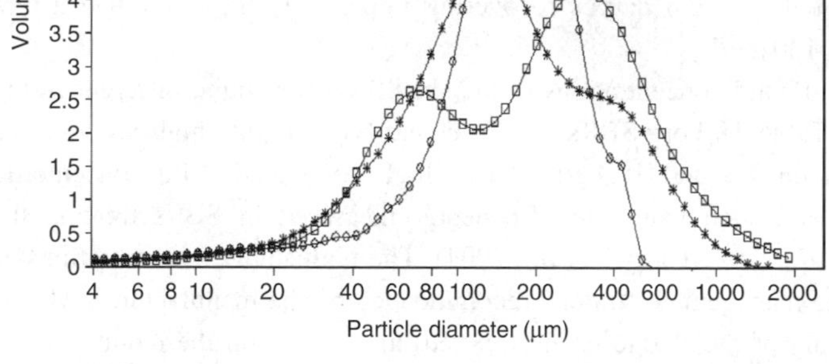

Figure 1 Flocs size distribution of sludge A, F and O

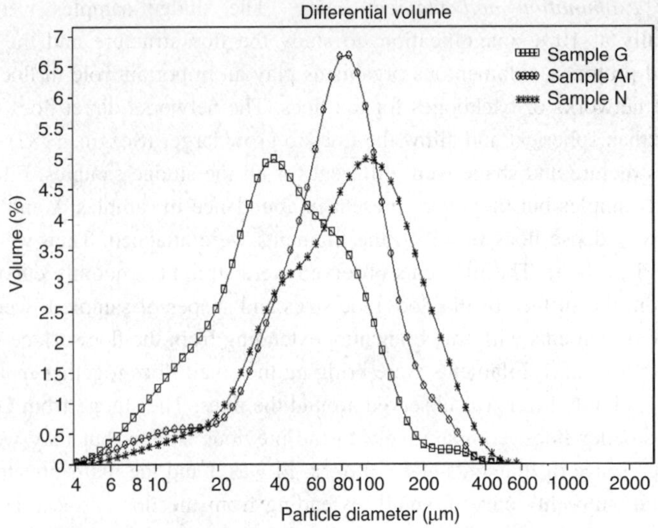

Figure 2 Flocs size distribution of sludge Ar, G and N

Discussion

In this study, the sludge of six SBRs treating the effluent of small cheese dairies and in operation for 3 to 9 years, were compared. These SBRs treat the effluent coming from the production of the same type of cheese (Comté), are located in the same geographical area, were built by the same manufacturer and were designed with the same criteria. Sludge characterization showed quite different results from one sludge to another, although the behavior of the reactors did not exhibit significant differences, except for sludge G.

Whey recovery is very carefully carried out in cheese dairies F, N and O for which average COD at inlet was less than 2 g/L. Whey recovery is less efficient in the three other dairies and the COD at inlet was between 4 and 5 g/L, indicating a loss of whey. Enough data on the volumes of effluent produced are not available to make any conclusion on the daily total quantity of COD discharged by the plants, nevertheless, for these last three plants, a risk of organic overloading of the SBRs exists due to a bad control of the discharge of whey.

COD at outlet showed that all SBRs except G were working correctly. The outlet of SBR G had a high soluble COD (312 mg/L) indicating insufficient purification efficiency. This result may be linked to a COD overloading of the reactor, due to the presence of whey in the effluent, and the quantity of COD loaded can then exceed the oxygen transfer capacity of the reactor. It is important to underline that SBRs from cheese dairies carrying out a good whey recovery had a very low COD at outlet. On the other hand, two SBRs out of three from dairies recovering imperfectly the whey had a COD at outlet greater than 130 mg/L.

Suspended solid concentrations in the six SBRs were quite different from one reactor to another (Table 3). Four SBRs (Ar, F, G and N) had quite high SS concentrations with values between 4.8 and 10.4 g/L. These high suspended solid concentrations are not uncommon and such values are frequently measured in SBRs treating dairy effluent (Torrijos *et al.*, 2001; Torrijos *et al.*, 2004). The high sludge concentrations are the result of the sludge management strategy recommended by the manufacturer, which is based on the monitoring of the sludge level after settling and not on the monitoring of the sludge concentration. Indeed, the following procedure is recommended: once a week, the operator must check the level of the sludge after 4 hours of settling with a measuring jar of 1 L and sludge removal is started manually, only when sludge level reaches a given level, generally 500 mL/1000 mL. As a result, sludge concentration can increase until the level of the sludge after settling reaches the maximum level. Ar, F and G have a low height of sludge after 4 hours (Table 3) and sludge concentration should then go on increasing as no sludge withdrawal was scheduled in the new future as the settled sludge levels in the SBRs were low. SBR N is an example for which the limit has been reached. Indeed, both the SS concentration and the level of the sludge after 4 hours of setting were high (10.4 g/L and 645 mL/L respectively), showing that sludge should be rapidly withdrawn from this reactor. On the other hand, 2 SBRs (A and O) had quite low SS concentrations. A different sludge management chosen by the cheese makers operating these two plants can explain this. In these two cases, a fixed volume of sludge is systematically withdrawn each week but the sludge level is not checked, which leads to an excessive sludge withdrawal.

Considering the SVI, the sludge can be separated into two groups, with 4 sludge (A, F, N and O) having a low SVI (<75 mL/g) indicating a very good settling and two sludge (Ar and G) with much higher SVIs (>140 mL/g) indicating a poorer settling. However, these values always remained in the range of the SVI generally measured with activated sludge processes treating municipal wastewater and working correctly. For example, Jin *et al.* (2003) reported SVI between 97 and 255 mL/g for municipal sludge, and during extensive pilot plant studies with municipal wastewater Daigger and Roper (1985) showed that the SVI values ranged between 36 and 402 mL/g.

Comparing the standard SVIs, the non-diluted SVIs and the sludge concentrations, the following comments can be made: (i) 4 sludge (A, Ar, F, and O) had a low height of sludge after 4 hours with less than 300 mL and then a low non-diluted SVI (<55mL/g) and these SBRs could be operated without any problem even for a SS concentration up to 6 g/L. Out of these four sludge, Ar had a high SVI but a low undiluted SVI. For this sludge, the discrete settling was poor but the settling due to compression was good; (ii) non-diluted SVI of sludge N was also good (62 mL/g) but the level of sludge was high due to a high SS concentration. Sludge should then be removed from this reactor to lower the SS concentration; (iii) SBR G was the only one where the sludge behavior was bad with a settling of half the volume of the reactor in 4 hours for a suspended solids concentration of only 5.7 g/L. Hence, the non-diluted SVI was significantly higher (89 mL/g) than for the five other sludges (39 to 62 mL/g) and, though SS was not very high, sludge should be removed from this reactor. For this sludge, both discrete settling and compression were low. The non-diluted SVI measured in this work were in accordance with the values normally observed for SBR sludge treating dairy effluent (Torrijos *et al.*, 2004), except for sludge G, which was the only one with a very poor settling. These results bring us to the conclusion that the standard diluted SVI after half an hour is not an adequate parameter to operate SBRs and the volume of the un-diluted sludge after 4 h of settling (non diluted SVI) should be used instead to fit to the real conditions of operation. Indeed, contradictory results can be obtained according to whether the SVI or the non-diluted SVI are considered. This happened for sludge Ar which had a high diluted SVI (153 mL/g) but a low undiluted SVI (42 mL/g) which indicated that the sludge behavior was good and that the reactor could be operated without any problem even at a solid concentration of 6.2 g/L.

Sludge G was the only one with a very low ZSV (1.19 m/h) confirming that this sludge had very bad settling properties. The five other sludges had quite close ZSV (3 to 4 m/h), which corresponded to the highest values found in domestic wastewater treatment (Daigger and Roper, 1985; Jin *et al.*, 2003).

Comparing SVI and ZSV, it is difficult to draw any correlation. Indeed, four sludges out of six had similar behavior with a high ZSV and a low SVI but the two other sludges had each a different behavior. Settling of G was bad with both a low ZSV and a high SVI, while sludge from Ar had a high ZSV, indicating a good discrete settling, but a high SVI, indicating a poor compression of the sludge.

Microscopic examination showed very different characteristics for the various sludges with quite different floc structures and shapes and quantities of filaments. Sludge G had highly disrupted flocs showing that this sludge was degraded. Two sludges (A and F) had quite a high quantity of filaments (FI of 4 to 5) but a very good settling (54 and 63 mL/g respectively). In this case, the presence of a high number of filaments did not disturb the settling of the sludge. Opposite results were reported by Jin *et al.* (2003) for sludge treating domestic wastewater with a low SVI for low to moderate numbers of filaments (SVI of 97 mL/g with FI of 2) and much higher SVI (235–255 mL/g) for higher FI (4 to 5).

No clear correlation appeared between SVI or ZSV and the presence of filaments. Indeed, two sludges (A and N), with two very different FI (4–5 for sludge A and 2 for sludge N) can have very close SVI or ZSV, and two sludges with the same FI (2 for sludge G and N) can have very different SVI and ZSV.

The present study showed that the size of the flocs might play an important role in the setteability of the sludge. Indeed, the sludge with the larger flocs had good settling properties as the samples (A, F, N and O) with floc size $>100\,\mu m$ had a diluted SVI <75 mL/g, even with a high filament index. On the other hand, the two sludge (Ar and G) with the lower average floc size had a high diluted SVI (140 and 153 mL/g). For municipal wastewater, Jin *et al.* (2003) obtained different results and showed that the samples with small and compact flocs with a low filament index had good settling properties and that the samples with larger diameter of flocs did not settle well. This may be due to the fact that the flocs with larger diameter were not dense but have a larger surface area. The same observations were also made by Andreadakis (1993).

Conclusion

The sludge from six industrial scale SBRs treating dairy effluent, designed and operated in close conditions were compared. The results obtained showed that the sludge which developed in each reactor had quite different characteristics.

One treatment plant out of six (SBR G) had a sludge completely different from the other ones with a very degraded structure (highly disrupted and very irregular flocs and almost no filaments) with high diluted and non-diluted SVI and a low ZSV, indicating that both discrete settling and compression were low. Furthermore, this reactor was not working fully satisfactorily with a too high COD at the outlet. One hypothesis to explain this result is that the reactor was undergoing repetitive overloading linked to a very bad recovery of the whey by the cheese maker. Nevertheless, the SBR can still be operated by maintaining a low sludge concentration and by reducing the COD input by improving the whey recovery at the cheese dairy to avoid any organic matter overloading.

The five other SBRs were working fully satisfactorily but the characteristics of the five sludges were quite different from one sludge to another. The most important results are: (i) very low SVI and undiluted SVI were measured with sludge treating dairy effluent; (ii) undiluted SVI appears to be the best parameter to estimate the settling of the sludge in the reactors; (iii) because of the low SVI, operating the SBRs according to the height of the sludge at the end of the settling period can lead to high suspended solid concentrations in the reactors; (iv) sludge with a high filament index (FI of 4 to 5) could have very good settling characteristics; (v) the size of the flocs seemed to be the only parameter measured which could be correlated to the settling characteristics of the sludge;

(vi) sludge treating dairy effluent was found to be very different from sludge treating domestic wastewater.

Finally, thanks to an easy-to-use process (no separate settling tank and no recirculation of the sludge) with an uncomplicated supervision, SBRs can be successfully operated by the cheese making staff, subject to the respect for the rated design data. The results of SBR G show also the necessity of the sensitizing of the cheese makers to the crucial importance of a good whey recovery during the cheese production to avoid any overloading of the reactors.

References

Andreadakis, A. (1993). Physical and chemical properties of activated sludge floc. *Wat. Sci. Tech.*, **27**(2), 1707–1714.

APHA (1995). *Standard Methods for the Examination of Water and Wastewater*, 19th edn., American Public Health Association, Baltimore, MD, USA.

Daigger, G. and Roper, R. (1985). The relationship between SVI and activated sludge settling characteristics. *Journal WPCF*, **57**, 859–866.

Govoreanu, R. (2004). Activated sludge flocculation dynamics: On-line measurement methodology and modeling. PhD thesis, Universiteit Gent, Belgium.

Jenkins, D., Richard, M. and Daigger, G. (1993). *Manual on the Causes and Control of Activated Sludge Bulking and Foaming*, Ridgeline Press, Lafayette, CA, USA.

Jin, B., Wilen, B.-M. and Lant, P. (2003). A comprehensive insight into floc characteristics and their impact on compressibility and settelability of activated sludge. *Chemical Engineering Journal*, **95**, 221–234.

Sezgin, M. (1982). Variation of sludge volume index with activated sludge characteristics. *Water Res.*, **16**, 83–88.

Torrijos, M., Vuitton, V. and Moletta, R. (2001). The SBR process: an efficient and economic solution for the treatment of wastewater at small cheese making dairies in the Jura mountains. *Wat. Sci. Tech.*, **43**(3), 373–380.

Torrijos, M., Sousbie, P., Moletta, R. and Delgenes, J.-P. (2004). High COD wastewater treatment in an aerobic SBR: treatment of effluent from a small farm goat's cheese dairy. *Wat. Sci. Tech.*, **50**(10), 259–267.

Potential of filter-vermicomposter for household wastewater pre-treatment and sludge sanitisation on site

D. Gajurel, S. Deegener, M. Shalabi and R. Otterpohl

TUHH-Hamburg University of Technology, Institute of Wastewater Management and Water Protection Eissendorfer Str. 42, 21071 Hamburg, Germany (E-mail: *gajurel@tuhh.de*; *otterpohl@tuhh.de*)

Abstract Septic tank systems have been widely used to separate and digest solid matter in the household wastewater for a long time. However, they contaminate groundwater with pathogens and nutrients and deprive agriculture of valuable nutrients and soil conditioner from human excreta. Compared with septic tank systems the filter–composter (Rottebehaelter), which usually consists of an underground monolithic concrete tank having two filter beds at its bottom or two filter bags that are hung side by side and used alternately at intervals of 6–12 months, is an efficient component for solid–liquid separation, pre-treatment and collection/storage of solid matter in household wastewater. The solids are retained and decompose in the filter bags or on the filter bed while the liquid filters through. However, because of the high moisture content of the retained solids decomposition is slow. Therefore, secondary treatment of the retained solids is required for sanitisation. The breakthrough was the combination of vermicomposting with the filter-composter system. Relatively dry and stable retained materials were obtained in the filter bags in about 3 months only. No secondary treatment is required as the human excreta will be converted to vermicastings, which are hygienically safe and can be reused as soil conditioner. Therefore, further development of the filter–composter with vermicomposting is worthwhile, especially the aspects of sanitisation of the faecal matter and its reuse as a soil conditioner.

Keywords Filter-composter; household wastewater; sludge sanitisation; vermicomposting

Introduction

The challenge in the 21st century remains to produce enough food to meet the needs of the increasing population while preserving and enhancing the natural resource. The extensive use of chemical fertiliser and overcropping cause land degradation. Also the production of chemical fertiliser is energy intensive, draws on very limited fossil resources and causes environmental problems. Therefore, nutrients and organic matter present in household wastewater should rather be reused instead of discharging them into the water body in order to minimise the production of mineral fertiliser as well as to avoid polluting the water body. Deficiencies of centralised wastewater management systems for nutrient recovery and public health have been well documented (Otterpohl *et al.*, 1997). Even the advanced technologies, which are not affordable for most of the population, discharge large amounts of nutrients present in wastewater to the aquatic environment, where they are lost for ever and cause severe problems. Those nutrients, which are captured in sludge are often contaminated with heavy metals such as cadmium (Cd) and organic compounds such as PCB (polychlorinated biphenyl), which pose potential toxic risks to plants, animals and humans (Metcalf and Eddy, 1991). Therefore, large amounts of sewage sludge are disposed of in landfills or incinerated. Only a smaller part is applied to agricultural land.

Compared to the centralised systems, decentralised wastewater management systems have many benefits (Wilderer, 2001; Otterpohl, 2001). The most important benefits are: there is no need to lay sewers for the transportation of sewage as in the centralised

doi: 10.2166/wst.2007.128

Water Science & Technology **Vol 55 No 7 pp 65–70** © IWA Publishing 2007

treatment plant; construction, maintenance and operation of sewers are very costly parts of sanitation systems; and there is far less dilution of sewage than in the centralised system, which creates possibilities to reuse treated wastewater and nutrients. Therefore, decentralised wastewater treatment technologies will play a significant role, if they are low-cost and allow reuse. In this paper the limitations and potential of some of these technologies are discussed.

Septic tank systems and their limitations

Septic tank systems have been used to collect household wastewater, settle out the solids and anaerobically digest them to some extent for a long time. Most of the people in urban and peri-urban areas of Asia, Africa and Latin America and peri-urban areas of industrialised countries use septic tank systems. Even in the USA, 25 per cent of the houses are served by septic tanks (Crites and Tchobanoglous, 1998). However, they cause pollution, i.e. nutrients and pathogens seeping from these systems contaminate groundwater and nearby surface water. They cannot destroy pathogens, and deprive agriculture of valuable nutrients and soil conditioner from human excreta. Moreover, they require expensive tanker-trucks to pump and transport the sludge deposited at the bottom of the tank off site, often far away. In large cities, transportation distances are normally long, since suitable sites for treatment and disposal can mostly only be found at the outskirts of cities. Transportation of relatively small faecal sludge volumes ($5-10\,m^3$ per truck) through congested roads over long distances in large urban agglomerations is not suitable, neither from an economical nor from an ecological point of view (Montangero and Strauss, 2000).

It has to be noted that occasionally problems with broken septic tanks occur leading to infiltration of nearly untreated wastewater. It is reported that there are 22 million septic system sites in the USA issuing contaminants such as bacteria, viruses, nitrate, phosphate, chloride, and organic compounds into the environment (Jenkins, 1994). According to the EPA, states of the USA reported septic tanks as a source of groundwater contamination more than any other source, with 46 states citing septic systems as sources of groundwater pollution (Figure 1) and nine of them to be the primary source of groundwater contamination in their state.

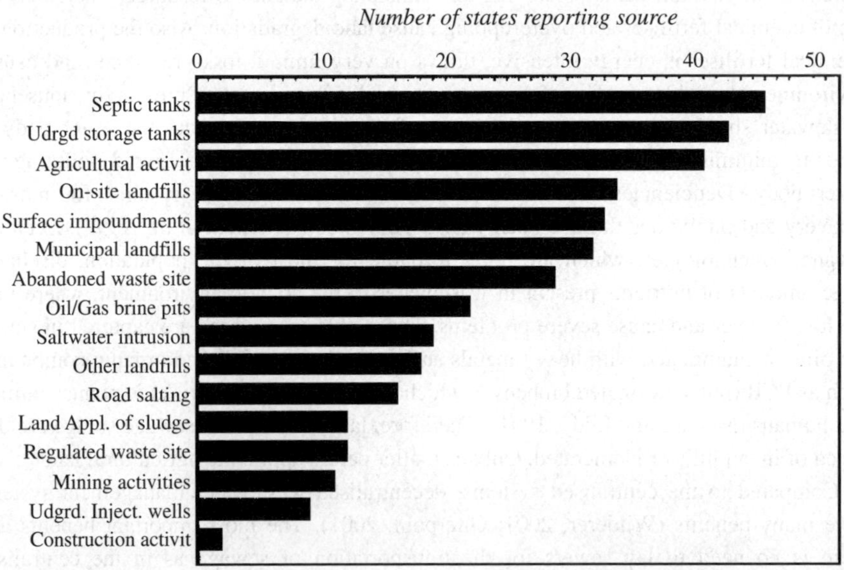

Figure 1 Reported sources of groundwater contamination in the United States (Jenkins, 1994)

The incomplete anaerobic decomposition in septic tanks results in unpleasant odour that spreads into the surroundings. Many households often add chemicals into septic tanks to reduce odour. These chemicals have adverse effects on the decomposition process and ultimately on the environment (Gray, 1989).

Because of the above mentioned drawbacks septic tank systems are not a suitable component for household wastewater pre-treatment. Compared to the septic tank systems, the conventional filter-composter (Rottebehaelter) that has been increasingly applied in Germany, Austria and Switzerland for domestic wastewater pre-treatment is an efficient component for solid–liquid separation, pre-treatment (dewatering to a certain extent) and collection/storage of solid matters in household wastewater (Otterpohl, 2001; Gajurel *et al.*, 2003a). In the following section its benefits and limitations are discussed.

Conventional filter-composter systems and their benefits and limitations

The conventional filter-composter consists of an underground monolithic concrete tank having two filter beds at its bottom or two filter bags that are hung side by side and used alternately at intervals of 6–12 months (see Figures 2 and 3).

It is watertight and structurally sound in order to avoid entering of extraneous groundwater into it and leakage of the filtrate into the groundwater, which would cause groundwater pollution. The top opening is covered by a prefabricated concrete slab and provided with ventilation so that air can enter. A shutter for changing the filter bag or emptying treated material, adding bulking agents such as straw, bark, etc. into the retained materials, inspection and cleaning is provided on the covering of the tank.

The influent is discharged into one of the two filter beds or filter bags retaining solid materials while draining the liquid. The principal role of the filter medium (filter bed or filter bag) is to cause a clean separation of particulate solids of the influent with no additional energy consumption. The filter medium is designed to recover a valuable solid product. Therefore, attempts are made to create a surface deposition of the solids in a recoverable form. When the influent is discharged into the filter bed or filter bag a filter cake is formed at the bottom. Its depth increases during the filtration due to deposition of solid material on its surface.

Results of the investigation in different conditions showed that the filter–composter system has a high potential for retaining and dewatering solid materials containing valuable nutrients and carbon. These retained materials are far more dewatered and less offensive than septic tank sludge that has to be treated in centralised treatment plants. However, the retained materials in the filter–composter still had a moisture content above 80% which is too high for composting of retained materials that have low structural stability (Gajurel *et al.*, 2003b). Therefore, retained materials were only slightly decomposed and still contained pathogens.

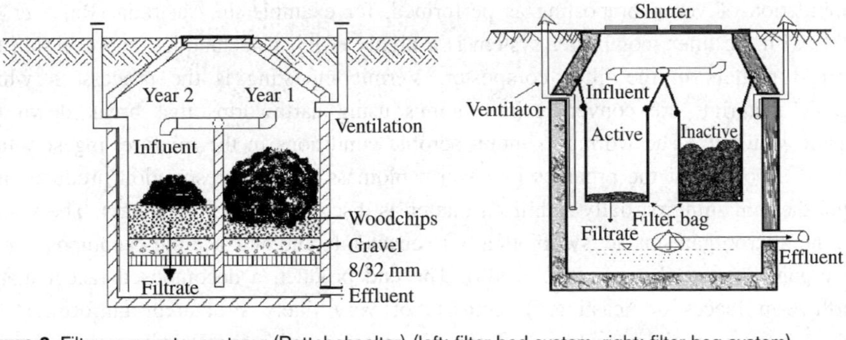

Figure 2 Filter-composter systems (Rottebehaelter) (left: filter bed system, right: filter bag system)

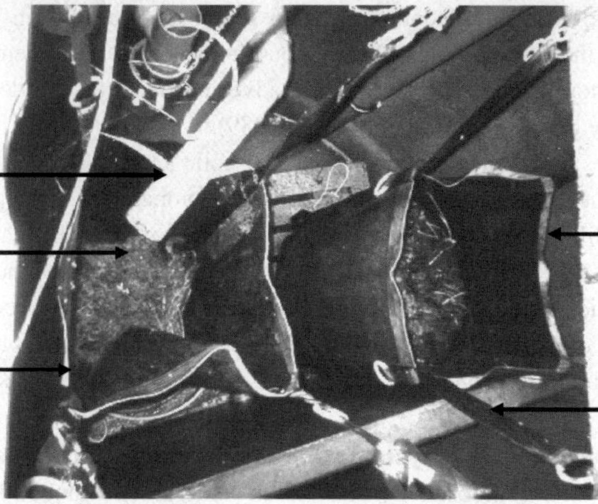

Figure 3 Filter-composter with filter bags applied in the Lambertsmuehle pilot project for faecal wastewater pre-treatment

The retained materials in the filter–composter, on the one hand have a high moisture content and, on the other hand have poor structure and a low carbon(C):nitrogen(N) ratio for composting. Therefore, the moisture content, structure and C:N ratio of retained materials should be maintained at an optimal level for composting. This can be achieved by adding straw or wood scrapes. Fibrous or bulky material such as straw or wood chips or wood scrapes can absorb relatively large quantities of water and still maintain their structural integrity and porosity as well as C:N ratio. Relatively dry and stable retained materials were obtained in the filter-composter by adding a sufficient amount of structural materials (barks or wood scrapes) after one year. However, temperature was low in the filter–composter, not more than 2–5 °C above the ambient temperature. Therefore, secondary treatment is required in order to eliminate pathogens further. It is also not very realistic to expect users of this system to add structural material every couple of weeks as our research has shown.

The filter–composter system in combination with sorting toilets, which separate urine and faeces at source, is more effective to recover high levels of nutrients as urine contains most of the soluble nutrients (Otterpohl, 2001). It was found in the Lambertsmuehle pilot project that considerably large amounts of nutrients were separated from the human excreta with this combination of sorting toilets and faecal matter pre-treatment in the filter–composter (Wupperverband, 2003).

Filter-vermicomposter and its potential

Compared to composting by adding bark or wood scrapes in the retained materials, combination of vermicomposting as performed, for example, in Australia (Bajsa *et al.*, 2003) with the filter–composter system is a highly efficient technique for processing the retained matters in the filter–composter. Vermicomposting is the process in which organic materials are converted into humus using earthworms that break down the organic materials. The worms maintain aerobic conditions in the mixture, ingest solids, convert a portion of the organics into worm biomass and into respiration products, and expel the remaining partially stabilised matter as discrete material (castings). The worms and the microorganisms act symbiotically to accelerate and enhance the decomposition of the organic matter (Loehr *et al.*, 1988). The end product, a decomposed faecal matter (earthworm faeces or "castings"), consists of very finely structured, uniform, stable and aggregated particles of humified organic material, with excellent porosity, aeration and water holding capacity, rich in available plant nutrients, hormones, enzymes and microbial populations (Appelhof, 1997). Thus this product is valuable, marketable and a superior plant growth medium (Aranda *et al.*, 1999). Moreover, pathogens cannot survive the vermicomposting process (Werner and Cuevas, 1996; Edwards *et al.*, 2002).

The processing of organic materials by worms occurs most rapidly at temperatures between 15 °C and 25 °C, at a moisture content of 70 to 90% and pH of 6.5–7.5 (Edwards, 1995). Outside these limits, worm activity and productivity, and thus the rate of waste processing, falls dramatically. The retained materials in the filter–composter offer an optimal range of moisture content, temperature and pH for worms. Results of the pilot scale experiments showed that relatively dry and stable retained materials can be obtained in the filter bags in about 3 months only, without adding structural material but including toilet paper at temperatures around 18 to 20 °C. Also, operation is far easier as no structural material has to be added, only once every half year the flow has to be directed to the other filter and worms have to be added to the now idle bag.

Vermicomposting is an effective process to sanitise faecal materials. The final product is hygienically relatively safe as the human excreta will be converted to vermicastings. The typical analysis of the vermicast of wastewater sludge in Australia revealed that pathogens such as enteric viruses, parasite eggs and bacteria such as *E. coli* have been reduced to safe levels for use in the garden (Bajsa *et al.*, 2003). However, there are very few data available so far. Therefore, further investigation is required to find out if pathogens can survive during the vermicomposting of retained materials in filter–composters.

Conclusion

In a well functioning filter–vermicomposter system, solid–liquid separation, sludge storage, dewatering and vermicomposting can be achieved in one unit with two chambers or bags. It can be implemented where people use flush toilets. Therefore, the septic tank systems can be replaced by the ecologically and hygienically advantageous filter–vermicomposter for household wastewater pre-treatment. In many cases there is no need to construct a new tank, since the existing tank can be modified, if the tank is watertight at least at the bottom where the filtrate is collected. The filter–vermicomposters are most beneficial when they are integrated in resources management sanitation (source control) for rural and peri-urban areas where local post treatment and reuse is possible. This has ecological, economic and hygienic benefits. Last but not least, the filter–vermicomposter system should be tested in pilot projects adapted to respective regional socio-economical conditions. For example, regional specific worms should be used that can easily adapt to the respective environment and are regionally available at low cost.

It has to be stated that because maintenance is a crucial factor, removal and handling of the retained material has to be improved. In addition, proper procedures of further treatment and usage should be established. The loss of water head does make it more appropriate for sites with a good gradient; otherwise an additional pump may be required to pump the filtrate collected at the bottom of the filter–vermicomposter for final treatment of the filtrate.

References

Appelhof, M. (1997). *Worms Eat My Garbage*, Second edition, Flower Press, Kalamazoo, MI, USA.

Aranada, E., Barois, I., Arellano, P., Irisson, S., Salazar, T., Rodriguez, J. and Patron, J. (1999). Vermicomposting in the tropics. In *Earthworms Management in Tropical Agroecosystems*, Lavelle, P., Brussaard, L. and Hendrix, P. (eds), CAB International 1999, pp. 253–387, ISBN 0 85199 270 6.

Bajsa, O., Nair, J., Mathew, K. and Ho, G.E. (2003). Vermiculture as a tool for domestic wastewater management. *Water Science and Technology*, **48**(11–12), 125–132.

Crites, R. and Tchobanoglous, G. (1998). *Small and Decentralised Wastewater Management Systems*, McGraw-Hill, USA.

Edwards, C.A. (1995). Commercial and environmental potential of vermicomposting: A historical overview. *BioCycle*, June 1995, 62–63.

Edwards, C.A., Eastman, B. and Bunandi, B. (2002). *The Fate of Human Pathogens During Vermicomposting*, Vermillennium Abstract S-5-3 Flowerfield Enterprises, Kalamazoo, MI, USA.

Gajurel, D.R., Benn, O., Li, Z., Behrendt, J. and Otterpohl, R. (2003a). Pre-treatment of domestic wastewater with pre-composting tanks: evaluation of existing systems. *Water Science and Technology*, **48**(11–12).

Gajurel, D.R., Li, Z. and Otterpohl, R. (2003b). Investigation of the effectiveness of source control sanitation concepts including pre-treatment with Rottebehaelter. *Water Science and Technology*, **48**(1), 111–118.

Gray, N. (1989). *Biology of Wastewater Treatment*, Oxford University Press, Oxford, UK, ISBN 0-19-859014-8.

Jenkins, J.C. (1994). *The Humanure Hand Book – a Guide to Composting Human Manure*, Jenkins Publ, PO Box 607, Grove City, PA 161227 USA.

Loehr, R.C., Martin, J.H. and Neuhauser, E.F. (1988). Stabilization of liquid municipal sludge using earthworms. In *Earthworms in Waste and Environmental Management*, Edwards, C.A. and Neuhauser, E.F. (eds), SPB Academic Publishing, The Hague, The Netherlands, pp. 95–110.

Metcalf and Eddy (1991). *Wastewater Engineering:Treatment Disposal Reuse*, McGraw-Hill, Inc.

Montangero, A. and Strauss, M. (2000). Faecal sludge treatment: Strategic Aspects and Treatment Options. Paper submitted to the *International Workshop on Biosolids Management and Utilization* at Nanjing and to the Forum on Biosolids Management and Utilization at Hong Kong, September 2000. Swiss Federal Institute for Environmental Science and technology (EAWAG) P.O. Box 611, CH-Duebendorf, Switzerland.

Otterpohl, R. (2001). Design of highly efficient source control sanitation and practical experiences. In *Decentralised Sanitation and Reuse*, Lens, P., Zeemann, G. and Lettinga, G. (eds), IWA Publishing, London, UK, pp. 164–179.

Otterpohl, R., Grottker, M. and Lange, J. (1997). Sustainable water and waste management in urban areas. *Wat. Sci. Tech.*, **35**(9), 121–133.

Werner, M.R. and Cuevas, J.R. (1996). Working with worms: Vermiculture in Cuba. *Biocycle Journal of Composting and Recycling*, **37**(6), 57–62.

Wilderer, P.A. (2001). Decentralized versus centralized wastewater management. In *Decentralised Sanitation and Reuse*, Lens, P., Zeemann, G. and Lettinga, G. (eds), IWA Publishing, London, UK, pp. 39–54.

Wupperverband (2003). Das Projekt Lambertsmühle:zukunftsfähiges Abwassermanagement Im ländlichen Raum (In German). Abschlussbericht des Projekts Lambertsmühle. Wupperverband(Hrsg.), Wuppertal (Hrsg.).

Optimization of subsurface vertical flow constructed wetlands for wastewater treatment

G. Langergraber*, Ch. Prandtstetten, A. Pressl*, R. Rohrhofer** and R. Haberl***

*Institute for Sanitary Engineering and Water Pollution Control, BOKU – University of Natural Resources and Applied Life Sciences, Muthgasse 18, 1190 Vienna, Austria (E-mail: *guenter.langergraber@boku.ac.at*)

**ÖKOREAL GmbH, Carl Reichert-Gasse 28, 1170 Vienna, Austria

Water Science & Technology **Vol 55 No 7 pp 71–78** © IWA Publishing 2007

Abstract Constructed wetlands (CWs) use the same processes that occur in natural wetlands to improve water quality and are used worldwide to treat different qualities of water. This paper shows the results of an Austrian research project having the main goals to optimize vertical flow beds in terms of surface area requirement and nutrient removal, respectively. It could be shown that a subsurface vertical flow constructed wetland (SSVFCW) operated with an organic load of 20 g COD.m^{-2}.d^{-1} (corresponding to a specific surface area demand of 4 m^2 per person) can fulfil the requirements of the Austrian standard regarding effluent concentrations and removal efficiencies. During the warmer months (May – October), when the temperature of the effluent is higher than 12 °C, the specific surface area might be further reduced. Even 2 m^2 per person have been proven to be adequate. Enhanced nitrogen removal of 58 % could be achieved with a two-stage system (first stage: grain size for main layer 1–4 mm, saturated drainage layer; and second stage: grain size for main layer 0.06–4 mm, free drainage) that was operated with an organic load of 80 g COD.m^{-2}.d^{-1} for the first stage (1 m^2 per person), i.e. 40 g COD.m^{-2}.d^{-1} for the two-stage system (2 m^2 per person). Although the two-stage system was operated with higher organic loads a higher effluent quality compared to a single-stage SSVFCW (grain size for main layer 0.06–4 mm, free drainage, organic load 20 g COD.m^{-2}.d^{-1}) could be reached.

Keywords Nitrogen removal; specific surface area requirement; subsurface vertical flow constructed wetlands

Introduction

Constructed wetlands (CWs) or treatment wetlands are wetlands designed to improve water quality. They use the same processes that occur in natural wetlands but have the flexibility of being constructed. CWs are used worldwide to treat different qualities of water (e.g. Kadlec *et al.*, 2000; Langergraber and Haberl, 2001).

Using subsurface vertical flow constructed wetlands (SSVFCWs) with intermittent loading it is possible to fulfil the stringent Austrian effluent standards regarding nitrification (Langergraber and Haberl, 2001). For small plants (less than 500 persons) the maximum effluent concentration allowed for ammonia nitrogen is 10 mg/L. This has to be met at water temperatures higher than 12 °C. For organic matter effluent concentrations (90 mg COD/L and 25 mg BOD_5/L) and treatment efficiencies (85 % and 95 % for COD and BOD_5, respectively) have to be met the whole year around. For small plants there is no standard for total nitrogen and phosphorus (1.AEVkA, 1996).

The required surface area for SSVFCWs treating wastewater is 5 m^2 per person according to the Austrian design standards (ÖNORM B 2505, 1997). Within the first part of the research project "*Optimization of subsurface vertical flow constructed wetlands (Bepflanzte Bodenfilter)*" the main goal was to optimize vertical flow beds in terms of surface area requirement. An additional goal was the optimisation of nutrient removal. The paper shows the results regarding the optimisation of the required specific surface area and nitrogen removal, respectively.

doi: 10.2166/wst.2007.129

Materials and methods

The experimental CW Ernsthofen

The investigations regarding the optimisation of the required filter surface area have been carried out at the experimental CW Ernsthofen. The experimental site is located at the wastewater treatment plant Ernsthofen (Lower Austria). A 3-chamber septic tank is used for mechanical pre-treatment. The experimental plant consists of three SSVFCWs with a surface area of $20\,m^2$ each, which are operated in parallel with intermittent loading (4 loadings a day). The organic loads applied are 20, 27 and $40\,g\,COD.m^{-2}.d^{-1}$, which corresponds to a specific area demand of 4, 3 and $2\,m^2$ per person, respectively (when considering a specific pollutant load for mechanically pre-treated wastewater of 80 g COD per person). The filter beds consist of a 50 cm main layer of sandy substrate (grain size 0.06–4 mm). An intermediate layer of 10 cm with a gravel size of 4–8 mm prevents fine particles being washed out into the drainage layer (15 cm thick; gravel size 16–32 cm) where the effluent is collected by means of tile drains. The beds are planted with common reed (*Phragmites australis*). Samples have been taken weekly and analysed in the lab of the wastewater treatment plant using Dr.Lange cuvette tests (Hach-Lange, Germany).

The indoor pilot-scale constructed wetlands (PSCWs)

Experiments regarding nutrient removal have been carried out at indoor pilot-scale constructed wetlands (PSCWs) in the Technical Laboratory Hall of the Institute. Each PSCW had a surface area of $1\,m^2$. The 50 cm main layer consists of sandy substrate with a grain size of 0.06–4 mm and 1–4 mm for 8 and 2 PSCWs, respectively. The PSCWs have been planted with *Miscanthus gigantea*. Samples have been taken on a monthly basis and analysed in the lab of the Institute according to DIN standards. For enhanced nitrogen removal a two-stage system was tested: the first stage consisting of a PSCW with a grain size of 1–4 mm for the main layer and a drainage layer operated under saturated conditions. The second stage was a PSCW with a grain size of 0.06–4 mm and free drainage. This two-stage system was operated with an organic load of $80\,g\,COD.m^{-2}.d^{-1}$ for the first stage corresponding to a specific surface area demand of $1\,m^2$ per person (i.e. $2\,m^2$ per person for the total two-stage system).

Statistical methods

STATGRAPHICS Plus for Windows 4.0 was used for statistical evaluation. Figures show "Box-and-Whisker Plots": the box encloses the middle 50% and the median value is drawn as a vertical line inside the box, the mean value is given as a cross. Horizontal lines (whiskers) extend from each end of the box. The lower/upper whisker is drawn from the lower/upper quartile to the smallest/largest value within 1.5 inter-quartile ranges. The dots represent outliers. The length of the notch around the median value represents a 95% confidence interval for the median. If the notches of 2 median values do not overlap there is a statistically significant difference amongst the median values at the 95 % confidence level.

Results and discussion

Optimisation of the required filter surface area

Influent concentrations and organic loads. Table 1 shows the concentrations of the mechanically pre-treated wastewater, i.e. the influent concentrations to the beds at the experimental CW Ernsthofen, during the investigation period (April 2004 – March 2005). The measured influent concentrations are high for mechanically pre-treated wastewater and have been very constant throughout the investigation period.

Table 1 Influent concentrations

Parameter	Unit	SS	BOD_5	COD	NH_4-N	NO_3-N	TN	TP
No. of analysis	–	49	46	49	55	49	6	49
Median value	mg/L	130	380	545	68.0	0.40	83.9	12.5
Mean value	mg/L	133	375	554	67.6	0.42	84.8	12.5
95% confidence interval	mg/L	10	11	17	1.5	0.02	3.2	0.3
Standard deviation	mg/L	34	37	61	5.8	0.07	4.0	1.0
Maximum	mg/L	270	460	728	81.7	0.63	89.6	15.2
Minimum	mg/L	70	280	402	52.9	0.28	81.0	9.5

Effluent concentrations. Figure 1 and Figure 2 show the COD and BOD_5 influent and effluent concentrations during the investigation period, respectively. The effluent standards in Austria have been met for bed 1 and bed 2 with a specific surface area of 4 and 3 m^2 per person, respectively. At higher loads, i.e. bed 3 with 2 m^2 per person, the

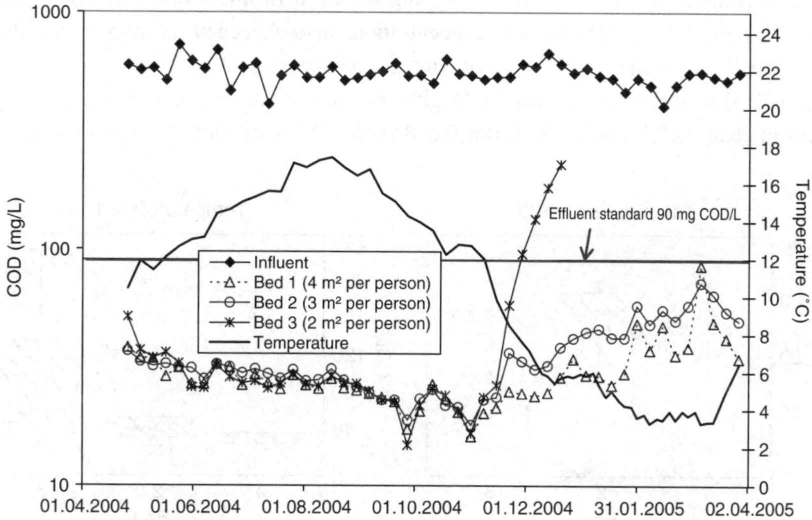

Figure 1 COD influent and effluent concentrations

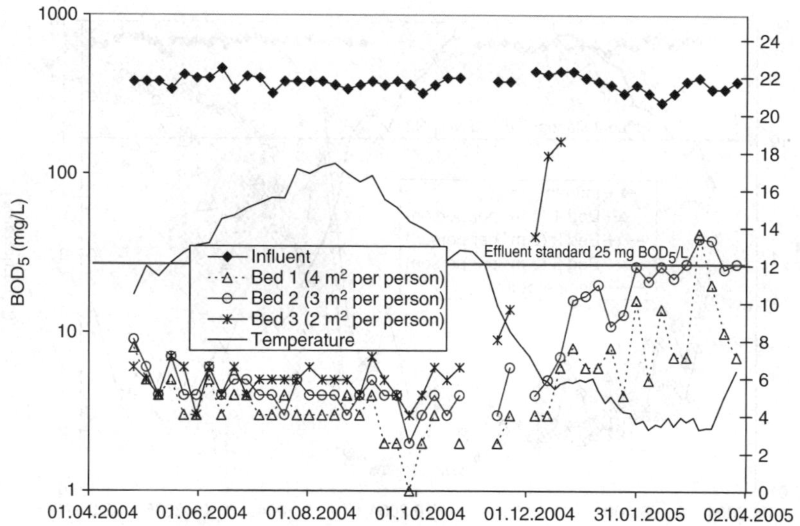

Figure 2 BOD_5 influent and effluent concentrations

Table 2 BOD_5 and COD effluent concentrations (mg/L)

Parameter	BOD_5						COD					
Temperature	>12 °C			<12 °C			>12 °C			<12 °C		
Bed	1	2	3*	1	2	3*	1	2	3*	1	2	3*
No. of analysis	26	27	27	19	19	8	29	29	29	20	20	9
Median value	3.0	4.0	5.0	7	20	135	27	28	27	32	45	182
Mean value	3.4	4.4	5.0	9	19	102	27	28	27	36	46	151
95% confidence interval	0.5	0.5	0.4	4	5	47	2	2	3	6	5	49
Standard deviation	1.4	1.4	1.1	9	11	68	5	5	7	14	12	75
Maximum	8.0	9.0	7.0	42	39	170	38	37	52	85	72	228
Minimum	1.0	2.0	3.0	2	3	9	16	18	15	21	24	27

*Bed 3 was only operated until mid of December 2004 when clogging occurred

effluent concentrations exceeded the effluent standard at water temperatures below 8 °C (end of November 2004) and finally clogging occurred in mid December 2004. During the winter period the BOD_5 effluent concentrations also exceeded the maximum allowed value for bed 2 (specific surface area of 3 m^2 per person).

Table 2 shows the BOD_5 and COD effluent concentrations for temperatures higher and lower than 12 °C, Figure 3 shows the Box-and-Whisker plots for the COD effluent

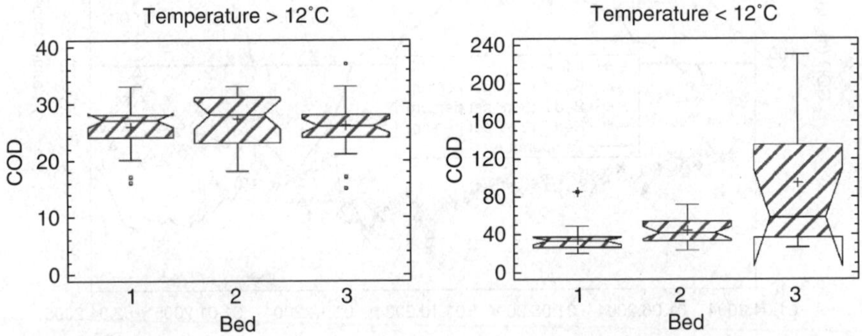

Figure 3 COD effluent concentration for >12 °C (left) and <12 °C (right)

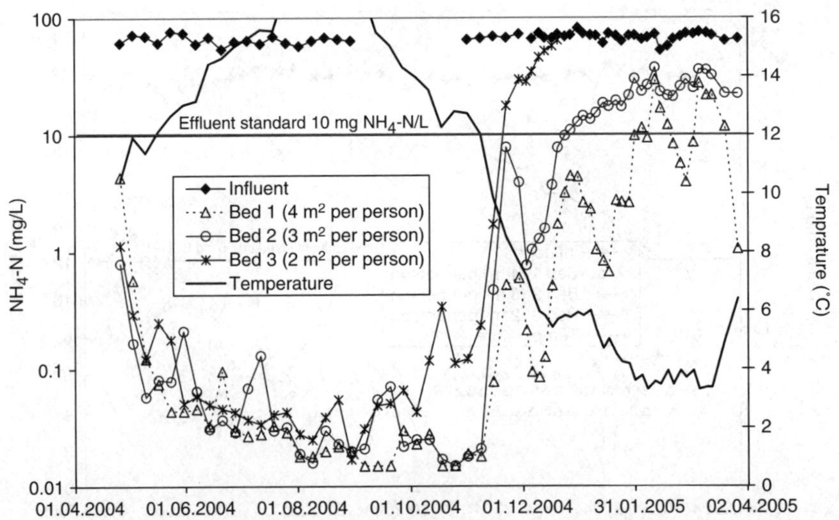

Figure 4 NH_4-N influent and effluent concentrations

Table 3 NH_4-N and NO_3-N effluent concentrations (mg/L)

Parameter	NH_4-N						NO_3-N					
Temperature	>12 °C			<12 °C			>2 °C			<12 °C		
Bed	1	2	3*	1	2	3*	1	2	3*	1	2	3*
No. of analysis	29	29	29	35	35	9	29	29	29	20	20	9
Median value	0.03	0.03	0.05	2.8	18.6	34.5	54.8	56.2	55.4	54.6	38.4	3.5
Mean value	0.20	0.08	0.13	6.6	17.5	36.9	53.8	55.5	55.8	49.4	39.0	11.1
95% confidence interval	0.29	0.05	0.08	2.7	3.6	13.1	2.8	2.9	3.3	5.5	6.0	9.7
Standard deviation	0.81	0.15	0.21	8.1	10.9	20.1	7.8	8.1	9.0	12.5	13.8	14.9
Maximum	4.37	0.80	1.14	29.5	37.4	64.4	69.4	69.0	76.2	63.8	59.4	47.2
Minimum	0.02	0.02	0.02	0.1	0.5	1.7	35.6	37.2	34.8	17.3	14.2	2.4

*Bed 3 was only operated until mid of December 2004 when clogging occurred

concentrations. During operation at temperatures higher than 12 °C no statistically significant difference occurred between the different organic loads. The median values of the BOD_5 and COD removal efficiencies at temperatures higher than 12 °C have been 99% and 96%, respectively. For water temperatures lower than 12 °C the median values of the removal efficiencies for bed 1, 2 and 3 have been 98%, 95% and 91% for BOD_5 and 94%, 91% and 80% for COD, respectively.

Figure 4 shows the NH_4-N influent and effluent concentrations during the investigation period. The effluent standards in Austria have been met for all beds, i.e. the effluent concentrations have been lower than 10 mg/l for effluent water temperatures higher than 12 °C. No statistically significant difference could be observed between the different organic loads for the NH_4-N and NO_3-N effluent concentrations (Table 3 and Figure 5) for water temperatures higher than 12 °C. For the highest loaded bed (bed 3, 40 g COD.m^{-2}.d^{-1}, 2 m^2 per person) the ammonia effluent concentration exceeded the effluent standard as soon as the temperatures dropped below 12 °C. For bed 2 (27 g COD.m^{-2}.d^{-1}, 3 m^2 per person) the ammonia effluent concentration exceeded the effluent standard at temperatures lower than 8 °C, for bed 1 (20 g COD.m^{-2}.d^{-1}, 4 m^2 per person) at temperatures lower than 6 °C.

Optimisation of nitrogen removal

Table 4 shows the influent concentrations, and the organic load applied and the resulting specific surface area for PSCWs with a main layer consisting of sand with a grain size of 0.06–4 mm and 1–4 mm, respectively. Although the concentrations varied the median values are close to the target loads. Table 5 shows the effluent concentrations and elimination rates for a standard single-stage system (single-stage SSVFCW, main layer with

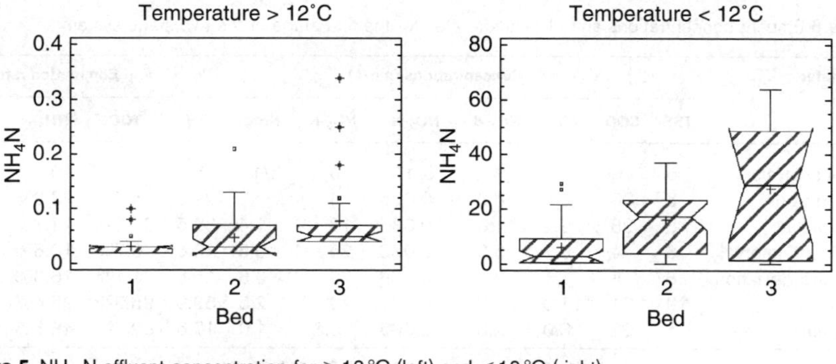

Figure 5 NH_4-N effluent concentration for >12 °C (left) and <12 °C (right)

Table 4 Influent concentrations, organic load and specific surface area

	Influent concentrations						Organic load		Specific surface area	
	mg/L						g $COD.m^{-2}.d^{-1}$		$m^2.PE^{-1}$	
	TSS	COD	TOC	NH_4-N	NO_3-N	TN	0.06–4 mm	1–4 mm	0.06–4 mm	1–4 mm
No. of analysis	7	15	7	7	7/7*	7	15	15	15	15
Median value	123	387	151	49.1	<0.1	56.3	19.3	77.1	3.85	0.96
Mean value	133	442	163	48.1	<0.1	58.5	18.9	75.4	3.77	0.94
95% conf. interval	33	77	22	2.7	–	4.0	3.1	12.3	0.62	0.15
Standard deviation	44	151	29	3.7	–	5.4	6.1	24.3	1.22	0.30
Maximum	223	814	204	53.8	<0.1	67.5	32.8	131.3	6.56	1.64
Minimum	95	253	134	41.9	<0.1	53.2	10.5	42.0	2.10	0.53

*Number of analysis below the limit of detection (0.1 mg NO_3-N/l)

Table 5 Effluent concentrations and elimination rates for the standard single-stage system

Parameter	Concentrations (mg/L)							Elimination rates		
	TSS	TOC	NH_4-N	NO_2-N	NO_3-N	Norg	TN	TOC	NH_4-N	TN
No. of analysis	11/4*	11	11/9*	11/5*	11	11/8*	11	11	11	11
Median value	2	4.3	0.030	0.003	52.3	1.0	52.8	96.8%	99.94%	2%
Mean value	2	4.7	0.032	0.005	51.7	1.1	52.5	96.9%	99.94%	8%
95% conf. interval	<1	0.4	0.002	0.002	5.4	0.1	5.3	0.3%	0.01%	8%
Standard deviation	<1	0.8	0.004	0.004	9.2	0.2	9.0	0.5%	0.02%	13%
Maximum	3	5.7	0.040	0.016	67.2	1.7	67.7	97.6%	99.98%	30%
Minimum	<1	3.8	<0.030	<0.003	38.9	<1.0	39.4	96.1%	99.92%	-9%

*Number of analysis below the limit of detection (1 mg TSS/l, 0.03 mg NH_4-N/l, 0.003 mg NO_2-N/l, and 1 mg Norg/l)

grain size of 0.06–4 mm, free drainage; organic load 20 mg $COD.m^{-2}.d^{-1}$), Table 6 for the effluent of the first stage of the two-stage system and Table 7 for the effluent of the second stage, i.e. the total effluent of the two-stage system.

Figure 6 compares the effluent concentrations for TOC and the influent and effluent concentrations for TN. Compared to a standard single-stage system the two-stage system showed a significantly higher mean nitrogen removal of about 58% (compared to about 2% for the standard single-stage system). Regarding the effluent concentrations of organic matter and ammonia nitrogen both systems showed similar results. However, due to the higher organic load the eliminated loads of the two-stage system are higher. In the effluent of the first stage the maximum COD concentrations has been 31 mg/l resulting in a maximum organic load of the second stage of $< 8\,g\,COD.m^{-2}.d^{-1}$. This organic load

Table 6 Effluent concentrations and elimination rates for the first stage of the two-stage system

Parameter	Concentrations (mg/L)								Elimination rates		
	TSS	COD	TOC	NH_4-N	NO_2-N	NO_3-N	Norg	TN	TOC	NH_4-N	TN
No. of analysis	5	5	9	9	5/1*	5	5/1*	5	9	9	5
Median value	5	30	9.4	8.0	0.026	4	1.2	29.0	94.5%	82.9%	57%
Mean value	8	28	9.2	11.0	0.035	18	1.4	32.5	94.3%	77.2%	46%
95% conf. interval	6	4	0.7	5.3	0.033	20	0.5	13.6	0.7%	10.6%	22%
Standard deviation	6	4	1.1	8.1	0.038	23	0.6	15.5	1.1%	16.3%	25%
Maximum	19	31	11.0	25.3	0.100	49	2.5	52.9	95.9%	95.4%	67%
Minimum	4	22	8.0	2.5	<0.003	0.3	<1.0	17.5	92.5%	48.6%	15%

*Number of analysis below the limit of detection (0.003 mg NO_2-N/l, and 1 mg Norg/l)

Table 7 Effluent concentrations and elimination rates for the two-stage system

Parameter	Concentrations (mg/L)							Elimination rates		
	TSS	TOC	NH_4-N	NO_2-N	NO_3-N	Norg	TN	TOC	NH_4-N	TN
No. of analysis	6/1*	6	6/5*	6/4*	6	6/4*	6	6	6	6
Median value	2	4.8	0.03	0.003	26.1	1.0	26.9	96.7%	99.94%	58%
Mean value	2	5.5	0.10	0.019	30.5	1.0	31.9	96.4%	99.78%	46%
95% conf. interval	<1	1.1	0.13	0.032	11.9	0.1	12.5	1.1%	0.31%	21%
Standard deviation	<1	1.4	0.16	0.040	14.9	0.1	15.6	1.4%	0.38%	27%
Maximum	3	8.1	0.42	0.100	55.5	1.2	56.0	97.7%	99.94%	69%
Minimum	<1	4.6	<0.03	<0.003	15.9	<1.0	16.4	94.0%	99.00%	10%

*Number of analysis below the limit of detection (1 mg TSS/l, 0.03 mg NH_4-N/l, 0.003 mg NO_2-N/l, and 1 mg Norg/l)

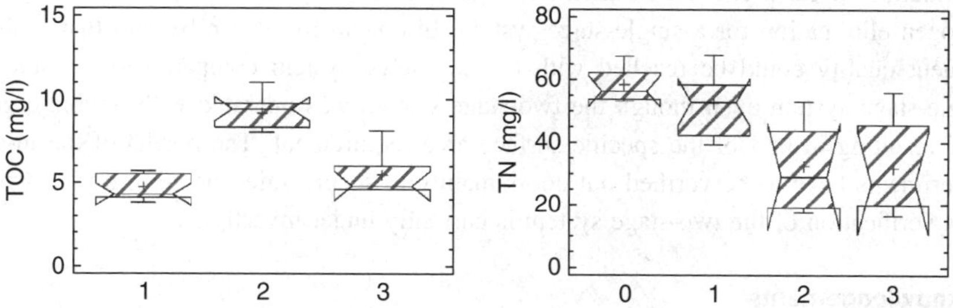

Figure 6 TOC effluent and TN influent and effluent concentrations (0…influent; 1…effluent of the standard system; 2… effluent of the 1st stage of the 2-stage system; 3… effluent of the 2nd stage of the 2-stage system)

is far below the maximum organic load of 20 g COD.m^{-2}.d^{-1} that was suggested by Winter and Goetz (2003) for operation of vertical flow beds without clogging.

Summary and conclusions

The results of the experiments regarding the optimisation of the required filter surface area at the experimental CW Ernsthofen can be summarised as follows.

- For water temperatures higher than 12 °C the effluent concentrations were far below the maximum allowed values for all three beds. No statistically significant differences between the beds with different organic loadings could be observed. Beds with higher organic loading respond more sensitively to fluctuations of the influent concentrations.
- The bed with an organic load of 20 g COD.m^{-2}.d^{-1} (4 m^2 per person) could meet the required effluent standards and removal efficiencies the whole year around. Ammonia nitrogen concentrations still could be met when effluent water temperatures were around 6 °C (only required above 12 °C).
- For the bed with an organic load of 27 g COD.m^{-2}.d^{-1} (3 m^2 per person) the effluent concentrations for BOD_5 were higher than the maximum allowed concentration for a short period in winter. Ammonia nitrogen concentrations exceeded the maximum concentration allowed at water temperatures lower than 10 °C.
- The bed with an organic load of 40 g COD.m^{-2}.d^{-1} (2 m^2 per person) failed during operation in winter. Despite the high organic load the filter worked well as long as the effluent water temperatures were higher than 12 °C. High organic loads are therefore applicable for systems that are only operated during the warm season when effluent water temperatures are above 12 °C.

It can be concluded that a subsurface vertical flow constructed wetland with intermittent loading operated with an organic load of 20 g COD.m^{-2}.d^{-1} (i.e. 4 m^2 per person) can fulfil the requirements of the Austrian standard regarding effluent concentrations and elimination efficiencies. Therefore a specific area demand of 4 m^2 per person is suitable to be included in the revision of the Austrian design standard. Additionally it was shown that during the warmer months (May – October) when the temperature of the effluent is higher than 12 °C the specific surface area might be further reduced, even 2 m^2 per person have been proven to be adequate.

A two-stage system (first stage: grain size for main layer 1–4 mm, saturated drainage layer; and second stage: grain size for main layer 0.06–4 mm, free drainage) was operated with an organic load of 80 g COD $m^{-2}.d^{-1}$ for the first stage (1 m^2 per person), i.e. 40 g COD $m^{-2}.d^{-1}$ for the whole system (2 m^2 per person). Organic matter removal and ammonia nitrogen reduction are the same as for a single-stage system (grain size for main layer 0.06–4 mm, free drainage, organic load 20 g COD $m^{-2}.d^{-1}$). A nitrogen elimination of 58% could be reached for the two-stage system compared to almost no nitrogen elimination for a single-stage system. In conclusion it can be said that a higher effluent quality could be reached with the two-stage system compared to the standard single-stage system even though the two-stage system was operated with double organic load resulting in half of the specific surface area requirement. The results of the in-door experiments have to be verified out-doors mainly to secure safe operation during winter. This verification of the two-stage system is currently under investigation.

Acknowledgements

The experiments were carried out within the project "*Optimization of subsurface vertical flow constructed wetlands (Bepflanzte Bodenfilter)*" funded by the Austrian Ministry for Agriculture, Forestry, Environment and Water Management. The authors are grateful for support and especially thank the mayor of the Ernsthofen municipality, Mr. Karl Huber, for making the construction of the experimental plant at the WWTP Ernsthofen possible as well as the staff of the WWTP Ernsthofen, Mr Franz Eglseer and Mr Karl Hiebl, for operating the experimental plant.

References

1.AEVkA (1996). *1. Abwasseremissionsverordnung für kommunales Abwasser (Austrian regulation for emissions from domestic wastewater)*. BGBl.210/1996, Vienna, Austria [*in German*].

Kadlec, R.H., Knight, R.L., Vymazal, J., Brix, H., Cooper, P. and Haberl, R. (2000). Constructed wetlands for pollution control – processes, performance, design and operation. In *IWA Scientific and Technical Report No.8*, IWA Publishing, London, UK.

Langergraber, G. and Haberl, R. (2001). Constructed wetlands for water treatment. *Minerva Biotecnologica*, **13**(2), 123–134.

ÖNORM B 2505 (1997). *Bepflanzte Bodenfilter (Pflanzenkläranlagen) – Anwendung, Bemessung, Bau und Betrieb* (Subsurface-flow constructed wetlands – Application, dimensioning, installation and operation). Österreichisches Normungsinstitut, Vienna, Austria [*in German*].

Winter, K.-J. and Goetz, D. (2003). The impact of sewage composition on the soil clogging phenomena of vertical flow constructed wetlands. *Water Science & Technology*, **48**(5), 9–14.

Anaerobic membrane bioreactor for treatment of synthetic municipal wastewater at ambient temperature

J.H. Ho, S.K. Khanal and S. Sung

Department of Civil, Construction, and Environmental Engineering, Iowa State University, 394 Town Engineering Building, Ames, IA 50011, USA (E-mail: *sung@iastate.edu*)

Abstract Non-woven fabric filter and poly-tetrafluoroethylene (PTFE) composite membrane were investigated to determine their applicability to treat low strength wastewater in an anaerobic membrane bioreactor (AMBR). Sludge cake resistance of the membrane was quantified using pure water flux of anaerobic sludge cake accumulated on the glass fiber filter of similar pore size. It is hypothesized that the formation of thin cake layer on the porous medium, e.g. non-woven and PTFE acts as a dynamic membrane. Thus, the capture of thin sludge cake inside the non-woven fabric matrix and accumulation on the PTFE membrane surface forms a membrane system equivalent to a commercial membrane system. The permeate quality was found to improve as the cake became more dense with filtration time. The PTFE composite membrane coated with thin PTFE film on the non-woven fabric filter enhanced the filtration performance by improving flux and minimizing the propensity of bio-fouling. The membrane flux was restored by back-flushing with permeate. The AMBR coupled with PTFE laminated membrane was operated continuously during the experiment at a cross flow velocity (CFV) of 0.1–0.2 m/sec and a transmembrane pressure (TMP) of 0.5–3 psi. Although about a month of acclimation was required to reach steady state, the effluent chemical oxygen demand (COD), volatile fatty acids (VFAs) as acetic acid, and suspended solids (SS) concentrations were below 30, 20 and 10 mg/L, respectively, during 90 days of operation with intermittent back washing. The lower operation TMP and CFV were subjected to less shear stress on the microbial community during continuous AMBR operation. In addition, thin sludge film accumulated on the membrane surface also acted as a biofilm bioreactor to remove additional COD in this study

Keywords Ambient temperature; anaerobic membrane bioreactor (AMBR); municipal wastewater; non-woven filter; PTFE

Introduction

The anaerobic process has been widely employed for municipal and industrial sludge digestion because of several inherent merits, including high loading rate, generation of valuable methane gas, less sludge production and less energy consumption. Application of the aerobic process, as distinct from the anaerobic system, is primarily confined to wastewater with low organic loading rate due to the oxygen transfer limitation and high bacterial sludge production. Therefore, the low strength wastewater such as municipal sewage has been historically treated by an aerobic system. However, anaerobic treatment could be a good alternative for small wastewater treatment systems due to its simplicity and long-term sustainability. It requires no aeration and produces relatively less sludge, which could result in significant reduction in capital and operation costs. In addition, it is a net energy producer. However, the anaerobic process has been has been primarily applied to treat high strength wastewater due to its poor effluent quality and solid–liquid separation. As a result, many high-rate anaerobic processes have been developed to expand their application to medium and low strength wastewater treatment. Some of the high rate anaerobic processes include the anaerobic contact process, anaerobic filter, anaerobic fluidized bed, upflow anaerobic sludge blanket (UASB), and expanded granular

doi: 10.2166/wst.2007.130

Water Science & Technology **Vol 55 No 7 pp 79–86** © IWA Publishing 2007

sludge bed (EGSB) (Rittmann and McCarty, 2001). These processes essentially maintain long solids retention time (SRT) irrespective of hydraulic retention time (HRT) through sludge granulation, biofilm formation or internal/external solid–liquid separation. UASB and EGSB have been given much attention for sewage treatment. Lettinga (1995) reported that EGSB could be a sustainable alternative for a high rate anaerobic process under psychrophilic conditions due to its very high substrate affinity. Many full scale UASB/EGSB plants are currently in operation in tropical countries like India, Colombia and Brazil. However, the full-scale did not show good organic removal efficiency as compared to lab scale results (Seghezzo *et al.*, 1998). Thus, often, anaerobic treatment is followed by a polishing step such as a trickling filter or stabilization pond.

Faced with these shortcomings, the membrane coupled anaerobic process has started to attract much attention for the treatment of low to medium strength wastewater (Baek and Pagilla, 2006; Hu and Stuckey, 2006). However, the high cost of the membrane is aggravating the wide spread application of membrane processes in wastewater treatment. Several researchers have tried to develop a cost effective membrane using low cost material such as non-woven filters (Pillay *et al.*, 1994; Muhammad *et al.*, 1997; Nomura *et al.*, 1997; Chang *et al.*, 2001; Seo *et al.*, 2003). The goals of this research are to evaluate the AMBR performance to treat low strength wastewater at ambient temperature using poly-tetrafluoroethylene (PTFE) laminated non-woven filter and to develop a suitable operation strategy to control membrane fouling.

Methods

Anaerobic membrane bioreactors

Two completely stirred anaerobic reactors coupled with external membrane separation units were used. The schematic diagram of the AMBR system is shown in Figure 1. Non-woven filter made of polypropylene with a pore size of 12 μm and PTFE membrane laminated on the non-woven filter with pore size of 10 μm (KNH Enterprise Co., Ltd., Taiwan) were used. The dimensions of the membrane were 600 mm (length) and 8 mm (inner diameter) and the surface area of each membrane was 0.015 m^2. Anaerobic reactors were equipped with pH and oxidation–reduction potential (ORP) control units, and maintained at 25 °C using a heating and cooling loop. Flux and fouling tests were conducted

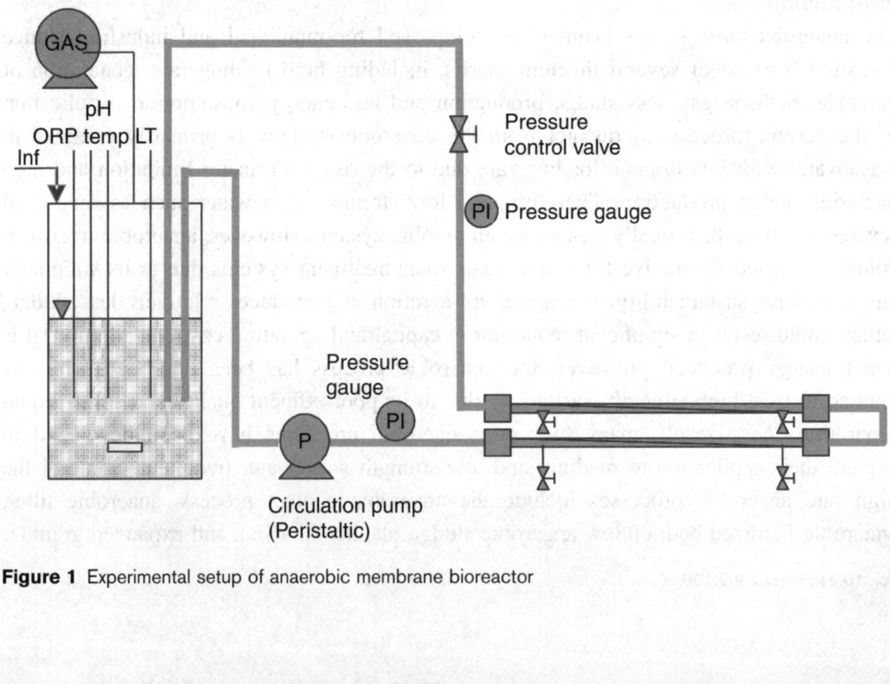

Figure 1 Experimental setup of anaerobic membrane bioreactor

using the flux test module. The seed sludge for AMBR start-up was obtained from a local municipal anaerobic digester located at Ames, IA, USA.

Synthetic municipal wastewater

Synthetic wastewater was prepared to represent a typical municipal sewage. The characteristics of the wastewater were modified from Syntho, which was developed to represent a settled domestic wastewater (Nopens *et al.*, 2001). Table 1 shows the characteristics of influent used in this research. The major component for chemical oxygen demand (COD) was non-fat dry milk (NFDM). The properties of non fat dry milk and required trace elements were provided by Dague *et al.* (1998).

Analysis

COD, volatile fatty acids (VFA), suspended solids (SS), and mixed liquor suspended solids (MLSS) were measured as per *Standard Methods* (APHA, 1998). Soluble bioreactor COD was measured after filtration using a 0.22 μm filter. The microscopic observation of the surface of the PTFE laminated non-woven filter was carried out using a scanning electron microscope (SEM). Total resistance was calculated using the following equation:

$$J = \frac{\Delta P}{\mu \cdot R_t}$$

where J is the flux ($L \cdot T^{-1}$, m/sec); ΔP is the trans-membrane pressure ($M \cdot L^{-1} \cdot T^{-2}$, N/m^2); μ is the dynamic viscosity ($M \cdot L^{-1} \cdot T^{-1}$, $N \cdot sec/m^2$); R_t is total membrane resistance (L^{-1}, m^{-1}). A flux test of the anaerobic sludge cake was conducted to evaluate the cake resistance. GF/C filters with pore size of 1.2 μm were used as a medium to attach a given amount of anaerobic sludge. Fine wire mesh with 50 μm opening was used to protect the flexible filter from breaking. A glass syringe was connected to the filter medium to apply positive pressure. Flux tests were carried-out until the three consecutive flux data varied less than 5%. The membrane modules were dissembled to collect the biomass accumulated on the membrane surface using mechanical brushing on 2, 3, 4 and 8 weeks of operation.

Results and discussion

Sludge cake accumulation and resistance

Anaerobic sludge cake resistance at different TMPs was measured to quantify dynamic membrane resistance. Controlling the dynamic membrane or secondary membrane thickness is a key issue in this research due to relatively large membrane pore size. Figure 2 (a) shows the attached sludge mass on the membrane surface at different operation times. The cake became more dense with filtration time which improved the permeate quality.

Table 1 Characteristics of synthetic municipal wastewater

	COD = 500 mg/L				COD = 1000 mg/L			
	COD mg/L	N mg/L	P mg/L	$CaCO_3$ mg/L	COD mg/L	N mg/L	P mg/L	$CaCO_3$ mg/L
Milk	250				540			
Starch	130				260			
Acetate	80				160			
Yeast	40				40			
NH_4Cl		20				20		
Urea		20				20		
KH_2PO_4			10				10	
$KHCO_3$				300				600
Total	500	40	10	300	1000	40	10	600

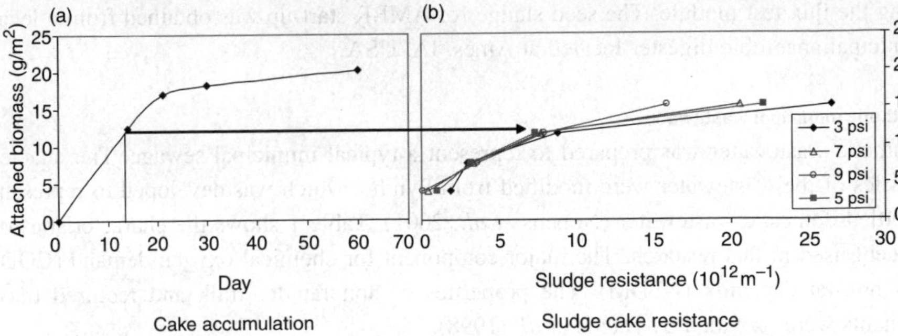

Figure 2 Anaerobic sludge cake growth and pure water flux on sludge cake

On the other hand, the cake thickness or accumulated sludge layer increased with filtration time resulting decline in flux decline increase in resistance. According to Murkes and Carson (1998), the secondary membrane in this research can be classified as Class 2 which occurs when the pore size of the membrane is much larger than the particle size. The short initial pore filling was followed by thin cake deposition. Sludge cake grew rapidly at the beginning stage and then slowed down after about 20 days of operation. Thus, it can be hypothesized that the cake may not grow further after reaching steady state which is governed by the membrane operation conditions. The amount of sludge cake accumulated on the membrane surface increased from 12.5 g/m^2 on day 14 to 20 g/m^2 on 60 days. Choo and Lee (1996) also found a significant amount of biomass accumulation within the membrane system.

Figure 2 (b) shows anaerobic sludge resistance at different sludge densities and TMPs. The sludge cake resistance increased with an increase in attached cake density at a given TMP. One interesting fact is that there was no significant difference in resistance at $7 \sim 12$ g/m^2 of sludge cake density. The sludge cake was broken and detached at low sludge cake density due to the insufficient cake layer development. The cake resistance rapidly increased after 12 g/m^2 of cake density or 10×10^{12} m^{-1} of sludge resistance. Thus, it is essential to control the sludge cake density to less than 12 g/m^2, which occurred after about 12 days of continuous operation. In addition, lower CFV and TMP of 0.1 m/s and 0.5 psi, respectively required in the beginning especially with new membrane in order to develop sufficient thickness of secondary membrane. Thus, the class 2 type of secondary membrane cake captured inside of the non-woven fabric matrix and accumulated on the PTFE membrane surface could substitute the membrane and cake in the commercial membrane system with back washing every fortnight.

Figure 3 (a) shows the flux profile during a 20-day flux test with TMP variation. The flux of PTFE composite membrane varied from 4 to 12 L/m^2/hr at TMP of $6.9 \sim 20$ kPa

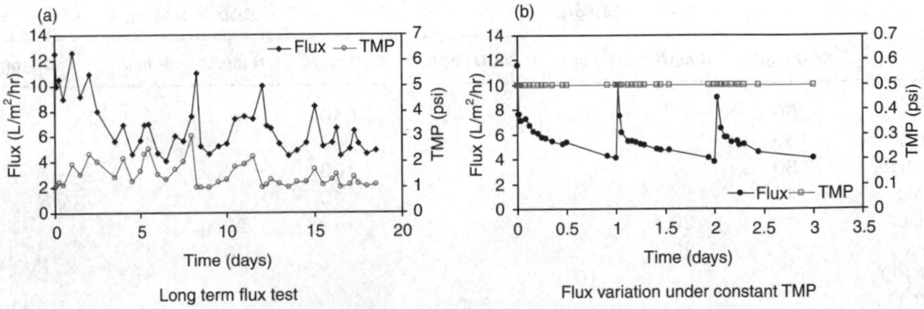

Figure 3 Flux profile of PTFE laminated non-woven filter

(1 ~ 3 psi) and CFV of 0.2 m/sec. Figure 3 (b) shows the flux decline under a constant TMP at the beginning of filtration. It was however possible to restored the flux to the original level with a daily back washing.

AMBR performance

A completely stirred anaerobic reactor coupled with PTFE composite membrane was operated for nearly 90 days. During the first 50 days of operation, influent COD was kept at 500 mg/L; whereas after day 50, it was increased to 1000 mg/L. It was essential to maintain a relatively low TMP and CFV due to the large pore size of the membrane. The operating TMP and CFV were 1 ~ 3 psi and 0.1 ~ 0.2 m/sec, respectively. The target flux was 5 L/m^2/hr. The initial flux of the PTFE composite membrane was 12 L/m^2/hr and it gradually decreased to 5 L/m^2/hr.

Once the flux declined below 5 L/m^2/hr, back flushing using permeate was carried-out to restore the flux. According to the sludge cake accumulation and resistance tests, back washing was done every 4 to 10 days depending on the flux. The flux was immediately restored right after the back flushing. However, after one day of operation, the flux again declined to about 5 L/m^2/hr, and remained nearly constant. It was possible to achieve a constant flux of 5 L/m^2/hr at 1.0 psi of TMP for more than 50 days of AMBR operation. After day 50, a higher TMP was needed to maintain the same flux. One of the reasons for the raised TMP was increased influent COD concentration.

Figure 4 (a) shows effluent COD, VFA, and SS levels at HRT of 18 hr and temperature of 25 ± 0.5 °C. The influent COD concentration was then increased to 1000 mg/L after day 50. A long acclimation time of more than 30 days was required to reach a steady state. Furthermore, the effluent quality deteriorated temporarily at the higher influent COD level of 1000 mg/L; but the acclimation time was shorter than the start-up. The effluent (permeate) COD, VFA, and SS concentrations were below 30, 20 and 10 mg/L, respectively. During 90 days of operation, the reactor mixed liquor and effluent pHs were ranged from 6.8–7.0 and 7.6–7.8, respectively. The ORP varied from −220 to −270 mV. Figure 4 (b) shows biogas composition. The average methane and carbon dioxide contents were about 60% and 5% (by volume), and the remaining 35% was nitrogen. After the influent COD was increased to 1000 mg/L, methane and carbon dioxide contents were increased to about 80% and 10%, respectively. The permeate side was exposed to atmosphere, which allowed air to come in contact with the retentate stream. This might have caused a slightly higher ORP in the reactor. The initial MLSS and MLVSS concentrations were 9600 and 5900 mg/L, respectively [Figure 4(c)]. Sludge was not withdrawn from the reactor which infers that theoretical SRT was infinity. After day 50, MLSS and MLVSS concentrations reached 12,500 and 8200 mg/L, respectively. The increase in MLSS concentration was mostly contributed by biomass synthesis; since the synthetic wastewater used in this research was completely soluble. Thus, the increase in influent COD led to further increase in MLSS/MLVSS concentration in the bioreactor. Based on MLVSS and COD data, the biomass yield was about 0.05 $gVSS/gCOD_{removed}$.

Figure 5 shows the SEM pictures of the surface of PTFE laminated non-woven filter before and after cake deposition. As apparent from the micrograph, the cake consisted of many different types of microorganisms such as rod, coccus, and spiral. It is hypothesized that the high MLSS concentration could compensate the low growth rate at low temperature. The biokinetic parameters of anaerobic sludge exposed to low strength wastewater would be different than that of high strength wastewater. Two distinct acetoclastic methanogens namely, *Methanosaeta* and *Methaosarcina*, have different substrate affinity and specific substrate utilization rates. It is expected that *Methanosaeta,* which has a higher substrate affinity and lower specific utilization rate than the *Methanosarcina*, would

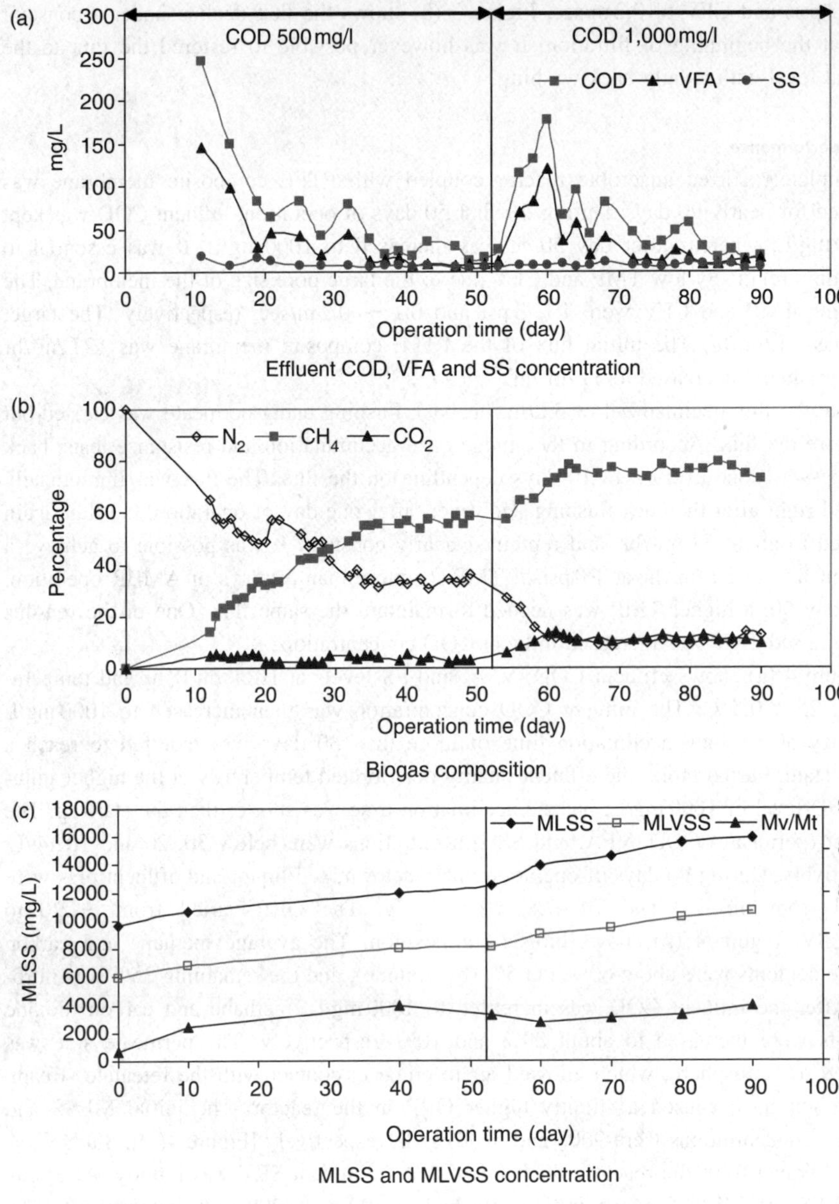

Figure 4 AMBR performance

become dominant over the latter under low substrate conditions. Thus, the mixed anaerobic sludge exposed to the low temperature and low strength wastewater as in this study, would have higher substrate affinity (and thus lower Ks value) and lower specific substrate utilization rate.

Challenge and opportunity of AMBR system

Anaerobic technology is getting attention as a sustainable technology due to its several inherent merits over the aerobic one. However, there are many uncertainties associated with membrane coupled anaerobic systems. For successful application, the AMBR system requires maintenance of optimal microbial activity in the bioreactor with minimal membrane fouling. AMBR has not been considered as a suitable process to treat low

Figure 5 SEM pictures on the surface of PTFE laminated non-woven filter before and after cake deposition

strength wastewater due to its high cost. Current aerobic membrane bioreactor (MBR) breakthroughs have also been somehow limiting the pace of AMBR development. However, if AMBR can take over the role of MBR, it would be a significant contribution for sustainable water quality.

There are two issues with membrane application in wastewater treatment and they are the cost of membrane and the cost of operation. The operation cost depends on the membrane performance and wastewater types. Stable flux and sustaining the time of membrane operation is more important than the initial flux with respect to the frequency of membrane cleaning and replacement. From this point of view, non-woven filter could be considered as an alternative for microfiltration for the anaerobic system. Non-woven fabric filter with PTFE surface modification can reduce the cake deposition effectively, which renders it suitable for operation at low pressure. It is important to point out that such a membrane system should be operated under low TMP and CFV in order to maintain a secondary membrane for particle free permeate. In most cases, anaerobic processes would not be applied as a sole treatment system due to the nutrient issues. However, it can be employed to remove organics and possibly inorganic nitrogen. Therefore, AMBR followed by biological aerated filter (BAF) for nitrification might be proposed for a small wastewater treatment system. Also, endocrine disrupting chemicals (EDCs) are becoming an issue for municipal sewage. There exists very little information on their fate and biodegradation in the AMBR system. Due to long SRT and secondary membrane, the AMBR system could enhance their removal more effectively. For example, an aerobic MF MBR with 10,000 mg/L of MLSS and 83.3 day SRT achieved an effluent estrogens of 2.2 ng/L as 17β-Estradiol (E2) and 10.3 ng/L as estrone (E1). On the other hand, the conventional activated sludge process with 1000 mg/L of MLSS and 4.8 day SRT could only achieve an effluent estrogen of 7.9 ng/L as E2 and 26.3 ng/L as E1 (Kikuta and Urase, 2003).

Conclusion

A completely stirred anaerobic reactor coupled with PTFE laminated non-woven filter was operated successfully to treat synthetic municipal wastewater at an HRT of 18 hrs without sludge withdrawal. A thin layer of anaerobic sludge cake accumulated on the porous medium, which acted as a dynamic membrane. This secondary membrane improved the permeate quality, although the membrane pore size was larger than the sludge particle size. Total amount of sludge attached on the membrane surface increased with time. Sludge cake grew rapidly at the early stage of filtration and then slowed down after about 20 days of operation when the sludge cake density reached 20 g/m^2. High biological activity on the membrane surface caused high biofouling potential. A lower CFV of 0.1 m/s and TMP of 0.5 psi were required in the beginning, especially with new membrane research in order to develop a secondary membrane of sufficient thickness. The effluent COD, VFA, and SS concentrations were below 30, 20 and 10 mg/L, respectively, during 90 days of operation with intermittent backwashing. The performance of an anaerobic CSTR coupled with PTFE composite membrane was comparable with granular sludge processes coupled with membrane module if operated with long SRT and would be ideal for a small wastewater treatment system.

Acknowledgements

The authors would like to thank KNH Co., Ltd. (Tainan, Taiwan) for providing non-woven filter and PTFE composite membrane. This study was supported in part by the Institute for Physical Research and Technology (IPRT) Seed Funding (Iowa State University) and in part by the National Science Foundation.

References

American Public Health Association/American Water Works Association/Water Environmental Federation (1998). *Standard Methods for the Examination of Water and Wastewater*, 20th edn, Washington DC, USA.

Baek, S.H. and Pagilla, K.R. (2006). Aerobic and anaerobic membrane bioreactors for municipal wastewater treatment. *Wat. Environ. Res.*, **78**, 133–140.

Chang, I.-S., Gander, M., Jefferson, B. and Judd, S.J. (2001). Low-cost membranes for use in asubmerged MBR. *Trans IChemE*, **79**, 183–188.

Choo, K.H. and Lee, C.H. (1996). Membrane fouling mechanisms in the membrane-coupled anaerobic bioreactor. *Wat. Res.*, **30**, 1771–1780.

Dague, R.R., Banik, G.C. and Ellis, T.G. (1998). Anaerobic sequencing batch reactor treatment of dilute wastewater at psychrophilic temperatures. *Wat. Environ. Res.*, **70**, 155–160.

Hu, A.H. and Stuckey, D.C. (2006). Treatment of dilute wastewaters using a novel submerged anaerobic membrane bioreactor. *J. Environ. Eng.*, **131**, 410–416.

Kikuta, T. and Urase, T. (2003): Removal of endocrine disruptors in membrane separation activated sludge, *Proc. of 5th International Membrane Science And Engineering Conference (IMSTEC-'03)*, Paper No. 27, Sydney, Australia.

Lettinga, G. (1995). Anaerobic digestion and wastewater treatment systems. *Antonie van Leeuwenhoek*, **67**, 3–28.

Muhammad, H., Al-Malack, and Anderson, G.K. (1997). Use of crossflow microfiltration in wastewater treatment. *Wat. Res.*, **31**, 3064–3071.

Murkes, J. and Carsson, C.G. (1998). *Crossflow Filtration*, John Wiley & Sons, UK.

Nomura, T., Fujii, T. and Suzuki, M. (1997). Application of the ceramic membrane with hydrophobic skin layer to separation of activated sludge. *Wat. Sci. Tech.*, **35**(8), 137–144.

Nopens, I., Capalozza, C. and Vanrolleghem, P.A. (2001). *Technical Report: Stability Analysis of a Systematic Municipal Wastewater*, Dept. Applied Mathematics, Biometrics and Process Control, Universiteit Gent, Gent, Belgium.

Pillay, V.L., Townsend, B. and Buckley, C.A. (1994). Improving the performance of anaerobic digesters at wastewater treatment works: The coupled cross-flow microfiltration/digester process. *Wat. Sci. Tech.*, **30**(12), 329–337.

Rittmann, B.E. and McCarty, P.L. (2001). *Environmental Biotechnology: Principles and Applications*, McGraw-Hill Book Co., Inc., New York, USA.

Seghezzo, L., Zeeman, G., van Lier, J.B., Hamelers, H.V.M. and Lettinga, G. (1998). A review: The anaerobic treatment of sewage in UASB and EGSB reactors. *Bioresource Tech.*, **65**, 175–190.

Seo, G.T., Moon, B.H., Lee, T.S., Lim, T.J. and Kim, I.S. (2003). Non-woven fabric filter separation activated sludge reactor for domestic wastewater reclamation. *Wat. Sci. Tech.*, **47**(1), 133–138.

Water Science & Technology **Vol 55 No 7 pp 87–93** © IWA Publishing 2007

Using molecular tools to assess biofilms in onsite systems

T.G. Flapper*, A. McKay, S. Hall** and P. Hill*****

*Ecowise Environmental, PO Box 1834, Fyshwick, ACT, Australia, 2609 (E-mail: *tflapper@ecowise.com.au*)

**University of Queensland, Advanced Wastewater Management Centre, Brisbane, QLD, Australia, 4072 (E-mail: *s.hall1@uq.edu.au*)

***AquaReuse, Unit 2, 7 Widemere Rd, Wetherill Park, NSW, Australia, 2164

Abstract Onsite sewage treatment is growing in diversity, and in regulatory control across Australia. This is occurring for both blackwater and greywater treatment, as the drought impact deepens and more of the community are exposed to options for 'managing' their own water. Regulators in each State are drafting and implementing Guidelines to cover a range of on-site system scenarios. In addition, more and more decentralised options are being tendered for sewage management in the commercial world. In this project we aim to use novel molecular tools, in combination with traditional physical/chemical/biological methods, to understand onsite treatment in a more detailed manner. The system tested is a new peat based biofilter which can be used for greywater or blackwater application, and can be retrofitted to current sewage systems. This project has been based on the AquaReuse greywater system for demonstration purposes, showing the strength of the information gained from the use of novel tools. The two systems investigated are installed at a caravan park in New South Wales (NSW) and a domestic residence in Tamborine, Queensland (QLD). A 20-week intensive sampling and analysis program was followed. The project monitored standard analytes such as biochemical oxygen demand (BOD_5), suspended solids (SS) and thermotolerant coliforms (TC). Additionally, we studied the biological community using fluorescent *in situ* hybridisation (FISH) on a monthly basis and full-cycle ribosomal RNA analysis (rRNA) for one sample to assess the biological community inhabitants. rRNA analysis at the NSW facility demonstrated a highly diverse biological community, in keeping with its long established operating period. In contrast, FISH analysis at the QLD installation showed a less diverse and younger community. rRNA and FISH identified organisms that are mostly associated with nutrient removing functions.

Keywords Biological; FISH; greywater; molecular; monitoring; rRNA

Introduction

Molecular based analytical procedures have to date mostly been used for pure culture, laboratory based, research studies. Rarely have they been applied to real environmental samples. In our study, we promote their use for industry study, demonstrating the valuable information obtained to understand, monitor and improve process performance.

The AquaReuse process includes peat based biological and synthetic membrane filters as primary and secondary series filters, a primary sump, intermittent regular feed to secondary filter, UV disinfection, and effluent storage (Flapper and Hill, 2005). This manufacturer provided access to test sites for monitoring, and thereby represents our demonstration process type. The system is treating domestic greywater, including kitchen sink, laundry, shower, bath and sink sources.

The test sites

AquaReuse has installed a number of test sites for accreditation and research and development purposes across Australia. These systems are located at a residence in Katoomba NSW, a caravan park in NSW, and a residential property in Tamborine, Queensland (Flapper and Hill, 2005; Flapper *et al.*, 2005).

doi: 10.2166/wst.2007.131

The caravan park test facility in NSW treats up to 900 Ld^{-1} of domestic character greywater including kitchen sink, laundry, bathwater, showers and sinks (Figure 1). The site was commissioned in October 2004 and has been routinely tested since that time. The biological community includes worms, bacteria, fungi, and macroinvertebrates (Figure 2).

The Tamborine facility is located at a domestic dwelling that services two adults, one child and one toddler. It treats around 300 to 500 Ld^{-1} and has been operating since March 2005. Figure 3 illustrates the installation, where the biological layers of the process units are clearly labelled as 1 to 4. The facility is enclosed in an insulated shed, and treated greywater is reused for irrigation purposes.

Sampling

Sampling of water was conducted by onsite operators at the NSW and QLD facilities, both of whom had been trained and audited by Ecowise staff. Each sample was placed into an appropriately preserved and labelled jar (500 mL plastic, thiosulphate). Samples were couriered on ice for delivery to the Ecowise laboratory for analysis within 24 hours.

Sampling of biofilm for molecular analysis was conducted by University of Queensland (UQ) resources, and at times by the onsite staff trained by UQ people. Figure 4 demonstrates this sampling method.

Analysis

Standard analysis

Thermotolerant coliforms (TC) were analysed using membrane filtration, APHA9222 D. *Escherichia coli* was analysed following TC, using the indole test. Results are reported as colony forming units per 100 mL sample (cfu/100 mL).

Figure 1 Caravan park test site installation (NSW)

Figure 2 Biological community within peat bed filter at test site installations

rRNA Phylogenetic analysis

rRNA analysis is used to access all genomic material available in a sample matrix. The steps below were followed.

1. Utilising FastDNA SPIN Kit for Soil, DNA extraction was performed on sample 31/3/05, Cell Two, Jervis Bay to determine if genomic material was present. Two different quantities of sample were used, 250 mg and 500 mg to ensure sufficient DNA volume was obtained. A favourable result was gained with a large quantity of DNA existing within both peat samples.
2. Polymerase chain reaction (PCR) amplification of the above DNA 250 mg sample was then performed using universal bacterial primers: 27f and 1492r in order to amplify 16S genes. Thermal profile for amplification included 30 cycles of denaturation at

Figure 3 Domestic residence test site installation (QLD)

Figure 4 Sampling biological community (QLD)

94 °C for one minute, primer annealing at 48 °C for one minute and primer extension at 72 °C for a period of two minutes.

3. Once screened for the 16S rRNA gene, clone ligation was then performed utilizing the pGEM-TEasy vector and subsequently transformed using *E. coli* cells.
4. PCR was then carried out using the plasmid specific primers of SP6 and T7. This allowed for the screening of clones in order to ascertain positive identification of 16S rRNA clones.
5. Digestion using the MspI restriction enzyme was then performed on clones positive for 16S that allowed for restricted fragment length polymorphism (RFLP) analysis.
6. Selected clones where then partially sequenced using primer 530f, and BLAST used to search for closely related sequenced relatives.

FISH analysis

Aliquots of samples were stored in 4% paraformaldehyde (PFA)/PBS mixture for approximately 1–3 hours. Following this, samples were resuspended and washed three times in PBS solution to finally be placed in a PBS/98% EtOH solution (1:1, v/v) and stored at −20 °C until hybridisation was performed. As samples proved to contain a large amount of auxiliary matter, it was required that a pipette tip be cut (approximately 10 mm from end) to allow adequate sample collection required for FISH.

FISH was then carried out utilising the oligonucleotide probes of EUBMix (EUB338, EUB338-II and EUB338-III, specific for all *Bacteria,* Amann *et al.*, 1990), BET42a (*Betaproteobacteria,* Manz *et al.*, 1992), GAM42a (*Gammaproteobacteria,* Manz *et al.*, 1992) and LGC354 (*Low G + C Gram positive bacteria,* Meier *et al.*, 1999). All probes were commercially synthesized with one of the sulfoindocyanine dyes, indocarboncyanine (Cy3) or indodicarbocyanine (Cy5) by ThermoHybaid (Interactivsa Division, Ulm, Germany).

After FISH, samples were observed and images collected using a BioRad Radiance 2000 confocal laser-scanning microscope (CLSM).

Findings

rRNA at Caravan Park Site, NSW

rRNA data is summarised in Table 1 for the NSW site (Plates A, B and C representing selected clones). Overall, a very diverse community was observed with 86 ordered taxonomic units (OTUs) being recognised. It was noted that a few dominant groups were present within the clone library, however, when sequenced and entered into BLAST, each of the selected clones within the particular OTU groups proved to be dissimilar

Table 1 rRNA Findings for the NSW Site

Clone Number, Plate	Organism	Origin OR Journal focus
PLATE A		
2A	*Betaproteobacterium, Zoogloea ramigera*	Batch reactor
9A	Uncultured unidentified bacterium	Equine contamination
10A	Unidenified *Betaproteobacterium*	PAO study
16A	Unidentified *Gammaproteobacterium*	Biological foaming, activated sludge
17A	Unidentified *Gammaproteobacterium*	Acidic uranium-contaminated aquifer
c	Unidentified *Deltaproteobacterium*	Coastal marine sediment
20A	Uncultured unidentified bacterium	Biological removal of P batch reactor
22A	*Gammaproteobacterium, Pseudomonas putida*	Equine contamination
26A	Uncultured unidentified bacterium	Arctic lake sediment
28A	*Betaproteobacterium, Variovorax* sp.	Soil
32A	Unidentified *Actinobacterium*	Subsurface microbial community structure
34A	*Betaproteobacterium, Comamonas testosteroni*	Plant pathogenic bacteria
36A	Uncultured *Eubacterium*	Travertine depositional facies
44A	Unidentified *Betaproteobacterium*	Underground oil storage cavity
49A	Uncultured unidentified bacterium	Bacterial community shifts in compost
51A	*Gammaproteobacterium, Pseudomonas putida*	Pseudomonas putida genome analysis
53A	Unidentified *Betaproteobacterium*	Paddy field soil
59A	Uncultured proteobacterium	Activated sludge
PLATE B		
1B	Uncultured *Eubacterium*	Travertine depositional facies
3B	*Betaproteobacterium*	Paddy field soil
13B	*Bacteroidetes*	Geothermal bacterial community
18B	*Deltaproteobacterium*	Marine communities
26B	Uncultured unidentified bacterium	Hawaiian archipelago
39B	Uncultured unidentified bacterium	Groundwater contaminated with uranium waste
PLATE C		
6C	*Betaproteobacterium, Zoogloea ramigera*	Molecular systematics of Zoogloea
7C	Unidentified *Alphaproteobacterium*	Acid and pristine sub alpine stream sediment
8C	Unidentified *Deltaproteobacterium*	Soil community in New Caledonia
16C	*Betaproteobacterium, Variovorax* sp.	Soil community
17C	*Betaproteobacterium, Ralstonia solanacearum*	Plant pathogen
24C	Unidentified *Planktomycete*	River biofilms

organisms. It is anticipated that these remaining clones within these dominant groups will be later sequenced and BLAST used to identify relatives.

Standard analysis at Tamborine site, QLD

TC and *E. coli* were routinely monitored for influent, secondary and effluent samples. Figure 5 shows the TC data over process units. It can be seen that successful removal of bacteria occurs, and that effluent quality is generally high.

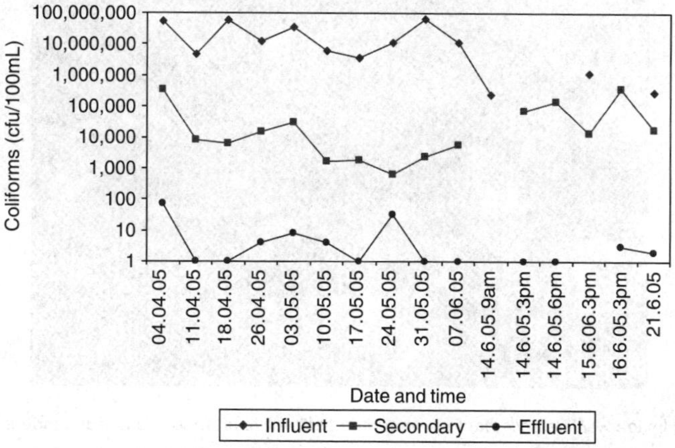

Figure 5 Thermotolerant coliform levels over process units (QLD)

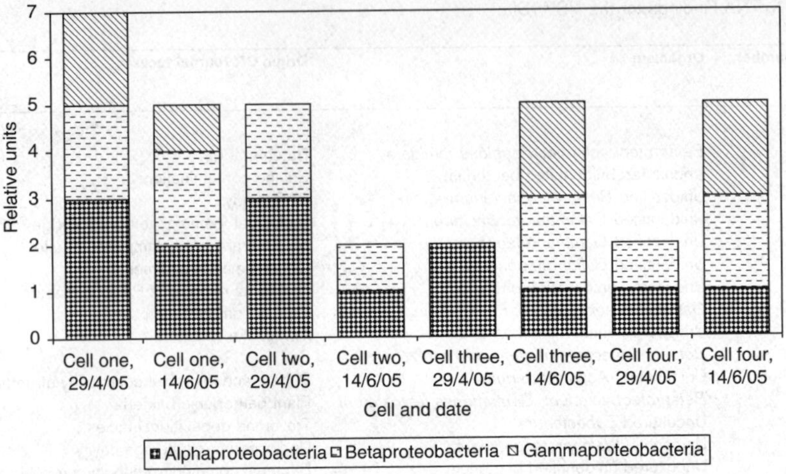

Figure 6 Biological community changes across cell unit and time, using FISH (QLD)

FISH at Tamborine site, QLD

Both FISH and FRNA phage work was conducted at the Tamborine site. It was noted that the dominant organisms observed in April 2005 were that of the *Alphaproteobacteria*. They were present within cells 1–3 in large abundance. However, alpha cell quantities decreased within cell 4. This was also the case for *Betaproteobacteria*, however, the latter were found not to be present in cell 3. Minimal numbers of *Gammaproteobacteria* were observed within cells 1–4, while no low G + C gram positive (LGC) classified bacteria were detected. This all suggests a very immature ecology is present.

Figure 6 graphically compares these data with that found for the more mature sample dated June 2005, showing bacteria type across time and cell unit. It can be seen that a more diverse ecology has developed during the six weeks between sampling events, with less

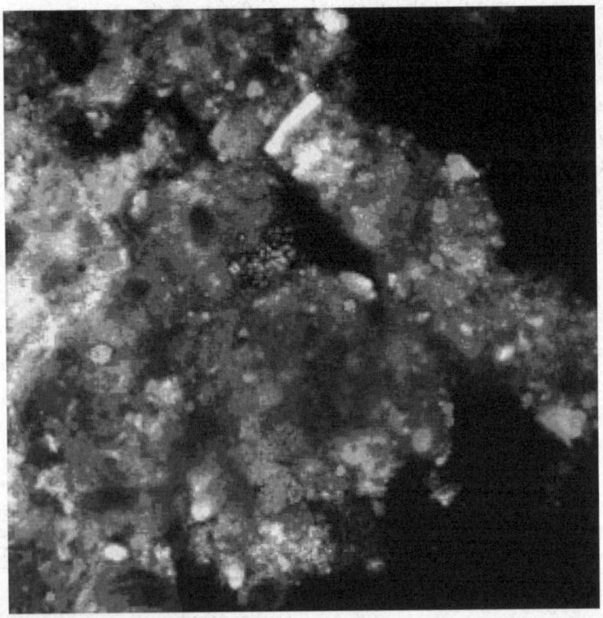

Figure 7 FISH Image – *Blue cells bacteria, magenta cells Alphaproteobacteria* Subscribers to the online version of *Water Science and Technology* can access the colour version of this figure from http://www.iwaponline.com/wst

Alphaproteobacteria and more *Betaproteobacteria* and *Gammaproteobacteria* present. Figure 7 shows a confocal microscope image using FISH, where blue cells are all bacteria, and magenta represent *Alphaproteobacteria*.

Conclusions

The rRNA phylogeny study of the NSW test site has indicated that the closest matches to the organisms identified are all involved in nutrient transformations in the environment. The most significant outcome is that the biological community identified by rRNA, has a tremendous diversity, with a domination from the *Betaproteobacteria* and *Gammaproteobacteria*. This is to be expected considering the facility has been in operation since October 2004, and is therefore a very mature ecology.

In contrast, the FISH results from Tamborine have shown a dominance of *Alphaproteobacteria*. This has highlighted a potential difference between a new community (operating since March 2005), and the more mature community at the NSW test site (operating since October 2004).

Acknowledgements

Ecowise wish to acknowledge that this project is supporting an Honours project at the University of Queensland via the Advanced Wastewater Management Centre. AquaReuse also provided the test facilities, and supported the project with routine staff sampling assistance and photographs. This project would not have been possible without their 'above and beyond' support.

References

APHA (1992). *Standard Methods for the Examination of Water and Wastewater*, 18th edn.

Amann, R.I., Binder, B.J., Olson, R.J., Chisholm, S.W., Devereux, R. and Stahl, D.A. (1990). Combination of 16S rRNA-targeted oligonucleotide probes with flow cytometry for analyzing mixed microbial populations. *Appl. Environ. Microbiol.*, **56**, 1919–1925.

Flapper, T.G. (September 2003), Meeting the NSW Health Accreditation Guidelines – a field perspective. *Proceedings of Onsite* 2003, Armidale, Australia.

Flapper, T.G. (September 2004), What is the onsite sewage system beast? *Proceedings of AWA NSW Operators Conference*, Penrith, Australia.

Flapper, T.G. and Hill, P. (2005). Assessing re-growth of organisms in stored treated greywater. *Onsite*, **2005**, Conference, September 2005.

Flapper, T.G., McKay, A. and Hill, P. (2005). Using molecular tools to assess biofilms in onsite systems. *Onsite*, **2005**, Conference, September 2005.

Flapper, T.G., Best, M., Squires, K. (2006). Using FRNA phage spikes to assess virus removal from onsite systems. Presented at IWA Small Water and Wastewater Systems conference, March 2006.

Manz, W., Amann, R., Ludwig, W., Wagner, M. and Schleifer, K.-H. (1992). Phylogenetic oligodeoxynucleotide probes for the major subclasses of Proteobacteria: problems and solutions. *Syst. Appl. Microbiol.*, **15**, 593–600.

Meier, H., Amann, R., Ludwig, W. and Schleifer, K.-H. (1999). Specific oligonucleotide probes for in situ detection of a major group of gram-positive bacteria with low DNA G + C content. *Syst. Appl. Microbiol.*, **22**, 186–196.

Development of an empirical mathematical model for describing and optimizing the hygiene potential of a thermophilic anaerobic bioreactor treating faeces

M. Lübken*, M. Wichern*, F. Bischof, S. Prechtl*** and H. Horn***

*Institute of Water Quality Control, Technical University of Munich, Am Coulombwall, 85748 Garching, Germany (E-mail: *m.luebken@bv.tum.de*)

**Hans Huber AG, Industriepark Erasbach A1, 92334 Berching, Germany

***ATZ Entwicklungszentrum, Kropfersrichterstraße 6-10, 92237 Sulzbach-Rosenberg, Germany

Water Science & Technology **Vol 55 No 7 pp 95–102** © IWA Publishing 2007

Abstract Poor sanitation and insufficient disposal of sewage and faeces are primarily responsible for water associated health problems in developing countries. Domestic sewage and faeces are prevalently discharged into surface waters which are used by the inhabitants as a source for drinking water. This paper presents a decentralized anaerobic process technique for handling of such domestic organic waste. Such an efficient and compact system for treating faeces and food waste may be of great benefit for developing countries. Besides a stable biogas production for energy generation, the reduction of bacterial pathogens is of particular importance. In our research we investigated the removal capacity of the reactor concerning pathogens, which has been operated under thermophilic conditions. Faecal coliforms and intestinal enterococci have been detected as indicator organisms for bacterial pathogens. By the multiple regression analysis technique an empirical mathematical model has been developed. The model shows a high correlation between removal efficiency and both, hydraulic retention time (HRT) and temperature. By this model an optimized HRT for defined bacterial pathogens effluent standards can be easily calculated. Thus, hygiene potential can be evaluated along with economic aspects. In this paper not only results for describing the hygiene potential of a thermophilic anaerobic bioreactor are presented, but also an exemplary method to draw the right conclusions out of biological tests with the aid of mathematical tools.

Keywords Bacterial pathogens; decentralized sanitation and reuse; faeces; multiple regression analysis; thermophilic anaerobic technology

Introduction

The International Drinking Water Supply and Sanitation Decade (1981–1990) and the Safe Water Supply Decade (1991–2000) with the goal to supply the population worldwide with drinking water and sanitation have just been finished. Up to the present, a significant part of the world population still has no access to clean potable water and suitable sanitation. Target 10 of the Millennium Development Goals demands to halve by 2015 the proportion of people without sustainable access to safe drinking water and basic sanitation (http://ddp-ext.worldbank.org). Concerning wastewater treatment, Asia has the biggest requirements for sanitation due to an increase in people to be supplied of about 320,000 per day, followed by Africa with 91,000 and Latin America with 32,000 (WHO, 2000; Mara and Faechem, 2001). In order to fulfil the goal "water and sanitation for all" alternatives have to be searched for, apart from the "classical" wastewater treatment systems consisting of a water-borne sewage system and a central wastewater treatment plant. An interesting alternative could be the DESAR-concept (Decentralized Sanitation and Reuse). A decentralized anaerobic process technique for treating domestic organic waste is presented in this work. Faecal sludge as the solid part of the single domestic wastewater streams is treated in an anaerobic bioreactor to produce energy by degrading

doi: 10.2166/wst.2007.132

organic compounds and to regain nutrients. The effluent is intended to be used for agricultural purposes like irrigation.

Anaerobic bioreactors as decentralized wastewater treatment systems are, especially for rural areas, economically and ecologically reasonable techniques. The main objective of applying thermophilic process conditions is the removal of pathogens and not an additional reduction of the organic compounds. If the effluent is used for irrigation, removal of pathogenic organisms is one of the most important tasks. Many pathogenic germs are excreted with faeces and so reach waterbodies, where they get access to the water cycle. Along with these epidemic germs, intestinal bacteria are also excreted in a high amount. These intestinal bacteria don't represent a hazard for consumers, but represent an indication that epidemic germs (in a much smaller amount) are also present.

In this work, intestinal enterococci and faecal coliforms have been detected as indicator organisms for bacterial pathogens to evaluate the hygiene potential of the thermophilic anaerobic reactor. Although the detection of these organisms is quite simple, the ongoing determination in the laboratory is time consuming. The overall objective was to develop a mathematical model, which can describe the reduction of the bacterial indicator organisms. Knowing the effluent contamination for different operational conditions and reducing laboratory effort can strongly enhance the benefit of the investigated decentralized anaerobic process technique.

Methodology

The anaerobic bioreactor (Figure 1) has been operated under thermophilic conditions (55 °C). The volume of 500 litres was separated into an outer chamber for hydrolysis and

Figure 1 Thermophilic anaerobic bioreactor

an inner part for methanogenesis. The experiments were started in October 2004 with faeces as sole substrate and were finished in March 2005.

Sampling and processing

Samples were collected from the inlet and outlet of the reactor once a week. Raw influent samples (0.5 litres) were collected in clean sampling bottles, transported to the laboratory and processed on the same day. Detection and enumeration of faecal coliforms was done in test tubes fitted with Durham tubes according to the "Most-Probable-Number-Process" (MPN-Process). Fluorocult-Lauryl-Sulfate-Bouillon (Merck 12588) was used as culture medium. The tubes were inoculated with 1 ml of faecal sludge and incubated for (48 ± 3) hours at (37 ± 1) °C. The presence of faecal coliforms is indicated by fluorescence under a long wavelength UV lamp (360 nm). The enumeration of the bacterial count was carried out with the aid of the "McCrady table" (deMan, 1983). The results are referred to 1 ml sample volume [MPN/1 ml]. Detection and enumeration of intestinal enterococci was performed according to ISO 7899-2. Cellulose nitrate filters with a pore size of 0.45 μm were placed on an enterococcus selective agar according to Slanetz and Bartley (1957) and inoculated for (48 ± 3) hours at (44 ± 2) °C. Pink to brown colonies with a diameter of 0.5 to 2 mm are usually enterococci. If characteristic colonies were detected, the filter was placed to a Bile Esculin Azide Agar and inoculated for 2 hours at (44 ± 0.5) °C. All colonies from tawny to black color were enumerated.

Multiple regression analysis

Simple regression and correlation are used when there are two variables multiple regression is used to express the relationship between one variable and two or more other variables. In general, regression is useful when the relationship between variables is of functional dependence. If so, the magnitude of one variable called the dependent variable, is assumed to be determined by the other variables, called the independent variables. When there is only one independent variable in addition to the dependent variable, the term simple regression is used. When there are two or more independent variables, the term multiple regression is used. To investigate the relationship between the independent variables x and the dependent variables y, a model of the form

$$y_i = \beta_0 + \beta_1 x_{1i} + \beta_2 x_{2i} + \ldots + \beta_p x_{pi} + \varepsilon_i \quad i = 1, 2, \ldots n$$

can be used, where $\beta_0, \beta_1, \ldots, \beta_p$ are called regression coefficients. Because there is a unique value ε_i corresponding to each value of y_i, the n data points will not all lie on a line (They would if ε_i was always zero). The n data points can be used, however, to estimate $\beta_0, \beta_1, \ldots, \beta_p$. The β are calculated using the method of least squares, which means to minimize the sum of the squared ε. This can be expressed by the following equation

$$S(\beta_0, \beta_1, \beta_2, \ldots, \beta_p) = \sum_{i=1}^{n} \varepsilon_i^2 = \sum_{i=1}^{n} (y_i - \beta_0 - \beta_1 x_{1i} - \ldots - \beta_p x_{pi})^2$$

Computing software

Multiple regression analysis as a complex mathematical method makes high demands on the computing software used. For multiple regression analysis only carefully tested professional mathematical software packages should be used. In our research we used *Mathematica* in its current version 5.0. *Mathematica*, first released in 1988 by Wolfram Research, Inc., combines symbolic manipulation, numerical mathematics, outstanding graphics, and a sophisticated programming language.

Results and discussion

The removal of faecal bacteria in anaerobic bioreactors depends on several mechanisms influenced by both, time and environmental conditions. The main factors in our research which have been found to have significant effects on faecal bacteria removal are the hydraulic retention time and the temperature within the reactor. During the investigation period the influent to the reactor was successively raised to gain information about removal efficiency under different HRT. Figure 2 and Figure 3 show the removal efficiency for intestinal enterococci and faecal coliforms depending on the hydraulic retention time.

The thermophilic reactor should reach high removal efficiencies ($E > 99.9$), so that the effluent can be used for irrigation. Both figures demonstrate that the hydraulic regime of the reactor has a great influence on the removal efficiency of the indicators of faecal contamination. An efficiency $E = 90\%$ corresponds to the removal of one logarithmic unit, an efficiency $E = 99.9999$ corresponds to the removal of 6 log units, according to equation 1.

$$\textit{removed}\ \log \textit{units} = -\log_{10}\left(\frac{100 - E}{100}\right) \qquad (1)$$

If the effluent is to be used for unrestricted irrigation, the World Health Organization (WHO, 1989) recommends values of less than 1000 faecal coliforms per 100 ml. Considering that the faecal coliforms count in the raw influent is in the order from 10^6 to 10^9 per 100 ml, the removal efficiency of the reactor should be very high. To comply with the WHO criteria, faecal coliforms removal efficiency in the order of 3 to 5 log units (99.9 to 99.999) are necessary to be reached.

To obtain efficiencies higher than 99.9% (3 log units removed) for faecal coliforms removal, the hydraulic retention time has to be above 2 days. The same efficiency for intestinal enterococci can be reached by a hydraulic retention time above three days. A hydraulic retention time above 2.5 days showed to be sufficient for a complete reduction of faecal coliforms. Intestinal enterococci, however, couldn't be removed completely, but were always less than 10/ml when a HRT of 5 days was applied. To remove at least two log units of both intestinal enterococci and faecal coliforms, the HRT must be higher than 1.5 days.

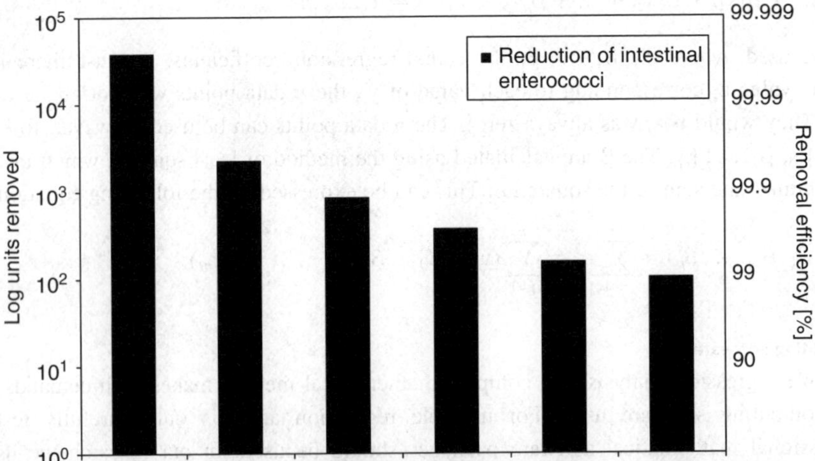

Figure 2 Removal efficiency of intestinal enterococci depending on the hydraulic retention time

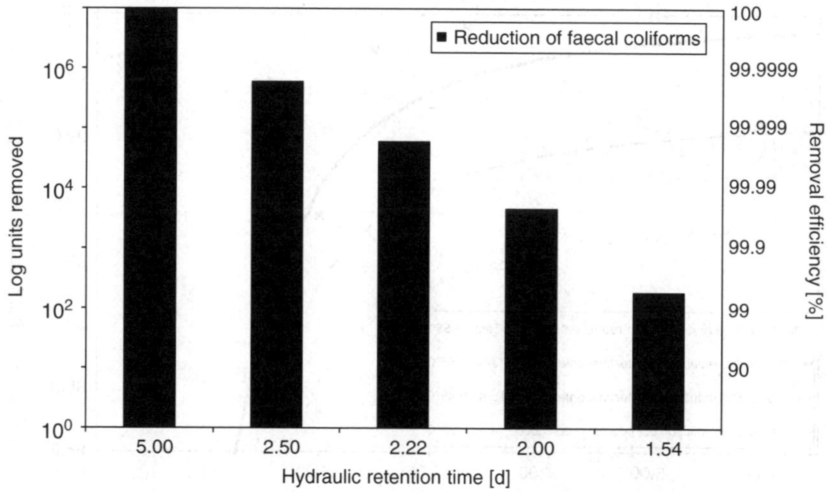

Figure 3 Removal efficiency of faecal coliforms depending on the hydraulic retention time

For more detailed information about the optimal hydraulic retention time, a regression term can be derived describing removal efficiency in terms of hydraulic retention time. Besides the hydraulic retention time, however, temperature is the main factor influencing removal efficiency. Figure 2 and Figure 3 show results only under constant thermophilic conditions (55 °C). A regression term is useless if the temperature hasn't been taken into account.

Temperature drop as a result of power failure delivered valuable information about removal efficiency under different temperature conditions. Using multiple regression analysis, intestinal enterococci and faecal coliforms removal can be represented in terms of HRT and temperature by the following equations:

Equation 2 shows the multiple regression term for intestinal enterococci removal.

$$\eta_{IE} = 98.29 - 2.2\left(\frac{1}{HRT}\right)^2 + 0.031T \qquad (2)$$

η_{IE} = intestinal enterococci removal [%]
HRT = hydraulic retention time [d]
T = temperature [°C]

Equation 3 shows the multiple regression term for faecal coliforms removal.

$$\eta_{FC} = 98.29 - 1\left(\frac{1}{HRT}\right)^2 + 0.031T \qquad (3)$$

η_{FC} = faecal coliforms removal [%]
HRT = hydraulic retention time [d]
T = temperature [°C]

Equation 2 has a correlation coefficient of 0.93 with $p = 0$. Equation 3 has a correlation coefficient of 0.99 with $p = 0$. This shows that the removal efficiency corresponds very well with the hydraulic retention time and the temperature.

Figure 4 depicts the comparison between measured and modelled reduction of intestinal enterococci, Figure 5 for faecal coliforms.

Both figures demonstrate a good concordance between measured and modelled removal efficiencies. According to the Equations 2 and 3 and as demonstrated in Figures 4 and 5, removal of intestinal enterococci is more sensitive to the applied hydraulic

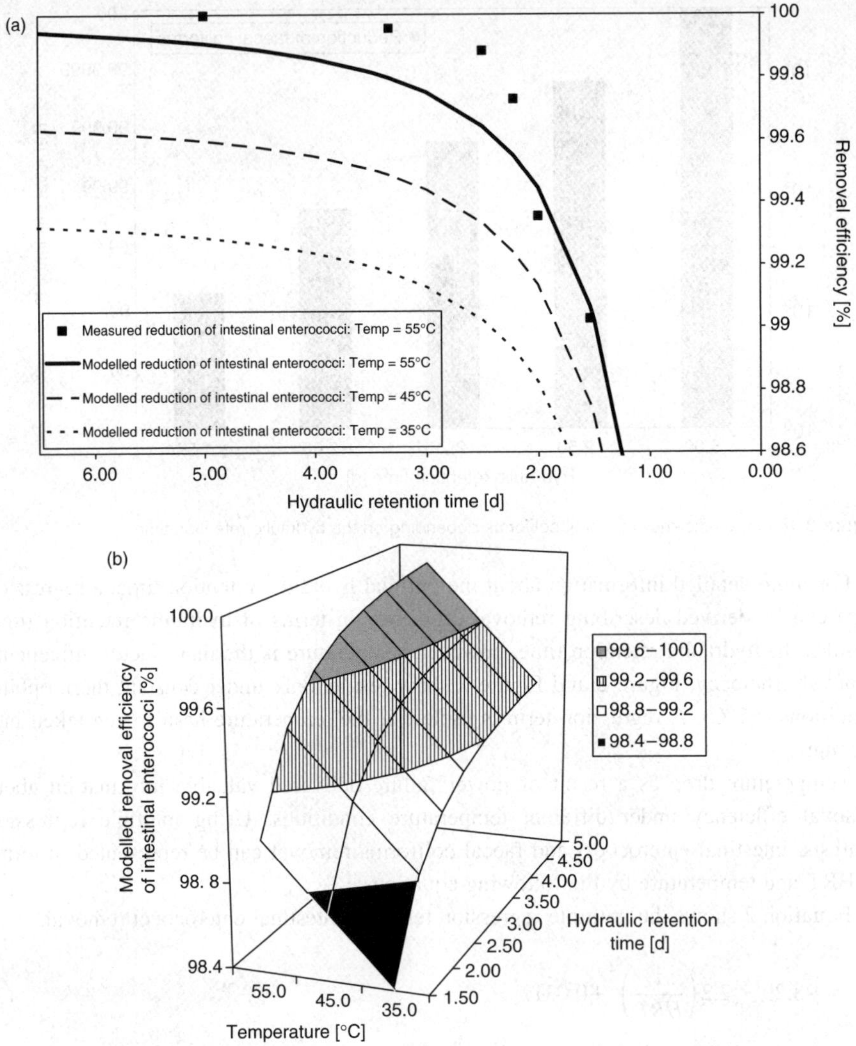

Figure 4 Modelled and measured removal efficiencies of intestinal enterococci in terms of hydraulic retention time and temperature: (a) modelled and measured, (b) only modelled shown as surface graphic

retention time than for faecal coliforms. Both models show the same linear influence of temperature, but the coefficient describing the influence of HRT is 2.2 times higher for intestinal enterococci than for faecal coliforms. As indicated in Figure 4 and Figure 5, the lower the HRT the higher the influence of HRT on the achievable removal efficiency. At higher HRT temperature is the main factor for increasing removal efficiency. Figure 4 explains that until a HRT of 2 days a stepwise increase of 0.5 days regarding the HRT is more efficient in intestinal enterococci removal than increasing temperature by 10 °C. Above 2 days, raising the temperature is more efficient. Referring to faecal coliforms, the same threshold can be set to 1.5 days.

The area of validity of both models is the range of data used for regression analysis. In praxis, the HRT varied between 1.5 and 5 days, the temperature between 35 and 55 °C. It is not possible to widely expand the models out of the data ranges from which the models have been derived (Draper and Smith, 1998). However, the range of data is expanded enough to give valuable information about an optimized HRT in case of an effluent standard of interest.

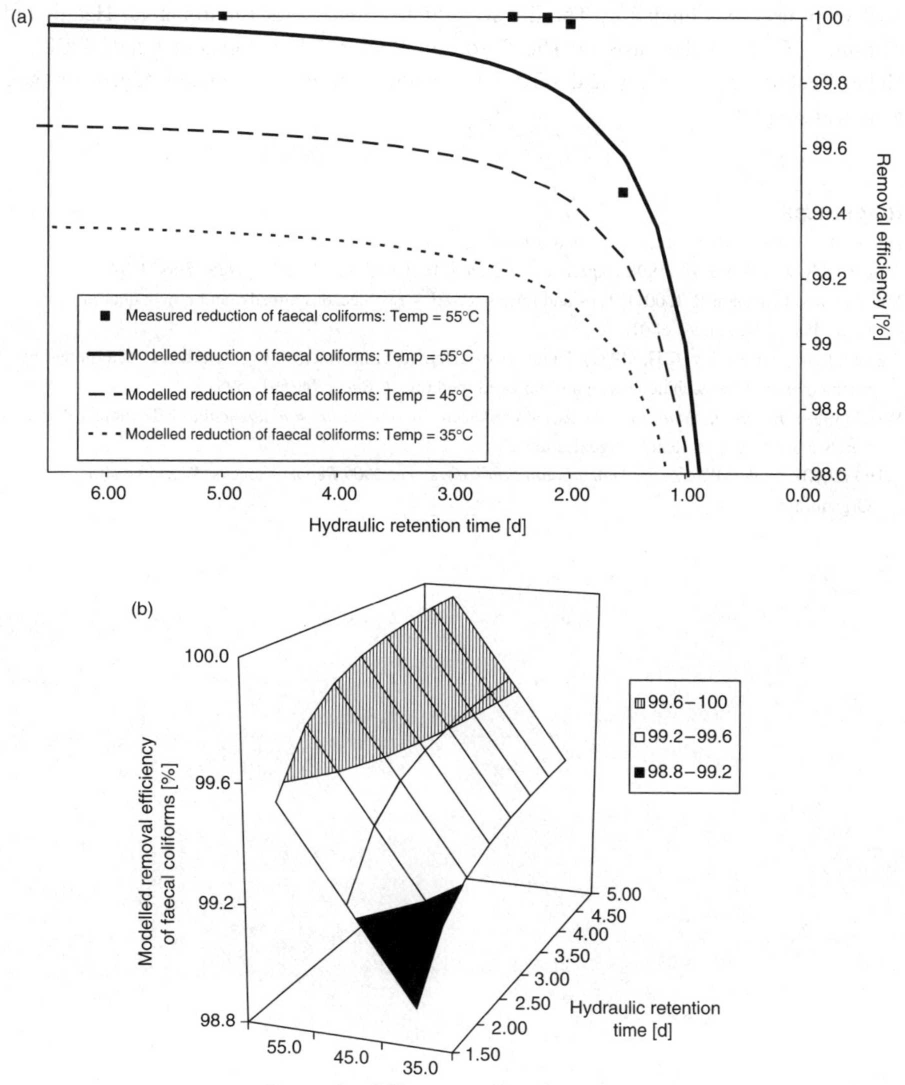

Figure 5 Modelled and measured removal efficiencies of faecal coliforms in terms of hydraulic retention time and temperature: (a) modelled and measured, (b) only modelled shown as surface graphic

Conclusions

Results obtained in our study indicated that both HRT and temperature have the most significant effect on pathogen removal, measured in terms of faecal coliforms and intestinal enterococci numbers. The derived models indicated that intestinal enterococci removal is more sensitive to the applied HRT than faecal coliforms removal. Secondarily, the models show an identical linear influence of temperature. Until a threshold of 2 days for intestinal enterococci and 1.5 days for faecal coliforms, increasing HRT by 0.5 days is more efficient than increasing temperature by 10 °C.

The models represent a mathematical tool to optimize both, hygiene potential and operation, for different effluent standards. The derived empirical models are expected to apply to faecal sludge treatment in decentralized thermophilic anaerobic bioreactors. The objective is to give planners a mathematical tool to optimize hygiene potential and operation for different effluent standards.

This project was funded by The Bavarian State Ministry for Environment, Health and Consumer Care on the basis of The European Regional Development Fund – ERDF. Huber Technology Germany and ATZ Entwicklungszentrum are acknowledged for their technical support.

References

deMan, J.C. (1983). MPN tables, corrected. *Eur. J. Appl. Biotechnol.*, **17**, 301–305.

Draper, N.R. and Smith, H. (1998). *Applied Regression Analysis*, 3rd ed. Wiley, New York, USA.

Mara, D. and Feachem, R. (2001). Taps and toilets for all – two decades already, and now a quarter century more. *Water 21*, August 2001.

Slanetz, L.W. and Bartley, C.H. (1957). Numbers of enterococci in water, sewage and faeces determined by the membrane filter technique with an improved medium. *J. Bact.*, **74**, 591–595.

WHO (1989). *Health guidelines for the use of wastewater in agriculture and aquaculture*. Technical Report Series No.778. World Health Organisation.

WHO (2000). *Global Water Supply and Sanitation Assessment 2000 Report*. Geneva, World Health Organisation.

Waste stabilization ponds and rock filters: solutions for small communities

D.D. Mara and M.L. Johnson

School of Civil Engineering, University of Leeds, Leeds LS2 9JT, UK (E-mail: *cen6ddm@leeds.ac.uk*; *cenmlj@leeds.ac.uk*)

Abstract In temperate climates facultative ponds and rock filters (either unaerated or, if ammonia removal is required, aerated) are a low-cost but high-performance treatment system for small rural communities. Effluent quality is suitable for surface water discharge or, in summer, for restricted crop irrigation. In tropical climates anaerobic and facultative ponds and either unaerated rock filters or, if ammonia reduction is required, subsurface horizontal-flow or vertical-flow constructed wetland, can be used if the effluents are discharged to surface waters. However, if the treated wastewater is to be used for crop irrigation, then a 3-log unit pathogen reduction by treatment in anaerobic, facultative and single maturation ponds is required for both restricted and unrestricted irrigation, provided that, in the case of unrestricted irrigation, there are in place post-treatment health-protection control measures that together provide a further 4-log unit pathogen reduction.

Keywords Constructed wetlands; pathogen reduction; rock filters; waste stabilization ponds; wastewater reuse

Water Science & Technology **Vol 55 No 7 pp 103–107** © IWA Publishing 2007

Introduction

It is now well established in many countries, both industrialized and developing, that waste stabilization ponds (WSP) are a very suitable method of wastewater treatment for small sewered communities. What is not so well established, despite many years experience in the United States (O'Brien *et al.*, 1973; Swanson and Williamson, 1980; Middlebrooks, 1988, 1995; US EPA, 2002), is that rock filters are equally suitable for 'polishing' WSP effluents (i.e. to remove algal solids and associated BOD). In contrast, constructed wetlands have been strongly advocated in the United States for this purpose (Kadlec, 2004).

WSP for small communities are designed in much the same way as those for larger populations. To illustrate this WSP systems are now designed for a typical small community in a temperate climate and for one in a tropical climate, using design temperatures of 8 and 20 °C, respectively.

Temperate climates

A facultative pond loaded at 80 kg BOD per ha per day (i.e., 8 $g/m^2 d$) would be appropriate (Abis and Mara, 2003). Assuming a BOD contribution of 40 g per person per day in rural areas (Pujol and Liénard, 1990), the pond mid-depth area would be 5 m^2 per person. Such a pond produces an effluent complying with the UWWTD (Abis and Mara, 2003).

Effluent discharge to surface water

To produce a final effluent with a low suspended solids (SS) concentration (e.g. ≤30 mg/l) the facultative pond effluent is best treated in a rock filter. Early work in the US has shown that rock filters are as good as other 'pond polishing' technologies (such as constructed wetlands or intermittent sand filters) but much cheaper (Middlebrooks, 1995). Our work in the UK shows that a hydraulic loading rate (HLR) of 0.3 m^3 of facultative pond effluent per m^3 of gross rock filter volume day per day is suitable for the

doi: 10.2166/wst.2007.133

production of ≤40 mg unfiltered BOD/l and ≤60 mg SS/l (95-percentile values) (Johnson and Mara, 2005; Mara and Johnson, 2006).

If the environmental regulator specifies a maximum ammonia concentration for the final effluent of 10 mg N/l or less, then the rock filter should be aerated (Johnson and Mara, 2005; Mara and Johnson, 2006) to provide the conditions for ammonia removal by nitrification. At an HLR of $0.3\,d^{-1}$ and an air flow rate of 20 l/min an aerated rock filter can produce an effluent with ~9 mg BOD/l, ~9 mg SS/l and ~7 mg ammonia-N/l (95-percentile values) throughout the year (Figure 1), consistently better than the effluent quality of an unaerated rock filter. Our experience with subsurface horizontal-flow constructed wetlands is that they produce a good quality effluent in summer, but not in winter (Figure 2).

The area of both aerated and unaerated rock filters (A_{rf}) is given by:

$$A_{rf} = \frac{Q_i}{(HLR) \times D_{rf}} = \frac{0.2}{0.3 \times 0.6} = 1.1\,m^2 \text{ per person}$$

where Q_i is the wastewater flow (taken as $0.2\,m^3$ per person per day; British Water, 2005) and D_{rf} is the wastewater depth in the rock filter (taken as 0.6 m).

Thus the area of a primary facultative pond and a rock filter is $6.1\,m^2$ per person (i.e. around $8\,m^2$ per person overall).

Crop irrigation in summer

The effluent from the aerated rock filter contains < 1000 faecal coliforms (FC) per 100 ml, and often in summer < 100 per 100 ml (Mara and Johnson, 2006). Thus the effluent would

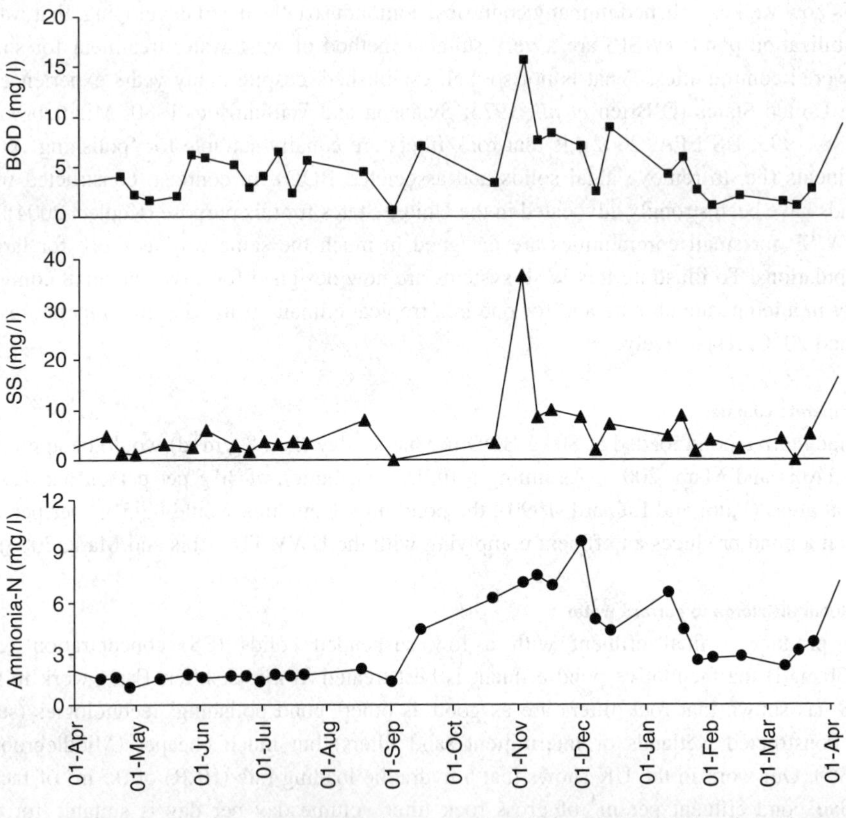

Figure 1 BOD, suspended solids and ammonia-N concentrations in the effluent from the pilot-scale aerated rock filter at Esholt, Bradford, during April 2004 – March 2005

Figure 2 Pilot-scale subsurface horizontal-flow constructed wetland and aerated and unaerated rock filters treating facultative pond effluent at a hydraulic loading rate of 0.3 d^{-1} at Esholt, Bradford, UK. Left, summer; right, winter

be suitable for restricted crop irrigation as the pathogen reduction, as indicated by this level of FC removal, would be more than the 2 − 3 log units required (WHO, 2006).

Tropical climates

In this case an anaerobic pond followed by a facultative pond would be generally used. At 20 °C the design loadings on anaerobic and facultative ponds are 300 g BOD/m^3 d and 250 kg BOD/ha d, respectively (Mara, 2004). Assuming a BOD contribution in the rural tropics of ~35 g/person d and a wastewater flow of 0.07 m^3 per person per day, the anaerobic pond volume is 0.12 m^3 per person (or, assuming a depth of 3 m, a mid-depth area of ~0.04 m^2 per person; the retention time is 1.7 days). With 60 per cent BOD removal in the anaerobic pond at 20 °C, the facultative pond retention time is 12 d and its mid-depth area is 0.56 m^2 per person – i.e. a total mid-depth area of ~0.6 m^2 per person.

Effluent discharge to surface water

Prior to surface water discharge facultative pond effluents can be treated in rock filters, subsurface horizontal-flow or vertical-flow constructed wetlands (the latter are preferred if ammonia removal is required). For rock filters an HLR of 0.5–1 m^3 of facultative pond effluent per m^3 of gross rock filter volume per day should be used (Saidam *et al.*, 1995; Mara *et al.*, 2001; Neder *et al.*, 2002). Assuming a wastewater flow of 0.07 m^3 per person per day, an HLR of 0.75 d^{-1} and a D_{rf} of 0.6 m, the rock filter area is 0.16 m^2 per person. Thus the area of the anaerobic and facultative ponds and the rock filter is 0.76 m^2 per person (i.e. around 1 m^2 per person overall). Effluent quality can be estimated on the basis of 80 per cent BOD removal in the anaerobic and facultative ponds and 50 per cent in the rock filter – i.e. a cumulative BOD reduction of 90 per cent.

Crop irrigation

For agricultural reuse a 2–3 log unit pathogen reduction is required for restricted irrigation and a 6–7 log unit pathogen reduction for unrestricted irrigation (WHO, 2006). The 2–3 log unit pathogen reduction for restricted irrigation is achieved solely by treatment, but the 6–7 log unit pathogen reduction for unrestricted irrigation is best partially achieved by treatment and partially post-treatment health-protection control measures (the most commonly employed of these are given in Table 1).

Restricted irrigation. The reduction of *Escherichia coli* (used as an indicator for bacterial, viral and protozoan pathogens) occurring in the anaerobic and facultative ponds

Table 1 Pathogen reductions achieved by health-protection control measures[a]

Control measure	Pathogen reduction (log_{10} units)	Notes
Wastewater treatment	1–7	The required pathogen reduction to be achieved by treatment depends on the combination of other health-protection control measures selected.
Drip irrigation	2 – 4	2-log reduction for low-growing crops and 4-log reduction for high-growing crops.
Pathogen die-off	0.5–2 d^{-1}	Die-off after last irrigation (value depends on climate, crop type, etc.).
Produce washing with water	1	Washing salad crops, vegetables and fruit with clean water.
Produce disinfection and rinsing with water	2	Washing salad crops, vegetables and fruit with a weak disinfectant solution and rinsing with clean water.
Produce peeling	2	Fruits, root crops.

[a]WHO (2006)

can be estimated by the equations of Marais (1974):

$$N_e = \frac{N_i}{(1 + k_{B(T)}\theta_A)(1 + k_{B(T)}\theta_F)}$$

where N_i and N_e are the number of *E. coli* per 100 ml of the raw wastewater and facultative pond effluent, respectively (N_i is taken as 10^7); $k_{B(T)}$, the first-order rate constant for *E. coli* removal, d^{-1} ($= 2.6(1.19)^{T-20}$; i.e. 2.6 d^{-1} for 20 °C); and θ_A and θ_F, the mean hydraulic retention time in the anaerobic and facultative pond, respectively, d (here, $\theta_A = 1.7$ d and $\theta_F = 12$ d). Thus $N_e = 6 \times 10^4$ per 100 ml. A single 3-d maturation pond would be required to reduce this to 7000 per 100 ml – i.e. to achieve a 3-log unit pathogen reduction (as indicated by the *E. coli* reduction from 10^7 to $<10^4$ per 100 ml), so making the effluent safe for restricted irrigation using labour-intensive agriculture (WHO, 2006). In this case this series of an anaerobic, a facultative and a single maturation pond would be the most cost-effective solution.

Unrestricted irrigation. The 3-log unit pathogen reduction by treatment in the pond series developed for restricted irrigation would also be satisfactory for unrestricted irrigation, provided that there were a combination of post-treatment health protection measures which together gave a 4-log unit pathogen reduction, such that the overall pathogen reduction would be 7 log units. This additional 4-log unit pathogen reduction could be realised by, for example, a 2-log unit reduction due to pathogen die-off and a 2-log unit reduction due to produce disinfection; by the drip irrigation of high-growing crops (4-log unit reduction); or by the drip irrigation of low-growing crops (2-log unit reduction), a 1-log unit reduction due to die-off and a 1-log unit reduction due to produce washing in clean water (Table 1). Wastewater treatment to achieve a pathogen reduction of >3 log units would not be cost-effective, especially as low-cost drip irrigation equipment is now available (Polak *et al.*, 1997; Intermediate Technology Consultants, 2003).

Conclusions

- In temperate climates facultative ponds and rock filters (either unaerated or, if ammonia removal is required, aerated) are a low-cost but high-performance treatment system for small rural communities.
- In tropical climates a similar system can be used if effluent disposal is to a surface water, although anaerobic ponds would generally be used ahead of the facultative

ponds, and a constructed wetland may often more appropriate than an aerated rock filter for nitrogen removal.

- If the treated wastewater is to be used for crop irrigation in tropical (or other warm) climates, then a 3-log unit pathogen reduction by treatment in a series comprising an anaerobic, a facultative and a single maturation pond is required for both restricted and unrestricted irrigation, provided that, in the case of unrestricted irrigation, there are in place post-treatment health-protection control measures that together provide a further 4-log unit pathogen reduction.

References

Abis, K. and Mara, D.D. (2003). Research on waste stabilization ponds in the United Kingdom: initial results from pilot-scale facultative ponds. *Water Science and Technology*, **48**(2), 1–7.

British Water (2005). *Flows and Loads – 2: Sizing Criteria, Treatment Capacity for Small Wastewater Treatment Systems (Package Plants)*, British Water, London, (Code of Practice No. BW COP 01/05, revised ed.).

Intermediate Technology Consultants (2003). *Low Cost Micro Irrigation Technologies for the Poor*, Intermediate Technology Consultants, Rugby.

Johnson, M.L. and Mara, D.D. (2005). Aerated rock filters for enhanced nitrogen and faecal coliform removal from facultative waste stabilization pond effluents. *Water Science and Technology*, **51**(12), 99–102.

Kadlec, R.H. (2004). Pond and wetland treatment. *Water Science and Technology*, **48**(5), 1–5.

Mara, D.D. and Johnson, M.L. (2006). Aerated rock filters for enhanced ammonia and fecal coliform removal from facultative pond effluents. *Journal of Environmental Engineering, American Society of Civil Engineers*, **132**(4) (in press).

Mara, D.D. (2004). *Domestic Wastewater Treatment in Developing Countries*, Earthscan Publications, London.

Mara, D.D., Pearson, H.W., Oragui, J.I., Arridge, A. and Silva, S.A. (2001). *Development of a New Approach to Waste Stabilization Pond Design*, School of Civil Engineering, University of Leeds, Leeds, (Tropical Public Health Engineering Research Monograph No. 12).

Marais, G.v.R. (1974). Faecal bacterial kinetics in waste stabilization ponds. *Journal of the Environmental Engineering Division, American Society of Civil Engineers*, **100**(EE1), 119–139.

Middlebrooks, E.J. (1988). Review of rock filters for the upgrade of lagoon effluents. *Journal of the Water Pollution Control Federation*, **60**(9), 1657–1662.

Middlebrooks, E.J. (1995). Upgrading pond effluents: an overview. *Water Science and Technology*, **31**(12), 353–368.

Neder, K.D., Carneiro, G.A., Queiroz, T.R. and de Souza, M.A.A. (2002). Selection of natural treatment processes for algae removal from stabilisation pond effluents in Brasília, using multi-criterion methods. *Water Science and Technology*, **46**(4–5), 347–354.

O'Brien, W.J., McKinney, R.E., Turvey, M.D. and Martin, D.M. (1973). Two methods for algae removal from wastewater stabilization ponds. *Water and Sewage Works Journal*, **120**(3), 66–73.

Polak, P., Nanes, B. and Adhikari, D. (1997). A low cost drip irrigation system for small farmers in developing countries. *Journal of the American Water Resources Association*, **33**(1), 119–124.

Pujol, R. and Liénard, A. (1990). Qualitative and quantitative characterization of waste water for small communities. *Water Science and Technology*, **22**(3–4), 253–260.

Saidam, M.Y., Ramadan, S.A. and Butler, D. (1995). Upgrading waste stabilization pond effluent by rock filters. *Water Science and Technology*, **31**(12), 369–378.

Swanson, G.R. and Williamson, K.J. (1980). Upgrading lagoon effluents with rock filters. *Journal of the Environmental Engineering Division, American Society of Civil Engineers*, **106**(EE6), 1111–1129.

US EPA (2002). *Rock Media Polishing Filter for Lagoons* (Wastewater Technology Fact Sheet No. EPA 832-F-02-023). Washington, DC: Office of Water, US Environmental Protection Agency.

WHO (2006). *Guidelines for the Safe Use of Wastewater, Excreta and Greywater: Volume 2, Wastewater Use in Agriculture*, World Health Organization, Geneva.

Water Science & Technology Vol 55 No 7 pp 109–117 © IWA Publishing 2007

Comparison of the nutrient removal efficiency of onsite wastewater treatments systems: applications of conventional sand filters and sequencing batch reactors (SBR)

R. Vilpas* and E. Santala**

*Research Programme for Production and Consumption, Finnish Environment Institute, P.O.Box 140, 00251 Helsinki, Finland (E-mail: *riikka.vilpas@ymparisto.fi*)

**Water Resources Management Division, Finnish Environment Institute, P.O.Box 140, 00251 Helsinki, Finland (E-mail: *erkki.santala@ymparisto.fi*)

Abstract When domestic wastewater was treated with different onsite applications of buried sand filters and sequencing batch reactors, good organic matter removal was common and effluent BOD_7 concentrations from 5 to 20 mg/l were easily achievable. For total nitrogen, effluent concentrations were usually between 20 and 80 mg/l. Good phosphorus removal, even using special adsorption or precipitation materials, was difficult to achieve and large variations occurred. The median effluent concentration of total phosphorus in the most successful sand filter application was less than 0.1 mg/l and other sand filters and SBRs had the median concentrations varying from 1.7 to 6.7 mg/l. These results are based on one year *in situ* monitoring of 2 conventional buried sand filters, 6 sand filter applications with special phosphorus adsorbing media within the filter bed, 5 sand filters with separate tertiary phosphorus filtration and 11 small SBRs of three different types. The study was carried out in southern Finland during 2003–05. The whole project included monitoring of more than 60 plants of 20 different treatment types or methods, used in normal conditions to treat domestic wastewater. Evaluation of the different systems was made by comparing the measured effluent concentrations. In addition the effluent concentrations were compared to the discharge limits calculated according to the new Finnish regulation.

Keywords Domestic wastewater; nitrogen; nutrients; performance; phosphorus; rural area; wastewater treatment

Introduction

In Finland the number of inhabitants not connected to public sewer networks is about one million, 19% of the total population. The traditional onsite wastewater treatment method has been a two- or three-chamber septic tank, with no further treatment. At present, more advanced methods are required. The Government Decree on Treating Domestic Wastewater in Areas outside Sewer Networks (Government Decree, 2003; Santala and Kaloinen, 2004) came into force in the beginning of the year 2004. It sets maximum requirements for the discharges and minimum requirements for construction, operation and maintenance for all onsite wastewater systems, with a population equivalent up to 100. The new Decree has created a lot of pressure especially to houseowners and to municipal environmental authorities. The manufacturers have become active in developing more efficient and cost-effective systems. This has been enhanced by the implementation of the CE-marking for small wastewater treatment plants according to the European standard, EN 12566-3 (CEN, 2005; Nonet, 2005). This product standard gives general requirements and test methods for the plants. Especially the testing of wastewater treatment efficiency according to the standard is an effective tool when evaluating the performance of different systems.

doi: 10.2166/wst.2007.134

In order to get up to date information of *in situ* constructed treatment methods and pre-fabricated (packaged) treatment plants, especially on those capable of reducing also nutrients, the Finnish Environment Institute (SYKE) launched a monitoring project in winter 2003. During the project "Enhancing nutrient removal efficiency of onsite wastewater treatment systems – Ravinnesampo", which was carried out jointly with West Finland Regional Environment Centre and Tampere University of Technology, more than twenty different types of onsite wastewater treatment systems were monitored *in situ* over a year. In total over sixty treatment plants were selected for evaluation in normal conditions to treat all the wastewater from single family homes, including 15 treatment plants which treated also wastewater from milking rooms of dairy farms. This paper describes the performance of different applications of conventional buried sand filters and sequencing batch reactors (SBR), mostly equipped with phosphorus adsorption or precipitation.

Method

The operation of 24 onsite wastewater treatment plants was evaluated, representing nine different treatment systems. The plants were treating wastewater from ordinary single family homes. The aim was to monitor as many identical treatment systems as possible to improve the reliability of the results. The wastewater flows varied between 350 and 1000 l/d. Almost all the treatment plants were recently constructed but a couple of them had already been used for several years. Geographically the treatment plants were located mainly in southern and south-western Finland.

Sampling and monitoring

Composite samples from the effluent of each plant were taken regularly 3 to 8 times during a period of one year starting in June 2003 and finishing in August 2004. Influent samples were taken only on a few occasions. Representative sampling from the influent was found unreliable due to huge variation of the concentrations (Kujala-Räty, 2004; Kujala-Räty and Santala, 2001). Thus the exact influent loadings to the treatment plants were not determined. Composite samples were collected from the effluent by combining 2 to 4 individual grab samples taken during the same day. Any automatic sampling device was not used because available devices were found unreliable for this purpose. However, the chosen sampling method was quite laborious and demanded careful planning in advance, in order to minimize traveling. Sampling was known to play a significant role in this kind of experimental *in situ* research where it is the basis for all the conclusions (Kujala-Räty, 2004). The competence of the persons responsible for the sampling was taken into special consideration and their possible sampling education and experience were evaluated when recruiting them. All observations concerning the performance of the treatment plants were recorded during the sampling period including any maintenance works and unusual conditions.

Treatment plants

The evaluated treatment plants were 2 conventional buried sand filters (SF), 6 applications of buried sand filters with phosphorus adsorbing media within the filterbed (SFP), 5 conventional buried sand filters with a separate tertiary phosphorus filtration (SF + P) and 11 sequencing batch reactors (SBRs) of three different types. The treatment plants with the process diagrams and the abbreviations used in this paper are shown in Figure 1.

Analysis and information

Effluent samples were analyzed for total phosphorus (TP), dissolved total phosphorus, total nitrogen (TN), nitrite/nitrate (NO_{2+3}-N), ammonium (NH_4-N), biochemical oxygen

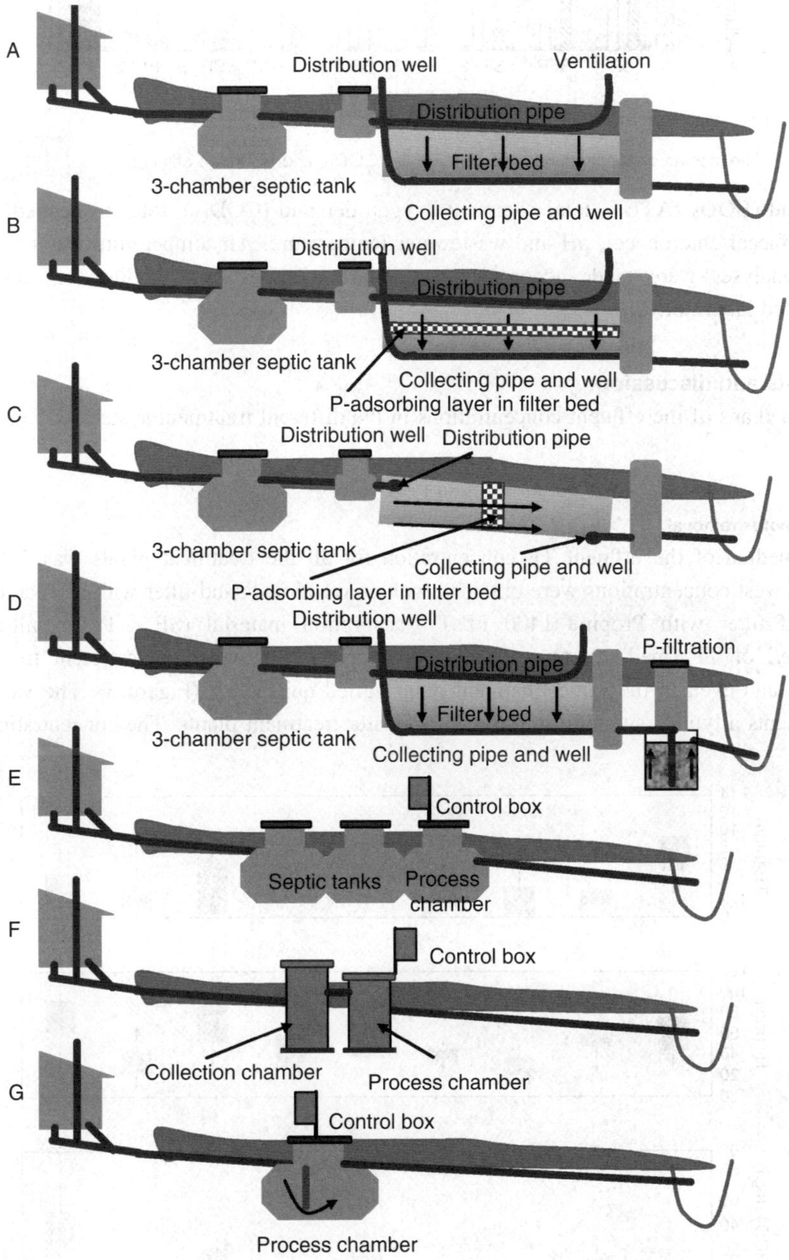

Figure 1 The evaluated treatment systems

Code	Treatment system	Abbreviation	Number of systems	Amount of samples
A	Conventional buried sand filter	SF	2	14
B	Buried sand filter with P-adsorbing media (Fosfilt, Kemira Ltd)	SFP1	1	7
B	Buried sand filter with P-adsorbing media (Biotite, Kemira Ltd)	SFP2	3	15
C	Horizontal flow buried sand filter with P-adsorbing media (Fe(III)+Ca(OH)2)	SFP3	2	16
D	Conventional buried sand filter with tertiary P-filter (Biotite, Kemira Ltd)	SF+P1	2	10
D	Conventional buried sand filter with tertiary P-filter (1400 FILT, Propipe Ltd)	SF+P2	3	19
E	Upoclean 5 –SBR with P-precipitation (Uponor Ltd)	SBR1	5	35
F	Wehoputs 6 –SBR with P-precipitation (KWH Pipe Ltd)	SBR2	3	15
G	BioKem –SBR with P-precipitation (Wavin-Labko Ltd)	SBR3	3	20

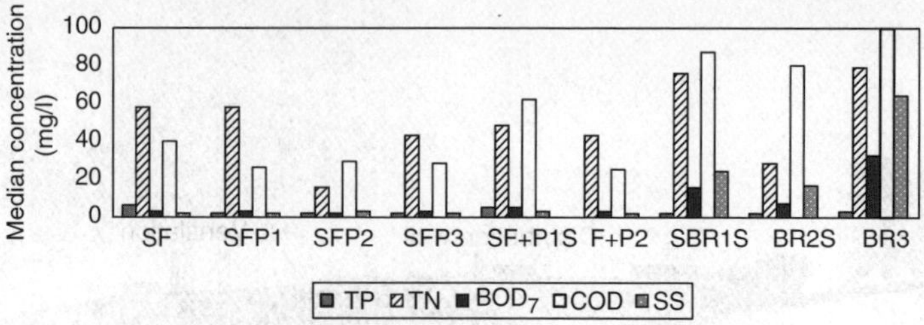

Figure 2 The median concentrations of TP, TN, BOD_7, COD_{Cr} and SS in the effluent

demand (BOD_7 (ATU)), total chemical oxygen demand (COD_{Cr}), total suspended solids (SS), faecal enterococci, pH and wastewater temperature. Air temperature was recorded. The analyses were made according to standard methods in accredited or otherwise accepted laboratories.

Results and discussion

The medians of the effluent concentrations in the different treatment systems are given in Figure 2.

Phosphorus removal

The median of the effluent TP concentration for all the treatment plants was 2.2 mg/l. The lowest concentrations were obtained with conventional sand filter with a separate tertiary P-filter with Propipe 1400 FILT –adsorption material (SF + P2) resulting in 0.1 mg/l of TP (median). Overall, the concentrations between the different treatment plants and even in the same treatment plant varied quite a lot (Figure 3). The variation represents a typical situation in onsite wastewater treatment plants. The concentrations of

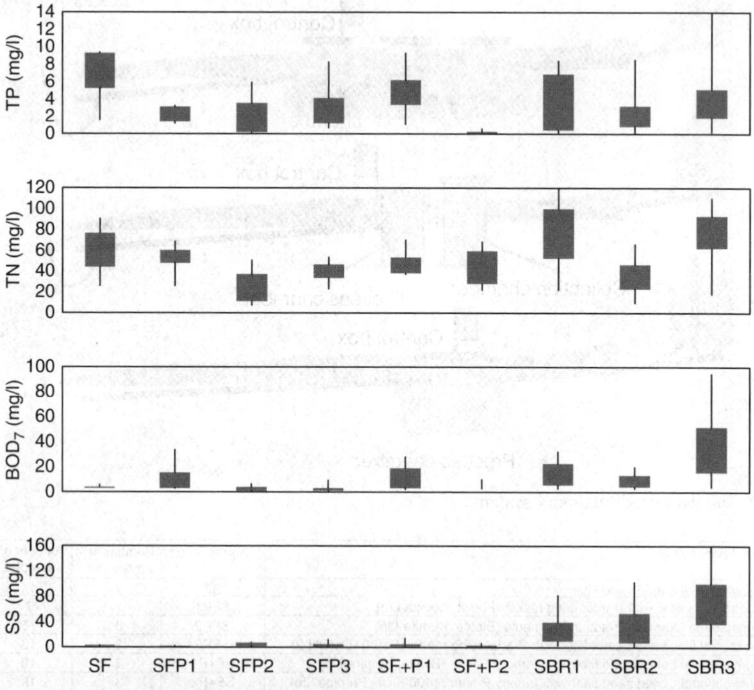

Figure 3 The variation of the effluent TP, TN, BOD_7 and SS concentrations. The middle 50% of the results is represented by the grey boxes and the lines describe the lowest and highest 25% of the results

Table 1 The median concentrations of dissolved TP, suspended TP and pH in the effluent and the amount of faecal enterococci and the ratio of COD_{Cr}/BOD_7 in the effluent

Treatment system	Unit	SF	SFP1	SFP2	SFP3	SF + P1	SF + P2	SBR1	SBR2	SBR3
Suspended TP (mg/l)	mg/l	0.1	0.1	0.02	0.01	0.3	0.01	0.9	0.7	1.7
Dissolved TP (mg/l)	mg/l	6.6	1.8	2.2	1.8	5.2	0.1	0.8	0.2	0.9
pH	–	6.1	6.3	6.8	6.9	7.2	9.4	6.9	6.4	7.7
COD/BOD	–	11	8	19	15	12	7	6	12	6
Faecal enterococci	cfu/100 ml	195	460	10	2	16	10	270	1400	8400

TP in the effluent of the conventional buried sand filters (SF) were somewhat higher than in the SBRs and in the more advanced sand filter applications with phosphorus precipitation within or after the filter bed (SFP and SF + P). The concentration difference of the advanced sand filter applications and SBRs was not significant. The proportion of suspended TP was higher in the effluent of the SBRs than in the sand filter applications (Table 1). This is a consequence of the higher effluent SS concentrations in the SBRs. Still in many cases there were rather high amounts of dissolved TP in the effluents of the SBRs indicating that the conditions for precipitation had not been optimal. Possibly the chemical dose or the maintenance of the plant was not adequate. The SBRs with chemical precipitation require regular addition of the chemical, which usually is the responsibility of the houseowner.

Nitrogen removal

The median of the effluent TN concentration for all the treatment plants was 49 mg/l. The lowest concentrations were obtained with a conventional buried sand filter with P-adsorbing media Biotite within the filter bed (SFP2) resulting in 16 mg/l of TN (median). Also the effluent TN concentrations varied a lot (Figure 3). The TN concentrations of the effluent were slightly lower in more advanced sand filter applications (SFP and SF + P) than in the conventional sand filters (SF). In the SBRs the effluent TN concentrations were higher (SBR1 and 3) or lower (SBR2) than in the sand filters. All applications of sand filters appeared to nitrify better than SBRs concluded from the NH_4-N concentrations in the effluent (Figure 4). In the sand filter applications only 20% (average) of the effluent TN was in the form of NH_4-N whereas in the SBRs the corresponding value was 63%. No correlation between air or wastewater temperature and nitrogen removal was observed.

Removal of organic matter

The median of the effluent BOD_7 and COD_{Cr} concentrations for all the treatment plants were 5 mg/l and 48 mg/l, respectively. The effluent BOD_7 concentrations were usually under 20 mg/l in all the treatment systems (Figure 3). In average all sand filter applications had lower effluent BOD_7 concentrations than the SBRs. However there was quite a big difference in concentrations between different types of SBRs. Between the effluent BOD_7 and COD_{Cr} concentrations a strong correlation ($R = 0.91$, df $= 147$) was observed. The ratio of COD_{Cr} and BOD_7 varied from 2 to 32, but 45% of the ratios were between 5 and 10. The median ratios are shown in Table 1. In the planning stage of the Finnish Decree it was discussed whether the COD_{Cr} could replace BOD_7 because the determination of COD_{Cr} is faster. Only according to this data the replacement can not be done and it requires further studies.

SS removal

The median effluent SS concentration for all the treatment plants was 7 mg/l. Usually the effluent SS concentrations were very low in the sand filter applications but in the SBRs

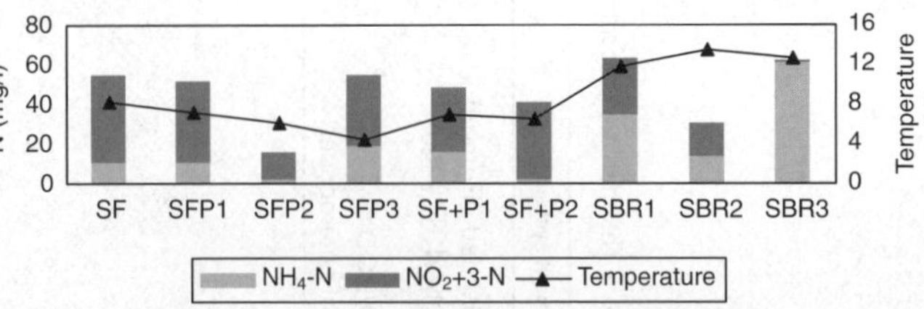

Figure 4 The effluent nitrogen parameters and wastewater temperatures

quite high concentrations were observed from time to time (Figure 3). Especially in the beginning of the effluent pumping a lot of SS escaped from the systems.

Comparison of the results with the treatment requirements of the new Finnish regulation

The influent wastewater concentrations were estimated by using the average values of untreated wastewater given in the Finnish Decree (Government Decree, 2003). Reliable influent concentrations were not obtainable as discussed above. Based on the Decree, the calculated influent concentrations are 20 mg TP/l, 130 mg TN/l and 450 mg BOD_7/l, when using the estimate of 110 l/person/d as the average water consumption determined in the Ravinnesampo project (Vilpas *et al.*, 2005). The reduction requirements given in the Decree are 85% for TP, 40% for TN and 90% for BOD_7. Consequently, the maximum allowable effluent concentrations are 3 mg TP/l, 78 mg TN/l and 45 mg BOD_7/l with the water consumption given above.

The conventional buried sand filters (SF) were not able to reduce TP as much as is required in the Finnish Decree (median TP 6.7 mg/l). All other studied treatment systems performed better in phosphorus removal and fulfilled the requirement (limit 3 mg TP/l). The effluent TN concentrations were usually under the limit value of 78 mg TN/l in all the investigated systems. All treatment systems removed organic matter well and the concentrations of effluent (limit 45 mg BOD_7/l) were not exceeded. All applications of the sand filters performed slightly better than the SBRs in this respect. As a conclusion all the treatment plants had the potential to achieve the treatment requirements when maintained properly.

Comparison of the results to the earlier studies

Several studies have been made about on-site wastewater treatment systems. There are many available technologies and for example Green and Ho (2005) have identified seventy individual small scale sanitation systems. Only Upoclean 5 (SBR1) was

Table 2 The effluent concentrations of Ravinnesampo project compared to some earlier studies

	BOD mg/l	TP mg/l	TN mg/l	SS mg/l
Ravinnesampo project (Vilpas *et al.*, 2005):				
Conventional buried sand filters (SF)	3	6.7	58	2
Buried sand filters with P-adsorbing media (SFP1 and 2)	2	2.3	31	2
Horizontal flow buried sand filters with P-adsorbing media (SFP3)	3	1.8	44	2
Conventional buried sand filters with tertiary P-filter (SF + P)	3	0.3	43	2
All SBRs	16	2.2	68	35
Upoclean 5 -SBR	16	2.0	76	24
Sweden (Hellström *et al.*, 2003; Hellström and Jonsson, 2003; Hellström *et al.*, 2000):				
Upoclean 5 – SBR (1 plant)	~20	< 1	10–35	
Package plants (5 plants)	4–18	0.7–5.7	15–65	15–38
Finland (Kujala-Räty, 2004; Kujala-Räty and Santala, 2001):				
Conventional buried sand filters (13 plants)	8	2.6	48	10
Buried sand filters with P-adsorbing media (9 plants)	5	7.9	78	11
Horizontal flow buried sand filters with P-adsorbing media (1 plant)	5	2.1	49	2
Australia (Charles *et al.*, 2005):				
Aerated wastewater treatment systems (AWTS) (140 plants)	11	–	–	18
Minnesota (Christopherson *et al.*, 2005)				
Recirculating sand filters (4 plants)	5–18	4–10	27–43	5–23

mentioned in their list. The differences e.g. in climatic conditions and design guidelines make it difficult to compare studies made in different countries. For example in the middle of Finland the mean air temperature is below zero for five months and the mean temperature of the warmest month is only 16.5 °C (Finnish Meteorological Institute, 2005).

In many countries outside Scandinavia nutrient removal is not even required. When it is required, e.g. in some areas of the USA and Australia, nitrogen is more of a concern than phosphorus (Lombardo, 2004; Beavers and Tully, 2005). This also makes it difficult to find comparable results concerning especially phosphorus removal. In Table 2 the effluent concentrations in the Ravinnesampo project are compared to earlier studies made in Finland, Sweden, Australia and Minnesota (USA). As can be seen from the table, the effluent concentrations based on the earlier studies are partly similar and partly different from the ones observed in the Ravinnesampo project. In all the studies concentrations of BOD_7 were similar compared to well operating large wastewater treatment plants (under 20 mg/l). Concentrations of TN were in the same level in all the studies. Concentrations of TP were usually lower in the systems aiming for phosphorus removal. Concentrations of SS deviated more and in sand filters the concentrations were usually lower.

Conclusions

1. In this study the plant specific median TP concentrations varied from 0.1 to 6.7 mg/l, TN from 16 to 79 mg/l, BOD_7 from 2 to 33 mg/l, COD_{Cr} from 25 to 160 mg/l, SS from 2 to 64 mg/l and faecal enterococci from 2 to 8400 cfu/100 ml. The study was executed in normal conditions to treat all the wastewater from single family homes. Consequently, the effluent concentrations of each plant varied a lot.
2. Good phosphorus removal, even using special adsorption or precipitation materials, was difficult to achieve. The sand filter applications with P-adsorbing media or tertiary P-filtration had smaller concentrations of phosphorus in the effluent than the conventional sand filters. Also all the SBRs performed well with phosphorus.
3. All types of sand filters nitrified better than the SBRs also reaching lower BOD_7 and SS concentrations in the effluent. Good nitrogen removal was obtained in one of the SBRs and most of the advanced sand filter applications. Also other systems performed moderately.
4. On average almost all the treatment systems had the potential to achieve the requirements given in the Finnish Decree reaching the effluent concentrations of $TP < 3$ mg/l, $TN < 78$ mg/l and $BOD_7 < 45$ mg/l. Proper maintenance was required to reach the limit value especially for phosphorus.
5. Considering the regulation, the performance differences between the advanced sand filter applications and SBRs were not remarkable. Thus the important factor when choosing the proper onsite wastewater treatment system is not only the performance of the system but also the maintenance it needs.

Acknowledgements

This research was financially supported by National Technology Agency of Finland (Tekes) and Ministry of Agriculture and Forestry. The authors also wish to thank Katriina Kujala-Räty, Timo Laaksonen and Taimi Määttä for their excellent work on the project and Pirjo Rantanen for her valuable advice in writing this paper. In addition, we highly appreciate the contribution of the plant manufacturers and private houseowners during this research.

References

Beavers, P.D. and Tully, I.K. (2005). Nutrient reduction evaluation of sewage effluent treatment options for small communities. *Wat. Sci. Tech.*, **51**(10), 221–229.

Charles, K.J., Ashbolt, N.J., Roser, D.J., McGuinness, R. and Deere, D.A. (2005). Effluent quality from 200 on-site sewage systems: design values for guidelines. *Wat. Sci. Tech.*, **51**(10), 163–169.

Christopherson, S.H., Anderson, J.L. and Gustafson, D.M. (2005). Evaluation of recirculating sand filter in a cold climate. *Wat. Sci. Tech.*, **51**(10), 267–272.

CEN (European Committee for Standardization) (2005). EN-12566-3: Small wastewater treatment systems for up to 50 PT – Part 3: Packaged and/or site assembled domestic wastewater treatment plants.

Finnish Meteorological Institute (2005). Ilmastotilastot (In Finnish). Available online, URL: http://www.fmi.fi (accessed 12/2005).

Government Decree (2003). The Government Decree on Treating Domestic Wastewater in Areas Outside Sewer Networks (542/2003). Finland. Available online, URL: http://www.finlex.fi/fi/laki/kaannokset/2003/en20030542.pdf (accessed 9/2006).

Green, W. and Ho, G. (2005). Small scale sanitation technologies. *Wat. Sci. Tech.*, **51**(10), 29–38.

Hellström, D. and Jonsson, L. (2003). Evaluation of small wastewater treatment systems. *Wat. Sci. Tech.*, **48**(11–12), 61–68.

Hellström, D., Qvarnström, L., Palm, O., Finnson, A. and Petterson, C.-M. (2000). Improvement of sanitary solutions for small wastewater treatment systems. *Vatten*, **56**(1), 15–19.

Hellström, D., Jonsson, L. and Sjöström, M. (2003). Bra Små Avlopp Slutrapport, Utvärdering av 15 enskilda avloppsanläggningar. Stockholm Vatten (In Swedish).

Kujala-Räty, K. (2004). On-site Wastewater Treatment Plants Functionality Study, Hajasampo Project. *Suomen ympäristö 654*, Helsinki (In Finnish). English summary available online, URL: http://www.ymparisto.fi/download.asp?contentid = 19089&lan = fi (accessed 9/2006).

Kujala-Räty, K. and Santala, E. (eds) (2001). Improvement of Wastewater Treatment in Rural Areas – Final Report of the Hajasampo Project. *Suomen ympäristö 491*, Helsinki (In Finnish).

Lombardo, P. (2004). Decentralized treatment removes nitrogen from septic effluent. *Water & Wastewater International*, **19**(2), 38–39.

Nonet, S. (2005). Requirements and test methods for on site domestic wastewater treatment plants: the European standard (prEN 12566-3) compared to other international standards. *Wat. Sci. Tech.*, **51**(10), 147–154.

Santala, E. and Kaloinen, J. (2004). New regulations enchance improvement of onsite wastewater treatment in Finland. *Proceedings of the IWA 6th International Specialised Conference on Small Water and Wastewater Systems and 1st International Specialised Conference on Onsite Wastewater Treatment and Recycling*, February 11–13, 2004, Fremantle, Australia.

Vilpas, R., Kujala-Räty, K., Laaksonen, T. and Santala, E. (2005). Enhancing nutrient removal efficiency of onsite wastewater treatment systems – Ravinnesampo, Part 1: Treatment of domestic wastewater. *Suomen ympäristö 762*, Helsinki (In Finnish). English summary available online, URL: http://www.ymparisto.fi/download.asp?contentid = 37356&lan = fi (accessed 9/2006).

Modification of submerged membrane bioreactors (MBRs) by inserting baffles: pilot scale study

K. Kimura*, M. Enomoto and Y. Watanabe***

*Department of Urban and Environmental Engineering, Hokkaido University, N13W8, Kita-ku, Sapporo 060-8628, Japan (E-mail: *kkatsu@eng.hokudai.ac.jp*; *yoshiw@eng.hokudai.ac.jp*)

**Ministry of Land, Infrastructure, and Transport of Japan, 2-1-3 Kasumigaseki, Chiyoda-ku, Tokyo 100-8918, Japan

Abstract Submerged membrane bioreactors (MBRs) have been gaining in popularity in various types of wastewater treatment. One drawback of submerged MBRs is difficulty in removing nitrogen as they are accompanied with intensive aeration inside the reactor and therefore principally operated under aerobic conditions. In order to address this problem, a simple modification for submerged MBRs, insertion of baffles to create alternative aerobic/anoxic conditions, was proposed. In this study, the performance of the proposed baffled membrane bioreactor (BMBR) was investigated based on a pilot-scale experiment using a real municipal wastewater. With appropriate operating conditions, the BMBR could remove more than 70% of total nitrogen contained in the feed water without any external carbon source. The BMBR demonstrated a good treatment performance in terms of TOC and phosphorus removal as well. Increase of trans-membrane pressure difference was subtle, which might be attributed to the alternative creation of aerobic/anoxic conditions.

Keywords Denitrification; membrane bioreactors; nitrification; wastewater treatment

Introduction

Membrane bioreactors (MBRs) are rapidly gaining in popularity in the field of municipal wastewater treatment (Stephenson *et al.*, 2000). Especially, submerged type MBRs (Yamamoto *et al.*, 1989) in which membrane modules are directly immersed in the bioreactors have been preferred due to their small footprint and low energy consumption compared with recirculated MBRs. Usually, intensive aeration is carried out in a submerged MBR. This accomplishes two things. First, supplement of oxygen to micro-organisms and second, cleaning of the membrane. As a result of the intensive aeration, one drawback associated with submerged MBRs is obvious: poor removal of nitrogen. Regarding reduction of NH_4^+ (i.e. nitrification), submerged MBRs generally show good performances due to high concentrations of nitrifiers and aerobic condition provided by the intensive aeration. However, anoxic conditions required us to promote denitrification cannot be created under such a high intensity of aeration. One possible option to overcome this drawback is to install an additional anoxic reactor where denitrification occurs besides the aerobic MBR (Côté *et al.*, 1997; Ueda and Hata, 1999; Rosenberger *et al.*, 2002). Mixed liquor is circulated between the two reactors at a fixed ratio. This type of MBR, however, apparently impairs the advantages of the submerged MBRs such as small footprint or ease of operation.

In order to address the problems stated above, insertion of baffles into the membrane chamber was proposed by the authors. Although this approach proposed herein is rather simple, in combination with an appropriate method of feed water addition, it can significantly improve the performance of submerged MBRs by promoting simultaneous nitrification/denitrification. Preliminary experiments (Kimura and Watanabe, 2005)

Water Science & Technology **Vol 55 No 7 pp 119–126** © IWA Publishing 2007

doi: 10.2166/wst.2007.135

demonstrated the proposed baffled membrane bioreactor (BMBR) certainly worked well. This paper will describe the results obtained in a pilot-scale experiment with a real wastewater to examine the feasibilities of the proposed reactor.

Materials and methods

Concept of the Baffled Membrane Bioreactor (BMBR)

Figure 1 shows the concept of the BMBR. An aeration bubbler is placed inside the inserted baffles. In the operation of the BMBR, membrane filtration is carried out in the constant flow rate mode and the flow rate of the raw wastewater feed is set at a higher value than that of filtration. When the water level is higher than the top of the baffles (Figure 1(a)), the whole reactor is vigorously mixed by the aeration and kept aerobic. Addition of the raw wastewater is stopped when the water level reaches the set highest level. Then, the water level goes down due to membrane filtration and eventually goes below the top of the inserted baffles. From this point, the reactor is separated by the baffles and discriminated into two zones. The outer zone should become anoxic due to oxygen consumption by biomass while the inner zone should be kept aerobic due to the aeration (Figure 1(b)). When the water level reaches the set lowest level, addition of the raw water is initiated and the liquid level is allowed to rise. The addition of raw wastewater must be done outside the baffles so that organic carbon contained in the raw wastewater can be utilized for denitrification. Eventually, the water level exceeds the top of the baffles and the outer zone should become aerobic again due to vigorous mixing provided by the aeration (Figure 1(a)). Thus, in the BMBR, aerobic and anoxic conditions are alternatively created in the outer zone at a constant interval and therefore the improvement in nitrogen removal is expected in comparison to normal submerged MBRs, which are principally aerobic.

Experimental apparatus

Continuous operation of a pilot-scale BMBR was conducted at an existing municipal wastewater treatment facility (Soseigawa Wastewater Treatment Plant, Sapporo, Japan), receiving the wastewater through combined sewer pipes. Temperature of the raw wastewater often decreased below 10 °C during the winter season, which makes biological treatment (especially nitrification) difficult. The BMBR used in this study was equipped with 6.0 m^2 of flat-sheet type micro-filtration (MF) membranes (Toray, Japan). Nominal pore size and material of the membrane were 0.1 μm and polyvinylidenfluoride (PVDF), respectively. Effective volume of the BMBR was 500 L, in which the ratio of the outer zone to the inner zone was approximately 2.2 when the water level was at the top of the inserted baffles. Aeration rate was fixed at 110 L/minute. As a result, dissolved oxygen

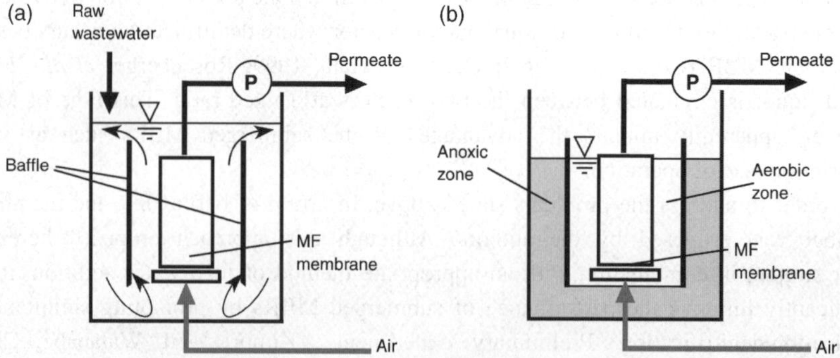

Figure 1 Concept of the baffled membrane bioreactor (BMBR)

concentration inside the baffles was always maintained above 5 mg/L. The authors had been examining "the hybrid wastewater treatment system" which is composed of pre-coagulation/sedimentation and a MBR (Itonaga and Watanabe, 2004). By carrying out the pre-treatment, enhanced removal of organic matter and phosphorus, and mitigation of membrane fouling can be achieved. The present study was conducted in the series of examination of the hybrid wastewater treatment system. Namely, the wastewater treated by coagulation and sedimentation processes was introduced to the BMBR as the feed water. Iron-based coagulant, poly-silicato iron (PSI) (Hasegawa *et al.*, 1991), was used as a coagulant. Dose of PSI was fixed at 10 mg-Fe/L.

Operating conditions

In this study, the constant flow rate mode of filtration was employed. Therefore, the required transmembrane pressure difference increased as the operation period became longer. Membrane flux was fixed at 0.4 $m^3/m^2/d$ in this study. Hydraulic retention time (HRT) in the reactor was approximately 5 hours. Intermittent operation of the suction pump (15 minutes operation and 1 minute pause) was also carried out. The continuous operation was initiated, and various water quality indices were monitored beginning on 4 November 2003 after acclimatization of biomass was confirmed. The operation was continued for approximately three months. The experimental period was divided into two phases (Runs 1 and 2). In Run 1, from 4 November 2003 to 5 December 2003, existence of the optimum OTPC (defined later) was explored with a fixed MLSS concentration of 10,000 mg/L at which concentration severe membrane fouling was not expected (Itonaga and Watanabe, 2004; Itonaga *et al.*, 2004). In Run 2, from 5 December 2003, MLSS concentration was allowed to rise by abandoning sludge extraction. As a result, MLSS concentration exceeded 20,000 mg/L in Run 2.

Results and discussion

Optimum Operation Time Per Cycle (OTPC) for the BMBR

In the operation of the proposed BMBR, the length of time that the system is aerobic/anoxic is obviously influential on denitrification efficiency. Operation time per cycle (OTPC) is defined herein as the time length that is required to complete one operation cycle. In the first phase of the continuous operation (Run 1), the optimum OTPC that could maximize nitrogen removal efficiency was explored. The OTPC was changed by adjusting the position of the water-level sensor that controlled the operation of the feed pump. Figure 2 shows the time course change in T-N concentrations in the feed water and the permeate in Run 1. The data shown in Figure 2 were obtained with various OTPCs. T-N concentration was obviously reduced by the BMBR process, demonstrating

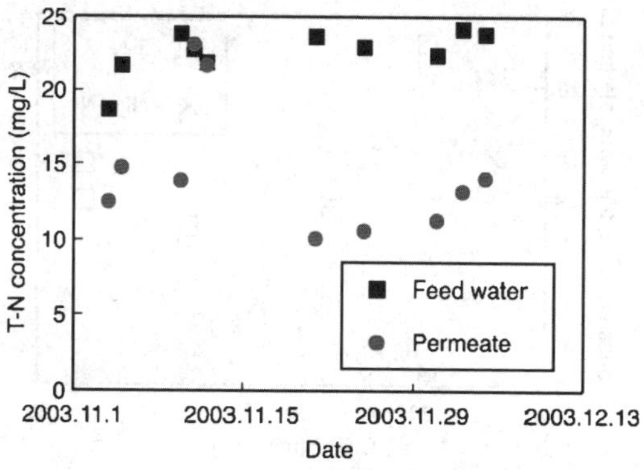

Figure 2 Removal of total nitrogen (T-N) by the BMBR in Run 1

that denitrification in the process was significant. Figure 3 shows the relationship between the OTPC and concentrations of T-N and NH_4^+-N in the permeate. Concentration of NH_4^+-N in the permeate was always maintained below 0.5 mg/L regardless of the length of the OTPC while T-N concentration in the permeate was apparently affected by the OTPC. Under the tested conditions, the optimum OTPC seemed to exist around 30 minutes.

Figure 4 shows time course changes of dissolved oxygen concentration measured outside the baffles at three different depths (numbers in Figure 4 are the distances from the top of the reactor). Figure 4(a) and (b) show data obtained with the OTPCs of 20 minutes and 29 minutes, respectively. It is clearly seen from Figure 4 that profiles of dissolved oxygen concentration measured outside the baffles significantly varied with the OTPC. Anoxic conditions were successively created outside the baffles with the OTPC of 29 minutes while it was not with the OTPC of 20 minutes. Sufficient anoxic conditions prepared by 29 minutes of the OTPC certainly promoted good denitrification (Figure 3). With the shorter OTPC (20 minutes), mixture between aerobic inside and part-time anoxic outside was made so often that dissolved oxygen could not be depressed enough to promote denitrification.

During Run 1, increase in trans-membrane pressure (TMP) was minimal (data shown later) and consequently membrane filtration could be stably continued. Creation of alternative aerobic/anoxic conditions might be beneficial for mitigation of membrane fouling in the operation of the BMBR.

Performance of the BMBR with high MLSS concentrations

In operations of MBRs, maintaining extremely high MLSS concentration is not necessarily good since it often causes severe membrane fouling. To avoid severe membrane fouling, MLSS concentration was kept around 10,000 mg/L in Run 1, according to the results of Itonaga *et al.* (2004) who used hollow-fiber submerged MBRs operated with the same feed as this study. However, judging from the extremely slow increase in TMP observed in Run 1, it was likely to be possible to continue a stable membrane filtration with the BMBR even at a higher MLSS concentration. In order to examine this point, extraction of sludge was abandoned on Dec. 1 in 2003 and MLSS concentration in the membrane chamber was increased to see the reactor performance at a higher MLSS concentration (Run 2). The OTPC was fixed at 29 minutes in Run 2.

Figure 5 shows changes in TMP. As a result of the abandonment of sludge extraction, MLSS concentration started to increase and eventually exceeded 20,000 mg/L at which the MBRs with hollow-fiber membrane expressed difficulty in filtering the mixed liquor (Itonaga and Watanabe, 2004; Itonaga *et al.*, 2004). In the present study, increase in TMP

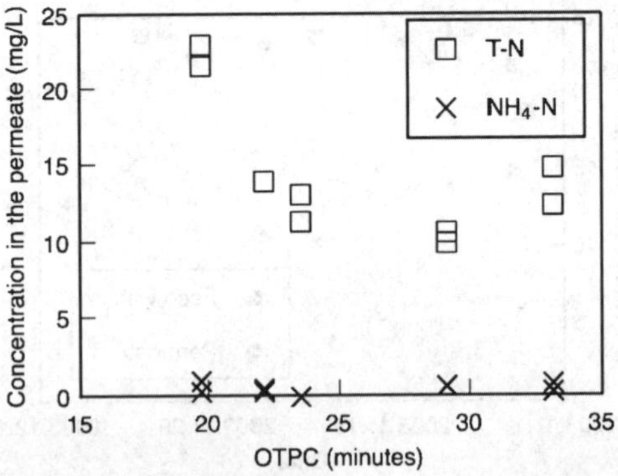

Figure 3 Relationship between the OTPC and nitrogen concentration in the permeate

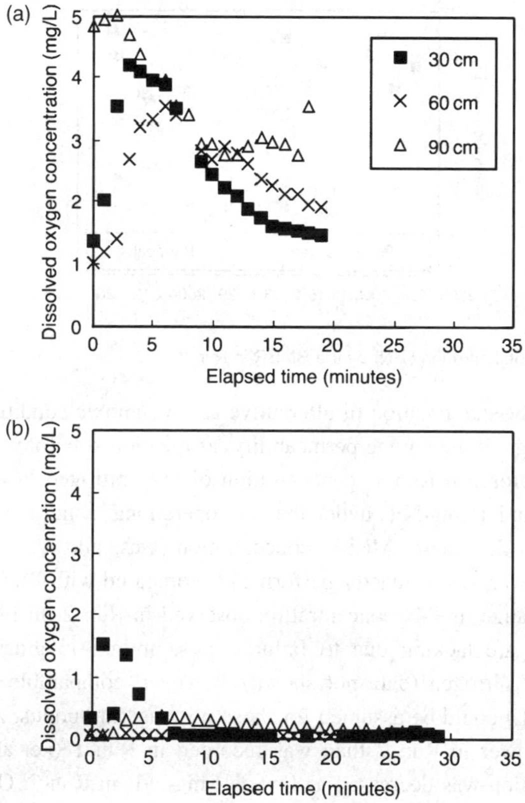

Figure 4 Time course change of dissolved oxygen concentration measured outside the inserted baffles: (a) OTPC of 20 minutes; (b) OTPC of 29 minutes

was not significant even after the abandonment of sludge extraction and membrane filtration could be continued stably. One exception can be found around 8 December 2003, when a sudden increase of TMP was observed. At that time, membrane fouling was so serious that chemical membrane cleaning (sodium hypochlorite and hydrochloric acid) was needed to restore the membrane permeability. This sudden membrane fouling might be attributed to a sudden drop of water temperature. After the membrane cleaning was carried out, increase of TMP was very slow although the MLSS concentration was very high. This stable membrane filtration could be partly attributed to the use of a flat sheet type of membrane, which is free from inter-fiber clogging. Another possible

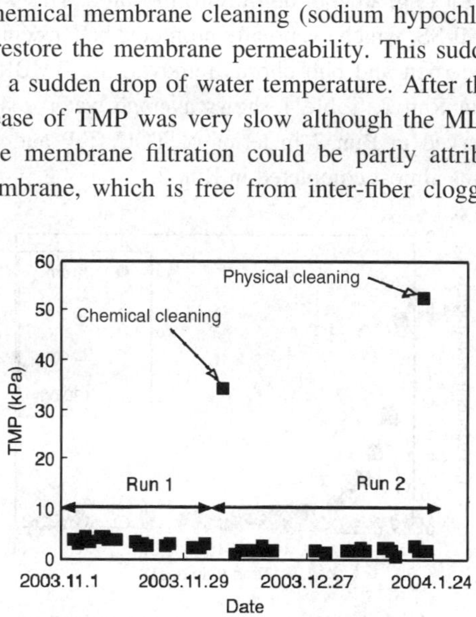

Figure 5 Change in trans-membrane pressure (TMP) in the BMBR

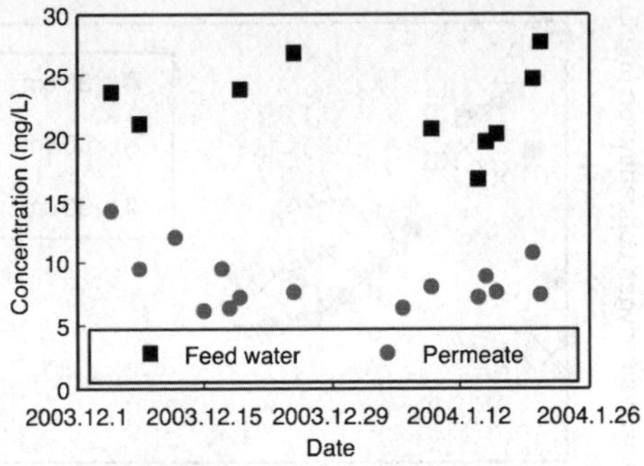

Figure 6 Removal of total nitrogen (T-N) by the BMBR in Run 2

explanation might be that creation of alternative aerobic/anoxic conditions would be beneficial to maintain good membrane permeability, as mentioned before.

As described before, in Run 1, concentration of total nitrogen in the permeate could be reduced to around 10 mg-N/L under the best operational conditions (i.e. OTPC of 29 minutes). In Run 2, where MLSS concentration was allowed to increase above 20,000 mg/L, however, better reactor performance compared with Run 1 was recognized. Figure 6 shows change in T-N concentration observed in Run 2. In Figure 6, a few data of the feed water are lacking due to failures in sampling. Judging from the data of inorganic forms of nitrogen (data not shown), however, comparable T-N concentration (e.g. 20 ~ 25 mg/L) could be assumed for those missing data points. Although the water temperature was lower in Run 2 than was recorded in Run 1 (i.e. around 15 °C), total nitrogen concentration was decreased as low as 6 mg-N/L in Run 2. One explanation for the better performance in Run 2 was the improvement in the creation of an anoxic environment. This was confirmed by dissolved oxygen measurement. Figure 7 shows variations of dissolved oxygen concentration at different depths measured in Run 2. In Run 1, the aerobic zone was recognized in the bottom part of the outside of the baffles (data not shown), which probably limited the denitrification performance of the proposed BMBR. In contrast, the creation of an anoxic environment could be achieved even in the bottom part of the reactor (e.g. a depth of 130 cm) in Run 2. This was probably due to the high concentration of MLSS, which eventually promoted better denitrification in Run 2.

Regarding organic carbon and phosphorus removal, the BMBR showed a very good performance throughout Run 2. Table 1 shows average water quality of the feed water and the permeate observed in Run 2 in terms of TOC, T-P and NH_4^+-N. As shown in Table 1, nitrification was almost completed in Run 2.

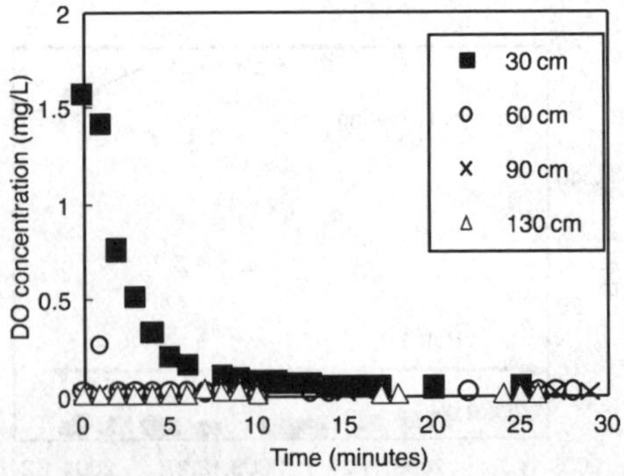

Figure 7 Time course change of dissolved oxygen concentration measured outside the inserted baffles at different depths (measurement was conducted on 4 January 2004)

Table 1 Reduction of TOC, T-P, and NH_4 by the BMBR in Run 2*

	Feed	Permeate
TOC (mg/L)	46.2	3.9
T-P (mg/L)	0.67	0.07
NH_4^+-N (mg/L)	16.1	0.57

*Averaged values

Influence of aeration rate inside the baffles on the reactor performance

For good denitrification with the BMBR, it is essential to create a sufficient anoxic condition outside the baffles. One approach to achieve this is to alter the OTPC, which was examined and described above. Reduction of aeration intensity can create good anoxic conditions as well. In order to preliminarily examine the influence of the aeration rate on the reactor performance, the aeration rate was reduced from 110 L/min to 55 L/min at the end of Run 2. This was expected to promote more significant denitrification. However, this did not enhance reduction of T-N, implying denitrification was already limited by the amount of available organic carbon. In this experiment, coagulation/sedimentation was carried out as pre-treatment for the BMBR, which certainly removed available organic carbon in the wastewater that could be used for denitrification. In Run 2, the average C/N ratio calculated for the feed water to the BMBR was 2.1, which was probably not enough for an ample biological denitrification. When the BMBR is used without such a pre-treatment, more significant reduction in T-N will probably be achieved while the advantages associated with the pre-treatment (e.g. phosphorus removal or mitigation of membrane fouling) might be impaired. This issue is currently under examination.

The reduction in the aeration rate brought about a sudden increase in TMP. However, the membrane permeability was almost completely restored by spraying pressurized water on the membrane surface and therefore chemical cleaning was not needed. Judging from this observation, reducing the air flow rate would cause a deposition of biomass on the membrane surface (reversible type of membrane fouling). Thus, caution must be taken when the aeration rate is reduced although it would enhance denitrification performance of the BMBR.

Summary

In this study, the performance of the proposed baffled membrane bioreactor (BMBR) was examined based on a pilot-scale study. Important findings obtained in this study can be summarized as follows:

The optimum OTPC that maximizes denitrification performance of the BMBR seems to exist. With such an OTPC, anoxic conditions can efficiently be created outside the inserted baffles, which leads to good denitrification. However, the optimum OTPC would be interrelated to other operating parameters (e.g. MLSS concentration) and therefore would have to be determined separately for each type of system configuration.

T-N concentration in the treated water could be reduced as low as 6 mg/L with the proposed BMBR operated under a high MLSS concentration (i.e. $>$20,000 mg/L) even when water temperature was around 12 °C. Considering the fact that the BMBR received the wastewater with a low C/N ratio due to the pretreatment in this study, it is certainly possible to achieve further reduction of T-N concentration in the treated water when wastewater without such a pre-treatment is fed.

Increase in TMP in the operation of the BMBR was generally quite slow. Alternative creation of aerobic/anoxic conditions and/or employment of a flat-sheet type of membrane might be beneficial for maintaining high membrane permeability. A sudden drop of water temperature would cause a severe irreversible membrane fouling. Also, reduction of the aeration rate inside the baffles, which would be desirable for creation of anoxic condition outside the baffles, brought about reversible membrane fouling via deposition of biomass on the membrane.

Our proposal for improvement of the performance of existing MBRs, just inserting baffles, can be applied to almost all of the submerged MBRs. However, there is still plenty of room for improvement in the reactor performance. For instance, as observed in this study, the aerobic zone might exist outside the baffles, especially in the bottom section of the reactor. This would be addressed by modification of the shape of the baffles or the position of the aeration bubbler.

References

Côté, P., Buisson, H., Pound, C. and Arakaki, G. (1997). Immersed membrane activated sludge for the reuse of municipal wastewater. *Desalination*, **113**, 189–196.

Hasegawa, T., Hashimoto, K. and Tambo, N. (1991). Characteristics of metal-polysilicate coagulants. *Water Sci. Tech.*, **23**(7–9), 1713–1722.

Itonaga, T. and Watanabe, Y. (2004). Performance of membrane bioreactor combined with pre-coagulation/sedimentation. *Water Sci. Tech.: Water Supply*, **4**(1), 143–149.

Itonaga, T., Kimura, K. and Watanabe, Y. (2004). Influence of suspension viscosity and colloidal particles on permeability of membrane used in membrane bioreactor (MBR). *Water Sci. Tech.*, **50**(12), 301–309.

Kimura, K. and Watanabe, Y. (2005) Baffled membrane bioreactor (BMBR) for advanced wastewater treatment: easy modification of existing MBRs for efficient nutrient removal. *Water Sci. Tech.*, **52**(10–11), 427–434.

Rosenberger, S., Kruger, U., Witzig, R., Manz, W., Szewzyk, U. and Krume, M. (2002). Performance of a bioreactor with submerged membranes for aerobic treatment of municipal waste water. *Water Res.*, **36**, 413–420.

Stephenson, T., Judd, S., Jefferson, B. and Brindle, K. (2000). *Membrane Bioreactors for Wastewater Treatment*, IWA Publishing, London.

Ueda, T. and Hata, K. (1999). Domestic wastewater treatment by a submerged membrane bioreactor with gravitational filtration. *Water Res.*, **33**, 2888–2892.

Yamamoto, K., Hiasa, M., Mahood, T. and Matsuo, T. (1989). Direct solid-liquid separation using hollow fiber membrane in an activated sludge aeration tank. *Water Sci. Tech.*, **21**(4/5), 43–54.

Vertical flow constructed wetland for textile effluent treatment

L.C. Davies, A. Vacas, J.M. Novais, F.G. Freire and S. Martins-Dias

Centre of Biological and Chemical Engineering, Instituto Superior Técnico, Av. Rovisco Pais, 1049-001 Lisboa, Portugal (E-mail: *susetedias@ist.utl.pt*)

Abstract A pulse feed vertical flow constructed wetland (VFCW) proved to be efficient in the treatment of a textile effluent being able to buffer, dilute and treat an Acid Orange (AO7) accidental discharge. The influence of the flooding level (FL) and pulse feed (PF) duration on the removal efficiencies of a VFCW was examined. Average AO7 removal efficiencies of 70% were achieved for an AO7 Inlet concentration of 700 mg l^{-1} applied during 15 min $cycle^{-1}$ (every three hours) at a hydraulic load of 13 l m^{-2} $cycle^{-1}$ and an FL of 21%. The VFCW was modelled by analogy with a combination of ideal reactors. The simplest combination that best reproduced the experimental results was an association of 2 reactors in series plus 1 reactor accounting the dead volumes. The model parameters helped to understand the hydrological and kinetic processes occurring in VFCW. Through the model simulation it was shown that 3 VFCW in series were enough to efficiently treat an organic mass load of 76 g_{AO7} m^{-2} day^{-1} in 9 hours and fulfil the discharge legislation. In this work it was possible to establish that the overall degradation kinetics was of first order.

Keywords Accidental discharge; acid orange 7 (AO7); flooding level; modelling; pulse feed

Introduction

Wetlands are complex, dynamic ecosystems. However, they can be engineered, designed and constructed for the purposes of waste treatment, taking advantage of natural processes within a controlled environment, involving the vegetation, soil and associated microorganisms (Haberl, 1999). Over the years, constructed wetlands (CW) have proved to be highly efficient wastewater treatment systems (in terms of BOD, COD, suspended solids, pathogens, nutrients and heavy metals removal) for municipal, industrial and agricultural effluents (Reddy and D'Angelo, 1997). CW are also attractive from an economic point of view, as their construction and running costs are low. Hydrology is recognized as the primary force influencing wetland development, survival and ecology, regulating the wetland structure and function. The hydrological regime of a wetland can be established by the frequency, timing, depth, duration of flooding and the amplitude of water level daily and/or seasonally (the latter is not applicable for CW as the water level is usually maintained uniform) (Gopal, 1999). There are three main basic types of wetland treatment systems based on their hydrology; natural wetlands; surface flow and subsurface-flow CW. Wastewater flows by gravity, vertically or horizontally, through the bed matrix.

In this work we studied the performance of a subsurface vertical flow (SSVF) system. Such systems generally use a bed of soil or gravel for the growth of rooted plants. VF systems with intermittent loading were developed to further increase oxygenation and ammonia nitrification (Cooper and Green, 1995; Ragusa *et al.*, 2004) as a consequence of the low removal efficiency of nutrients attained by the other hydraulic regimes.

The increasing application of wetlands to wastewater treatment together with progressively stricter water quality standards makes it desirable to predict the degree of treatment

doi: 10.2166/wst.2007.136

Water Science & Technology **Vol 55 No 7 pp 127–134** © IWA Publishing 2007

to be expected. This information is crucial to project the treatment potential of natural wetlands and for the design of CW (Kadlec and Knight, 1996; Rousseau *et al.*, 2004). However, the complexity of CW, which can be regarded as eco-reactors, makes them difficult to model and so their optimal design and operation is usually determined by the experience and intuition of the designer and/or operator (Pastor *et al.*, 2003). Nevertheless, several mathematical models of CW have been proposed, mainly for horizontal flow CW (Kadlec and Knight, 1996; Chazarenc *et al.*, 2003; Rousseau *et al.*, 2004).

Most modelling of wetlands and CW has been based on retention time distributions (RTD) and first order degradation kinetics (Kadlec, 2003). The RTD models are built on the supposition that each element of water passes through the wetland or CW bed matrix in different times (Rousseau *et al.*, 2004). Another approach that has been used involves mechanistic models such as continuous stirred tank reactors (CSTRs) in series (Wynn and Liehr, 2001) and plug flow reactors with dispersion (King *et al.*, 1997) involving a finite or infinite number of stages (Werner and Kadlec, 2000).

The aim of the present work was to develop a mathematical model useful to predict the behaviour of a pulse feed VFCW, in terms of an association of ideal reactors, choosing the simplest possible combination capable of reproducing the essential features of the experimental results. With this goal, an azo dye, acid orange 7 (AO7) was fed to a pilot scale VFCW, which is a dye widely used in the textile industries and often discharged into adjacent water bodies. It was used as a model molecule, due to its characteristic orange colour, which it is easily followed and because it is degraded within the CW. A well above average concentration was selected ($700\,mg_{AO7}\,l^{-1}$) representative of an accidental textile discharge compared to the typical concentration in wastewaters of less than $80\,mg_{AO7}\,l^{-1}$(Davies *et al.*, 2005), which is useful to validate the model predictions in terms of removal efficiency and dimensioning. Based on the model results obtained a prediction of the number of VFCW needed to overcome the AO7 accidental discharge is also discussed.

Methods

Operational conditions

The experimental work was carried out in a pilot VFCW located in Instituto Superior Técnico Campus, which has been extensively described in Davies *et al.* (2005). The layout of the experimental set-up is illustrated in Figure 1.

The influence of the flooding level (FL) and of the pulse feed duration (PF), on the VFCW performance was studied and modelled. The FL is regulated by the siphon height and the PF is regulated by the timer ON/OFF switch of the pump as Inlet flow remained constant. Before starting the trials, the bed matrix was already saturated with AO7 to

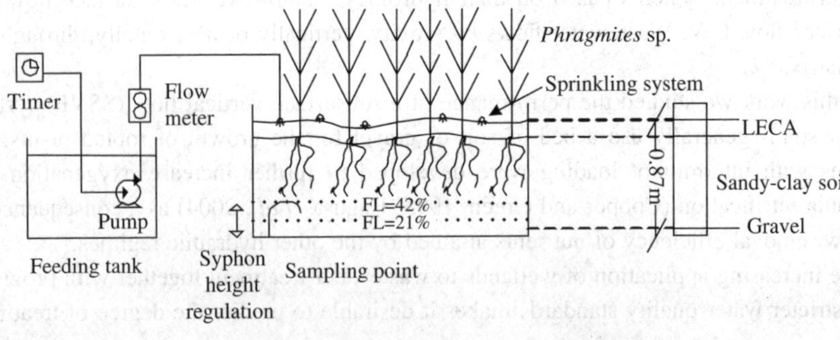

Figure 1 Schematic layout of the VFCW at pilot scale (1.60 × 0.60 × 0.87 m)

ensure that adsorption could be neglected. The system was stabilized for three weeks and three assays were then performed designated as A, B and C as shown in Table 1.

At the beginning of each assay the bed was allowed to stabilise (phase I). Afterwards, it was observed that even the results obtained during phase I were already near a pseudo-stationary state (phase II) and therefore they were included in the model development. In assay A and B, the PF was 13 minutes at $1\,l\,min^{-1}$every 3 hours, giving a hydraulic load (HL) of $108\,l\,m^{-2}day^{-1}$ and a organic mass load (OML) of $76\,g_{AO7}m^{-2}day^{-1}$. FLs of 21% and 42% (see Figure 1) were used, corresponding to water levels in the bottom of the bed of 18 and 36 cm. In assay C, the PF duration was of 26 minutes every 3 hours and the FL was 21% giving doubled HL and OML of the other assays. The assays were performed during the spring and summer time. A total of 1200 feeding pulses were applied in the VFCW and 16 cycles were sampled. The AO7 Outlet flow and AO7 concentration, [AO7], was characterised during each cycle by collecting up to 25 samples (including an inlet sample) of 50 mL at regular intervals, which were analysed for colour in terms of [AO7], COD and TOC. During trials air temperatures were monitored.

Analytical methods

COD was determined by a colorimetric method (600 nm) using a Digilab Hitachi, U-2810 UV-Vis spectrophotometer and TOC was determined using a high-temperature Dohrmann Apollo 9000 TOC analyser following the procedures described in *Standard Methods of Analysis*. [AO7] was quantified from the maximum absorbance wavelength in the visible range (490 nm).

Model development

The model for the VFCW was initially conceived by analogy with the hydrodynamics of pulsed systems and the RTD of a continuous system. As can be seen in Figure 2, the output flow has three distinct characteristics: (a) an initial delay between feeding and output, (b) a peak and (c) a long flat tail.

The initial delay could be modelled by a tubular reactor, but it was neglected since its volume was verified to be irrelevant in the overall treatment. The peak is reminiscent of the curves for RTD for two CSTRs in series. The long flat tail is suggestive of the existence of dead volumes. Therefore, the VFCW was modelled as 3 reactors, 2 CSTRs in series and 1 in parallel. Using a mechanistic approach over this model, we can consider that reactor 1 has variable volume and is located above the siphon height, reactor 2 has a fixed volume and is located near the siphon height and reactor 3 accounts for the dead volumes and is possibly distributed through the bed, the schematic representation is in Figure 3.

Table 1 Description of assays (FL-Flooding level; PF-Pulse feed)

Assay	FL (%)	PF (min)	Phase	Dates (dd/mm/2005)	Cycles	Sampling cycles
A	21	13	I	05/03–15/04	0–336	32nd, 88th, 304th
			II	16/04–26/04	336–424	360th, 424th
B	42	13	I	27/04–16/05	424–584	584th
			II	17/05–08/06	584–768	640th, 704th, 760th
C	21	26	I	09/06–07/07	768–1000	872nd, 976th, 992nd
			II	08/07–02/08	1000–1200	1032nd, 1048th, 1104th, 1200th

0-saturation; I-stabilisation; II-pseudo-stationary state

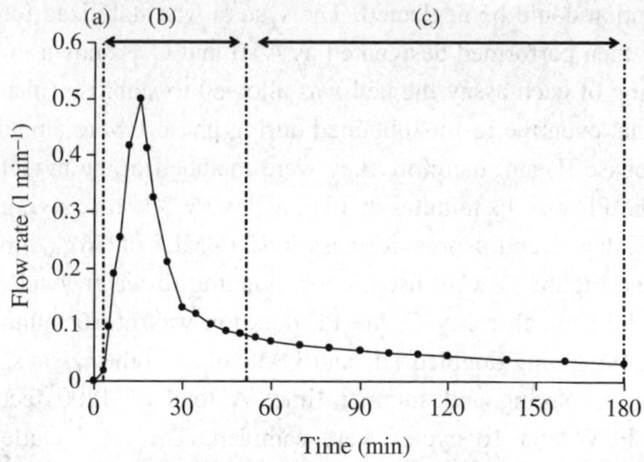

Figure 2 Typical output flow rate of a VFCW

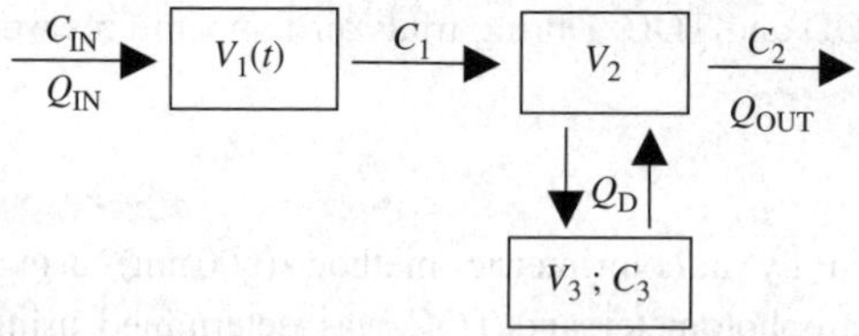

Figure 3 VFCW mechanistic modelling (C_{IN}-inlet concentration; Q_{IN}-inlet flow; V_1-volume of reactor 1; C_1-inlet concentration from reactor 1; Q_{OUT}-outlet flow; V_2-volume of reactor 2; C_2-outlet concentration; V_3-dead volume; C_3-concentration on dead volume reactor; Q_D-inlet and outlet flow of dead volume reactor)

Model assumption. The VFCW was pulse feed, so the inlet flow Q_{IN} for the model was defined by the following expression,

$$Q_{IN}(t) = \begin{cases} Q_{IN} & 0 \le t < PF_{time} \\ 0 & t \ge PF_{time} \end{cases} \quad \text{where } PF_{time} \text{ is the pulse feed time (min)} \tag{1}$$

The outlet flow was calculated using a simplification of the Carmen–Kozeny equation given by

$$Q_{OUT}(t) = \alpha \left(\frac{h(t)}{h_0} \right)^{\beta} \tag{2}$$

where α is a parameter related to the bed permeability, h_0 is the VFCW height and β is a parameter related with the flow regime. The liquid balance to reactor 1, which has variable volume, is

$$\frac{dV_1}{dt} = Q_{IN} - Q_{OUT} \tag{3}$$

Considering also that $V_1(t) = Ah(t)$ where A is the superficial area of the flooded bed. Equation (3) can be rewritten in the form of equation (4).

$$A\frac{dh(t)}{dt} = Q_{IN} - \alpha \left(\frac{h(t)}{h_0} \right)^{\beta} \tag{4}$$

For the model development the AO7 azo bond breakdown was only considered as first order irreversible reaction, thus equation (5) was considered for the kinetics.

$$(-r_{AO7}) = k\,C_{AO7} \tag{5}$$

r is the reaction rate, k is the apparent constant degradation rate and C_{AO7} is the AO7 concentration. The AO7 mass balance for reactor 1 is given by equation (6), for reactor 2 by equation (7) and for dead volume reactor by equation (8).

$$Q_{IN}C_{IN} - Q_{OUT}C_1 - kC_1Ah = A\frac{d(C_1h)}{dt} \tag{6}$$

$$Q_{OUT}(C_1 - C_2) - Q_DC_2 + Q_DC_3 - kC_2V_2 - KC_3V_3 = V_2\frac{dC_2}{dt} \tag{7}$$

$$Q_DC_2 - Q_DC_3 - kC_3V_3 = V_3\frac{dC_3}{dt} \tag{8}$$

The experimental flow rate and AO7 outlet concentration data were fitted using the Solver of MSExcel, performing non-linear optimisation of the least squares sum enabling the estimation of the overall parameters V_2, V_3, α, β, Q_D and k.

Results and discussion

VFCW hydraulic profile

The study of the influence of the FL increase from 21 to 42% (assay A and B), did not reveal significant changes in the outlet flow profiles as is presented in Figure 4 (a and b). On the other hand, the influence of the increase in the PF (assay C), from 13 to 26 min led to bed clogging. The clogging of the system was visually observed and was corroborated by the pronounced change in the outlet profiles, Figure 4 (c), which became low broad peaks denoting evident changes in the hydraulics of the bed. The FL affects the oxygen availability, with its increase, more bed is inundated and thus less interstitial oxygen is available. The FL also produces a buffer effect due to the existence of a permanent volume of effluent retained inside the bed, which dilutes the influent that enters in each pulse. The buffer effect took 25 and 35 min for assays A and B respectively. This is important information in the case of an accidental discharge as it smothers the impact of it in the system.

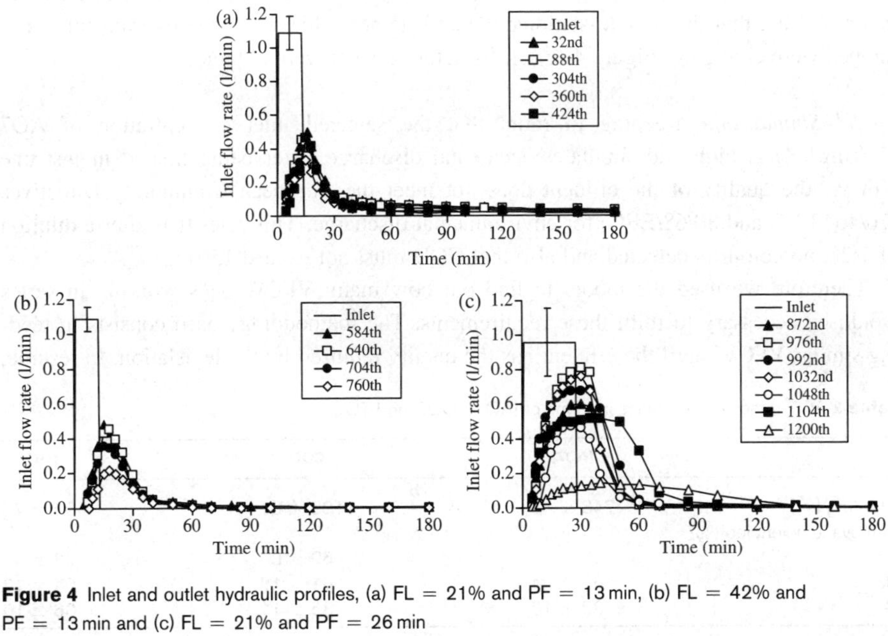

Figure 4 Inlet and outlet hydraulic profiles, (a) FL = 21% and PF = 13 min, (b) FL = 42% and PF = 13 min and (c) FL = 21% and PF = 26 min

Influence of the air temperature in the outlet flow

In Figure 4(b) a diminishing in the outlet flow quantity was noticed besides the fact that the inlet pulse remained constant. This effect was attributed to the severe changes in the evapotranspiration rate (ET), up to 70% of inlet flow, due to air temperature variation from 18 to 40 °C. In this respect, Kadlec (1997) suggests that the evapotranspiration has a double effect of lengthening the retention time and also concentrating the dissolved constituents. So, the AO7 concentration was calculated accordingly.

VFCW AO7 removal efficiency

The removal efficiencies were calculated by means of a mass balance, by the integration of the experimental data using equation (9) for each parameter.

$$\text{Removal Efficiency}(\%) = \frac{QC_{\text{IN}}\text{PF}_{\text{time}} - \sum_{n=1}^{N} Q_n C_{\text{OUT}_n} \Delta t_n}{QC_{\text{IN}}\text{PF}_{\text{time}}} \times 100 \qquad (9)$$

where Q is the flow rate, C is the concentration, N sample number and Δt is the sampling time interval. The obtained results can be seen in Table 2.

Similar values were obtained for [AO7], COD and TOC removals, which indicate that a mineralization process is occurring. Assay B, lower removal efficiencies could be attributed to the lesser free oxygen availability associated with higher FL. In assay C, the clogging of the system affected drastically the removal efficiencies that decreased to half of the previous values.

VFCW Model

The model was adjusted individually for each assay (A, B and C) and presented in Figure 5 and the estimated parameters can be found in Table 3.

With the increase in the FL to 42%, the model predicted that the dead volume was more isolated (lower Q_D). Increasing the PF_{time} three effects were noticed: (1) the k diminished that can be attributed to the decreasing oxygenation of the medium; (2) the V_3 double sized; (3) Q_D also doubled. The two last effects can be explained by overcoming the bed permeability capacity.

For assay A and B the model reproduces very accurately the results. For assay C, given the fact that the pseudo-stationary state was not achieved as the hydrodynamics of the bed were changing (Figure 4 (c)), satisfactory results were obtained.

Model Simulation. Keeping in mind that the selected inlet concentration of AO7 ($700\,\text{mg}\,\text{l}^{-1}$) is high and simulates accidental discharge, after being treated in just one VFCW, the quality of the effluent does not meet the European Community Directives (76/464/EEC and 80/68/EEC) for environmental discharge. This rules that after a dilution of 1:20, no colour is detected and also that COD must not exceed $150\,\text{mg}\,\text{l}^{-1}$.

Therefore we used the model to find out how many VFCW beds working in series would be necessary to fulfil these requirements. The methodology used consists in feeding similar VFCW until the effluent has the quality required by the legislation. In resume,

Table 2 AO7 removal efficiencies in terms of [AO7], COD and TOC

	[AO7]	COD	TOC
Inlet (mg l^{-1})	748 ± 42	1135 ± 61	452 ± 75
Removal efficiencies (%)			
A	68 ± 4	69 ± 8	67 ± 4
B	62 ± 24	61 ± 12	64 ± 12
C	32 ± 13	35 ± 11	36 ± 10

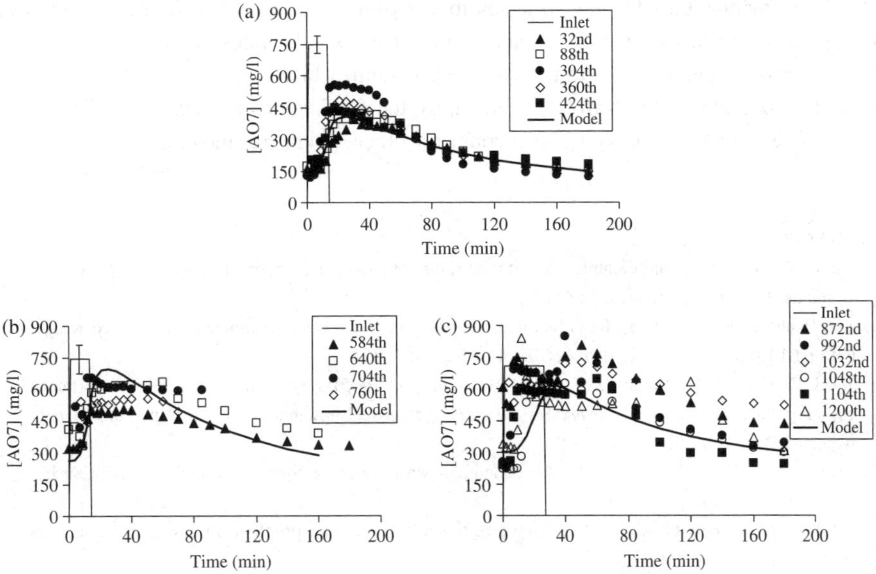

Figure 5 [AO7] Inlet and outlet profile for assays A, B, C and modelling predictions

Table 3 Parameters obtained through the VFCW model using the different operational conditions

Parameters	A	B	C
$k(s^{-1})$	8.2×10^{-4}	9.2×10^{-4}	5.7×10^{-4}
V_2 (m^3)	3.1×10^{-3}	37×10^{-3}	4.0×10^{-3}
Q_D ($m^3.s^{-1}$)	5.6×10^{-7}	1.8×10^{-7}	16×10^{-7}
V_3 (m^3)	4.9×10^{-2}	4.6×10^{-2}	7.9×10^{-2}
α ($m^3.s^{-1}$)	0.2		
β	10.76		

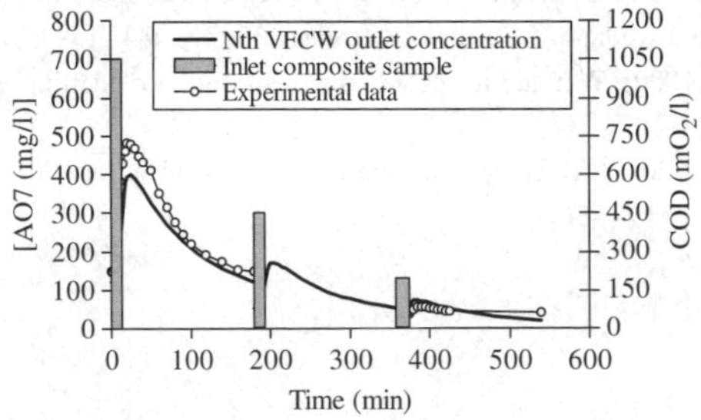

Figure 6 Simulation profiles in a 3 VFCW working in series for a 700 mg_{AO7} l^{-1} accidental discharge

it would be necessary to operate 3 VFCW in series with identical characteristics of size, porosity summary etc. In the third VFCW, a colourless composite sample with concentration of 50 mg l^{-1} would be expected. The experimental data validate this simulation as the VFCW was fed with an inlet concentration of 130 mg l^{-1} and the quality of the outlet effluent meet the legislation. This model validation also confirms the first order of the degradation mechanisms behaviour of VFCW as represented in Figure 6.

Conclusions

VFCW are efficient in enriched azo dye wastewater treatment. In this work it was indirectly shown that the degradation mechanisms or the limiting path is of first order. The parameter variation given by the model was shown to be essential to understand the

VFCW behaviour, thus leading to a possible optimisation. The VFCW model can easily be extended to other wastewater contaminants if the kinetic rates are known. The PF_{time} is an important operational parameter in a PF system, which can lead to the bed clogging rather than FL that did not affect significantly the AO7 removal efficiencies. The model is still under development trying to account for other design parameters.

References

Chazarenc, F., Merlin, G. and Gonthier, Y. (2003). Hydrodynamics of horizontal subsurface flow constructed wetlands. *Ecol. Eng.*, **21**(2–3), 165–173.

Cooper, P. and Green, B. (1995). Reed bed treatment systems for sewage treatment in the United Kingdom – the first 10 years experience. *Wat. Sci. Tech.*, **32**(3), 317–327.

Davies, L.C., Catia, C.C., Novais, J.M. and Martins-Dias, S. (2005). Phytoremediation of textile effluents containing azo dye by using *Phragmites australis* in a vertical flow intermittent feeding constructed wetland. *Ecol. Eng.*, **25**(5), 594–605.

Gopal, B. (1999). Natural and constructed wetlands for wastewater treatment: potentials and problems. *Wat. Sci. Tech.*, **40**(3), 27–35.

Haberl, R. (1999). Constructed wetlands: a chance to solve wastewater problems in developing countries. *Wat. Sci. Tech.*, **40**(3), 11–17.

Kadlec, R.H. (1997). Deterministic and stochastic aspects of constructed wetland performance and design. *Wat. Sci. Tech.*, **35**(5), 149–156.

Kadlec, R. (2003). Effects of pollutant speciation in treatment wetlands design. *Ecol. Eng.*, **20**(1), 1–16.

Kadlec, R.H. and Knight, R.L. (1996). *Treatment Wetlands*, CRC Press, Boca Raton, FL, USA.

King, A.C., Mitchell, C.A. and Howes, T. (1997). Hydraulic tracer studies in a pilot scale subsurface flow constructed wetland. *Wat. Sci. Tech.*, **35**(5), 189–196.

Pastor, R., Benqlilou, C., Paz, D., Cardenas, G., Espuña, A. and Puigjaner, L. (2003). Design optimisation of constructed wetlands for wastewater treatment. *Resou. Conserv. Rec.*, **37**(3), 193–204.

Ragusa, S., McNevin, D., Qasem, S. and Mitchell, C. (2004). Indicators of biofilm development and activity in constructed wetlands microcosms. *Wat. Res.*, **38**(11), 2865–2873.

Reddy, K.R. and D'Angelo, E.M. (1997). Biogeochemical indicators to evaluate pollutant removal efficiency in constructed wetland. *Wat. Sci.Tech.*, **35**(5), 1–10.

Rousseau, D.P.L., Vanrolleghem, P.A. and De Pauw, N. (2004). Model-based deign of horizontal subsurface flow constructed treatment wetlands: a review. *Wat. Res.*, **38**(6), 1484–1493.

Werner, T. and Kadlec, R. (2000). Wetland residence time distribution modelling. *Ecol. Eng.*, **15**(1–2), 77–90.

Wynn, T.M. and Liehr, S.K. (2001). Development of a constructed subsurface-flow wetland simulation model. *Ecol. Eng.*, **16**(4), 519–536.

Water Science & Technology **Vol 55 No 7 pp 135–141** © IWA Publishing 2007

Biological nutrient removal by applying SBR technology in small wastewater treatment plants: carbon source and C/N/P ratio effects

S. Puig, Ll. Corominas, M.D. Balaguer and J. Colprim

Lab. of Chemical and Environmental Engineering (LEQUIA). University of Girona, Facultat de Ciències, Campus Montilivi, E-17071 Girona, Catalonia, Spain (E-mail: *sebastia@lequia.udg.es*; *lluis@lequia.udg.es*; *marilos@lequia.udg.es*; *J.Colprim@lequia.udg.es*)

Abstract SBR technology is considered an alternative to conventional processes such as Phoredox, Five-stage Bardenpho, among the others for treating nutrients in wastewaters. It is especially applicable to small communities of a just few people to a population equivalent (p.e) up to 4000. In this paper, biological nutrient removal using SBR technology in a single reactor is presented. Biological nutrient removal requires a sequence of anaerobic–anoxic–aerobic phases with multiple feeding events over one cycle. This filling strategy was adapted to enhance denitrification and phosphate release, using the easily biodegradable organic matter from the wastewater. In spite of using this feeding strategy, the organic matter concentration can be insufficient. The results show that biological nutrient removal was successfully achieved by using only one reactor, working with a low organic matter concentration in the influent (C/N/P ratio of 100:12:1.8). Nevertheless, when the C/P ratio was lower than 36 g COD·g^{-1} P-PO_4, an accumulation of phosphate was observed. After that, the system responded quickly and returned to ideal conditions (C/P ratio of 67 g COD·g^{-1} P-PO_4), taking only 15 days to achieve the complete nutrient removal. Furthermore, the operational conditions and the synthetic wastewater used conferred a selective advantage to polyphosphate accumulating organisms (PAOs) over glycogen accumulating non-poly-P organisms (GAOs) as shown by the FISH analysis performed.

Keywords C/P and C/N/P ratios; nutrient removal; PAOs; SBRs; step-feed strategy

Introduction

Treatment of contaminated wastewater with biological processes has been widely used for classical urban and industrial wastewaters (Metcalf and Eddy, 2003). Since stricter legislation is being applied, more attention to phosphorus removal is needed. The traditional method for removing phosphate from wastewater is adding precipitating chemicals (iron or aluminium salts) to the wastewater. Besides the fact that using chemicals in waste treatment should be minimized, there are several negative aspects of this practice (Van Loosdrecht *et al.*, 1997). However, although there are many biological nutrient removal processes to deal with nutrient removal, some of their drawbacks must be considered, such as moderate levels of nitrogen removal (i.e. A^2/O^{TM}, VIP, UCT(University of Cape Town) processes), moderate to poor phosphorus removal (i.e. Five-stage Bardenpho process), large reactor volumes required (i.e. A^2/O^{TM} (Grady *et al.*, 1999)) or large investments in extra infrastructure (i.e. Phostrip process (Van Loosdrecht *et al.*, 1997)).

SBR technology is considered an alternative to these conventional processes for removing nutrients from wastewaters. It is especially applicable to small communities with a few inhabitants up to 4000 p.e. SBR technology has successfully been applied in small wastewater treatment plants treating urban (Lee *et al.*, 2004; Puig *et al.*, 2005) and industrial (Torrijos *et al.*, 2001; Vives *et al.*, 2003) wastewaters. It is especially

doi: 10.2166/wst.2007.137

appropriate for places where there is significant flow and load variability or where space problems are a restriction (Metcalf and Eddy, 2003).

Nutrient removal in SBRs requires a combination of anaerobic–anoxic–aerobic phases. During the anaerobic phase, the carbon source in the form of volatile fatty acids (VFAs) is taken up by polyphosphate-accumulating organisms (PAOs) releasing phosphate into the liquid phase. Within the cells, VFAs are stored basically as polyhydroxybutyrate (PHB) while intracellular glycogen is converted to polyhydroxyalcaonate (PHA). In the anoxic phases, if a soluble and biodegradable carbon source is present, a classic heterotrophic denitrification process takes place. Finally, under aerobic conditions, PAOs grow and accumulate phosphate in the cells, while autotrophic biomass is responsible for the nitrification process. Phosphorus removal is achieved through the wastage of excess sludge with high poly-P content.

Glycogen-Accumulating Organisms (GAOs) are a group of bacteria capable of competing with PAOs during the anaerobic VFA uptake. Like PAOs, GAOs take up VFAs anaerobically and convert them into PHAs via hydrolysis of glycogen as their sole source of energy for this process (Mino *et al.*, 1998). Nevertheless, PAOs have a clear competitive advantage in that they can accumulate VFAs in the cells without the necessity of an external electron acceptor because, under aerobic or anoxic conditions, PAOs can grow at the expense of their stored substrate.

When operating SBRs for nutrient removal, special attention has to be given to the availability and use of the easily biodegradable substrate. A step-feed strategy for nutrient removal means that the influent filling phases must be carried out under anaerobic or anoxic conditions in order to increase the denitrification and phosphate release efficiencies (Puig *et al.*, 2004). In spite of using this feeding strategy, the organic matter concentration can be insufficient. When an excessive C/P ratio is provided, PAOs can accumulate a high content of polyphosphate coupled with a higher and faster substrate uptake ability, successfully out-competing GAOs. In contrast, reducing the C/P ratio causes the depletion of the polyphosphate content in PAOs, eventually leading to them being replaced by GAOs as the majority (Liu *et al.*, 1997).

Puig *et al.* (2004) described a successful implementation in SBRs using 4 to 6 filling events with a sequence of anoxic–aerobic phases treating synthetic and urban wastewater for carbon and nitrogen removal. In this paper, the SBR cycle was adapted for biological nutrient removal (designed to treat 90 p.e.) in a single reactor. A filling strategy was adapted to enhance denitrification and phosphate release. Furthermore, the influence of C/P and C/N feeding ratios on the efficiency of nutrient removal using an SBR was investigated. In addition, biological phosphorus removal has been assessed by monitoring the evolution of PAOs versus GAOs.

Materials and methods

Synthetic feed

The synthetic wastewater was basically composed of a mixed carbon source, an ammonium solution, a phosphate buffer, an alkalinity control ($NaHCO_3$) and a microelements solution (adapted from Dangcong *et al.*, 2000)). The synthetic wastewater was prepared twice a week with a concentration of $502 \pm 175\,mg{\cdot}L^{-1}$ COD, $61.5 \pm 10.1\,mg{\cdot}L^{-1}$ N-TKN, $47.9 \pm 7.3\,mg{\cdot}L^{-1}$ N-NH_4^+ and 9.1 ± 3.8 mg P-$PO_4{\cdot}L^{-1}$ during stable-state operation, giving C/N and C/P ratios of 8 ± 2 mg COD·mg^{-1} N-TKN and 52 ± 16 mg COD·mg^{-1} P-PO_4, respectively. The composition contained per L: 0.27 ml EtOH, 0.56 g DME (Dehydrated meat extract), 0.4 mL milk, 13.3 mL leachate, 280 mg $NaHCO_3$, 183 mg NH_4Cl, a microelement solution (described in Vives, 2004) and a phosphate buffer (7.0 mg KH_2PO_4, 18 mg K_2HPO_4, 14 mg $Na_2HPO_4{\cdot}7H_2O$).

SBR pilot plant

The SBR was composed of a cylindrical reactor with maximum and minimum working volumes of 30 L and 20 L, respectively. To achieve nutrient removal in the SBR, the hydraulic retention time (HRT) and solids retention time (SRT) were maintained at 1 day and 20 days, respectively. The SBR plant was seeded with sludge from the Sils-Vidreres wastewater treatment plant (Girona, Spain) and it treated synthetic wastewater for nutrient removal.

An 8 hour cycle (Figure 1) with 3 feeding steps (the first one double the volume of the others) was implemented for nutrient removal. The wastewater was always introduced under anaerobic or anoxic conditions in order to enhance phosphate release and denitrification, using the easily biodegradable organic matter from the wastewater. The cycle was divided into two parts, one focussed on phosphorus removal and the other on nitrogen removal. The first anaerobic–aerobic pair phase was related to phosphorus removal. The anaerobic phase improved the readily biodegradable organic matter uptake by PAOs and was followed by an aerobic phase for the phosphate uptake. Phosphorus removal was achieved through the wastage of excess sludge with high poly-P content. The second part consisted of a sequence of anoxic–aerobic phases for nitrogen removal. The cycle was divided into a reaction phase (426 minutes; 48.8% and 20.2% under anaerobic and anoxic conditions respectively), a settling phase (39 minutes) and a draw phase (15 minutes).

FISH analysis

Fluorescence *in situ* hybridization (FISH) analysis was performed as specified by Amman (1995) using the Cy5-labeled-EUBMIX probe to target the entire bacterial community (Daims *et al.*, 1999), and the Accumulibacter and Competibacter probes were used to target Cy3-labelled PAOMIX (consisting of PAO462, PAO651, and PAO846) for Accumulibacter, the only currently identified PAO and Fluos-labelled GAOMIX (consisting of GAO431 and GAO989) was used to target Competibacter, the only currently identified GAO (Crocetti *et al.*, 2002). The probed sludge was examined using an Epifluorescence microscope. The area containing specific labelled probe cells (Cy3 and Fluos for PAOMIX and GAOMIX, respectively) was quantified as a percentage of the area of the entire Cy5-labeled bacterial population probe (EUBMIX).

Analytical methods

Total suspended solids (TSS), volatile suspended solids (VSS), total chemical oxygen demand (COD), ammonium (N-NH_4^+), total Kjeldahl nitrogen (N-TKN), nitrites (N-NO_2^-), nitrates (N-NO_3^-) and phosphate (P-PO_4^{3-}) were all measured according to *Standard Methods* (APHA, 1995). Total nitrogen (TN) was calculated as the sum of N-TKN, and nitrite and nitrate concentrations as mg N-TN·L^{-1}.

Results and discussion

SBR operation and performance

The SBR plant operated for three months using an 8-hour cycle (Figure 1) for nutrient removal. A filling strategy was adopted to enhance denitrification and phosphate release

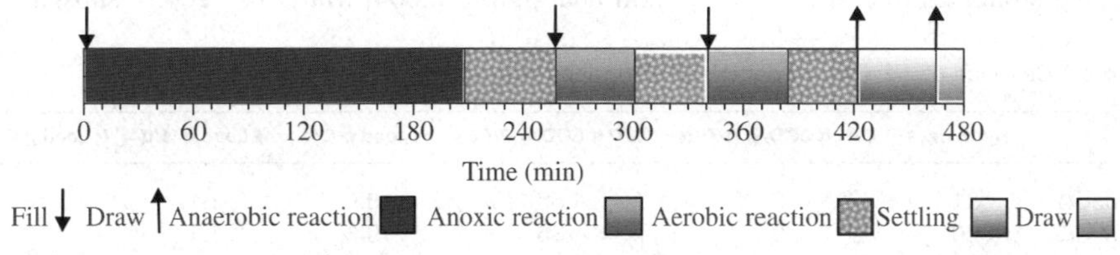

Figure 1 Definition of the SBR cycle used during the experimental study

using the easily biodegradable organic matter from the wastewater. This cycle operated with different load conditions divided into three periods according to the C/N and C/P ratios (Table 1).

Figure 2 shows the evolution of carbon, nitrogen and phosphorus components and the C/N and C/P feeding ratios during the experimental study.

Period 1: Start-up of the SBR. The reactor was seeded with sludge from urban wastewater (Sils-Vidreres WWTP) and fed with synthetic wastewater. A stable state (Figure 2) was reached after 2 weeks of operation achieving average COD, TN and phosphate effluent concentrations of 44, 2.2 and 0.15 $mg \cdot L^{-1}$, respectively, and leading to removal efficiencies of 94%, 95% and 98% for carbon, nitrogen and phosphorus, respectively.

In spite of the competition for the carbon source between PAOs and denitrifiers, C/P (55 g $COD \cdot g^{-1}$ $P\text{-}PO_4$) and C/N (8 g $COD \cdot g^{-1}$ N-TKN) ratios were enough to achieve complete nutrient removal.

Period 2: C/P ratio decrease. After achieving complete nutrient removal in period 1, an increase in phosphate concentration in the influent (17 mg $P\text{-}PO_4 \cdot L^{-1}$, on average) was applied in period 2 the carbon and nitrogen concentrations remained the same. Despite the small amount of carbon source available (Figure 2B), the nitrogen effluent concentrations (0.3 mg $N\text{-}NH_4^+ \cdot L^{-1}$ and 2.5 mg $N\text{-}NO_3^- \cdot L^{-1}$) were not affected and the nitrogen concentration was always lower than the European Directive values (D. level, Figure 2). However, the lower amount of available organic matter (C/P ratio of 36 g $COD \cdot g^{-1}$ $P\text{-}PO_4$) clearly affected the phosphorus removal performance (Figure 2D) as phosphate started to accumulate in the reactor.

Period 3: Recuperation time for the complete nutrient removal. The aim was to discover the time needed to achieve complete nutrient removal after a period of phosphate accumulation. For this reason, the C/P was kept higher (67 g $COD \cdot g^{-1}$ $P\text{-}PO_4$) than in period 2.

In spite of the high COD variability in the influent (Figure 2B, range between 102 and 596 mg $COD \cdot L^{-1}$), complete nutrient removal was achieved. Thus, the nitrogen removal efficiencies remained high in spite of the low influent COD concentration on days 66 and 69. Nevertheless, recovering the phosphorus removal took more time. The PAOs' response was slower and, after 15 days of operation, complete phosphorus removal was achieved again.

Table 2 compares the variation of C/N and C/P ratios and of nitrogen and phosphorus removal efficiencies in this study with the results reported by other researchers.

Comparing the data shown in Table 2, some variability can be observed between the C/N and C/P feeding ratios and nitrogen and phosphorus removal efficiencies in the different studies. The nitrogen efficiencies presented (periods 1 and 3) are higher than those obtained by other researchers working with the same or a higher C/N ratio (C/N of 7–10 g $COD \cdot g^{-1}$ TKN) and applying the step-feed strategy. In single-feed or continuous feeding strategies (Yu *et al.*, 1997; Akin and Ugurlu, 2004; Ma *et al.*, 2005), substantial

Table 1 Definition of the different periods of study according to the C/N and C/P ratios

	Length days	C/N g COD/g N-TKN	C/P g COD/g $P\text{-}PO_4$	C load g C/d	N Load g N/d	P Load g P/d
Period 1	31	8	55	8.8	1.1	0.16
Period 2	20	9	36	9.6	1.1	0.27
Period 3	37	7	67	6.2	0.9	0.09

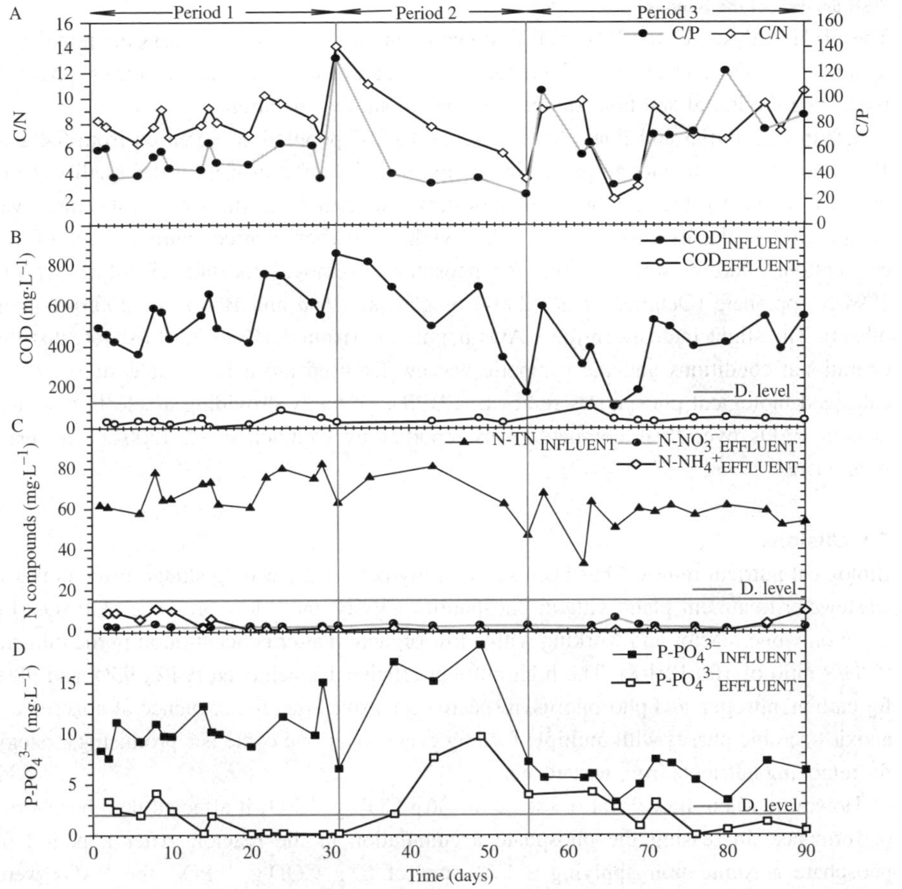

Figure 2 Evolution of carbon, nitrogen and phosphorus components during the experimental period applying an 8-hour cycle in the SBR pilot plant (D.level: Maximum discharge concentration admissible)

amounts of nitrate might remain in the effluent because endogenous denitrification cannot reduce nitrate efficiently in the anoxic reaction and settle period due to the lack of an organic electron donor (Chang and Hao, 1996).

However, the step-feed strategy also improved phosphorus removal in the experimental study working with the same C/P ratio (55–58 g $COD \cdot g^{-1}$ $P\text{-}PO_4$), as other researchers that used continuous feeding and feeding during the settle phase strategies have also reported (Yu *et al.*, 1997; Keller *et al.*, 2001). Akin and Ugurlu, 2004 showed that when the C/P ratio was too low (19 g $COD \cdot g^{-1}$ $P\text{-}PO_4$), the phosphorus removal efficiency decreased.

Table 2 Comparison of C/N and C/P ratios and of nitrogen and phosphorus removal efficiencies in different studies of nutrient removal

Publication	Kind of wastewater	Reactor configuration	C/N	C/P	C/N/P	% N removal	% P removal
This study (Period 1)	Synthetic	SBR	8	55	100:12:1.8	95	98
This study (Period 3)	Synthetic	SBR	7	67	100:14:1.5	96	90
Yu *et al.*, 1997	Synthetic	SBR	7	55	100:14:1.8	81	61
Keller *et al.*, 2001	Urban	SBR	10	58	100:10:1.7	96	84
Kargi and Uygur, 2003	Synthetic	SBR	10–50	40–250	100:2:0.54	94	99
Akin and Ugulu, 2004	Synthetic	SBR	8	19	100:13:5.3	37	80
Ma *et al.*, 2005	Synthetic	A^2/O	6	18 to 61	100:17: (1.6 to 5.6)	75–84	90–98

FISH analysis of the biomass

The FISH analysis for the PAO and GAO communities was done to assess the population dynamics. FISH analysis was done for the initial evolution of sludge during phase 1, presenting the initial and final sludge after one month of operation.

Comparing initial and final phase data, the PAOs' population increased from 9.4% to 17.5% in spite of the sludge presenting a significant change in wastewater quality (from the real plant to the synthetic wastewater). Nevertheless, influent wastewater was composed of ethanol and DME as the synthetic carbon source, without any of the conventional carbon sources used for phosphorus removal (acetate (Smolders *et al.*, 1994), propionate (Oehmen *et al.*, 2004) or glucose (Pala and Bölükbaş, 2005), among others). The slight increase in the GAOs population (from 3.1% to 3.5%) shows that the operational conditions and the synthetic wastewater used are a favourable substrate for enhanced biological phosphorus removal (EBPR), probably providing a selective advantage to PAOs over their competitors as reported by Oehmen *et al.* (2004) who used propionate.

Conclusions

Biological nutrient removal has been successfully achieved (seeding sludge from an urban wastewater treatment plant without phosphorus activity and a low presence of PAOs) by using only one reactor and working with a low organic matter concentration in the influent (C/N/P ratio of 100:12:1.8). The high nutrient efficiencies achieved (94%, 95% and 98% for carbon, nitrogen and phosphorus, respectively) proved that the sequence of anaerobic–anoxic–aerobic phases with multiple feeding events over one cycle is a promising strategy for removing nutrients from wastewater.

However, when the C/P ratio was below 36 g COD·g^{-1} PO_4 it affected the phosphorus performance, increasing the phosphate accumulation in the reactor. After a period of phosphate accumulation applying a C/P ratio of 67 g COD·g^{-1} PO_4, the PAOs were observed to respond quickly (15 days).

The selective enrichment of PAOs, working with a carbon source without any VFAs, shows that the operational conditions and the synthetic wastewater used are a favourable substrate for EBPR that probably provide a selective advantage to PAOs over their competitors. This topic will be further developed in future work.

Acknowledgements

The authors would like to thank the Spanish Government (MCYT-DPI-2002–04579-C02 and DPI2005-08922-C02) and the University of Girona (BR03/11) for their financial support of this study. In addition, the authors gratefully acknowledge Ariadna Cabezas and the Department of Biotechnology of TU Delft (Netherlands) for their valuable cooperation during the experimental study.

References

Akin, B.S. and Ugulu, A. (2004). The effect of anoxic zone on the biological phosphorus removal by a sequential batch reactor. *Bioresource Technology*, **94**(1), 1–7.

Amman, R.I. (1995). Fluorescently labelled, ribosomal-RNA-targeted oligonucleotide probes in the study of microbial ecology. *Molecular Ecology*, **4**, 543–553.

APHA, AWWA, WEF (1995). *Standard Methods for the Examination of Water and Wastewater*. 19th edn. American Public Health Association/American Water Works Association/Water Environment Federation, Washington DC, USA.

Chang, C.H. and Hao, O.J. (1996). Sequencing batch reactor system for nutrient removal: ORP and pH profiles. *J. Chem. Tech. Biotechnol.*, **67**, 27–38.

Crocetti, G.R., Banfieldm, J-F., Keller, J., Bond, P.L. and Blackball, L.L. (2002). Glycogenaccumulating organisms in laboratory-scale and full-scale wastewater treatment processes. *Microbiology*, **148**, 3353–3364.

Daims, H., Bruhl, A., Amann, R., Schleifer, K.H. and Wagner, M. (1999). The domain specific probe EUB338 is insufficient for the detection of all bacteria: development and evaluation of a more comprehensive probe set. *Syst. Appl. Microbiol.*, **22**(3), 434–444.

Dangcond, P., Bernet, N., Delgenes, J.P. and Moletta, R. (2000). Effects of oxygen supply methods on the performance of a sequencing batch reactor for high ammonium nitrification. *Water Environ. Res.*, **72**, 195.

Grady, J., Daigger, G. and Lim, H. (1999). *Biological Wastewater Treatment*, Marcel Dekker, Inc., New York.

Kargi, F. and Uygur, A. (2003). Nutrient removal performance of a five-step sequencing batch reactor as a function of wastewater composition. *Process Biochemistry*, **38**, 1039–1045.

Keller, J., Watts, S., Baytte-Smith, W. and Chong, R. (2001). Full-scale demonstration of biological nutrient removal single tank sequencing batch recators. *Water Sci. Technol*, **43**(3), 355–362.

Lee, H., Min, Y.M., Park, C.H. and Park, Y.H. (2004). Automatic control and remote monitoring system for biological nutrient removal on small wastewater treatment plants in Korea. *Water Sci. Technol*, **50**(6), 199–206.

Liu, W-T., Nakamura, K., Matsuo, T. and Mino, T. (1997). Internal energy-based competition between polyphosphate and glycogen-accumulating bacteria in biological phosphorus removal reactors- effect of C/P feeding ratio. *Water Res.*, **31**(6), 1430–1438.

Ma, Y., Peng, Y-Z., Wang, X-L. and Wang, S-Y. (2005). Nutrient removal performance of an anaerobic-anoxic-aerobic process as a function of influent C/P ratio. *J. Chem. Technol. Biotechnol.*, **80**, 1118–1124.

Metcalf and Eddy (2003). *Wastewater Engineering: Treatment and Reuse*, 4th edn, McGraw-Hill Higher Education, New York.

Mino, T., van Loosdrecht, M.C.M. and Heijnen, J.J. (1998). Microbiology and biochemistry of the enhanced biological phosphate removal process. *Water Res.*, **32**(11), 3193–3207.

Oehmen, A., Yuan, Z., Blackall, L.L. and Keller, J. (2004). Short-term effects of carbon source on the competition of polyphosphate accumulating organisms and glycogen accumulating organisms. *Water Sci. Technol.*, **50**(10), 139–144.

Pala, A. and Bölükbaş, Ŏ. (2005). Evaluation of kinetic parameters for biological CNP removal from a municipal wastewater through batch test. *Process Biochemistry*, **40**, 629–635.

Puig, S., Vives, M.T., Corominas, Ll., Balaguer, M.D. and Colprim, J. (2004). Wastewater nitrogen removal in SBRs, applying a step-feed strategy: from lab-scale to pilot-plant operation. *Water Sci. Technol*, **50**(10), 89–96.

Puig, S., Corominas, Ll., Vives, M.T., Colomer, J., Balaguer, M.D. and Colprim, J. (2005). Development and implementation of a real-time control system for nitrogen removal using OUR and ORP as end points. *Ind. Eng. Chem. Res.*, **44**, 3367–3373.

Smolders, G.J.F., Vandermeij, J., Van Loosdrecht, M.C.M. and Heijnen, J.J. (1994). Model of the anaerobic metabolism of the biological phosphorus removal process-stoichiometry and pH influence. *Biotechnol. Bioeng.*, **43**(6), 461–470.

Torrijos, M., Vuitton, V. and Moletta, R. (2001). The SBR process: an efficient and economic solution for the treatment of wastewater at small cheesemaking dairies in the Jura mountains. *Water Sci. Technol*, **43**(3), 373–380.

Van Loosdrecht, M.C.M., Hooijmans, C.M., Brdjanovic, D. and Heijnen, J.J. (1997). Biological phosphate removal processes. *Appl. Microbiol. Biotechnol.*, **48**, 289–296.

Vives, M.T (2004). SBR technology for wastewater treatment: suitable operational conditions for nutrient removal. Ph.D. thesis. University of Girona. Girona. Spain.

Vives, M.T., Balaguer, M.D., García, S., García, R. and Colprim, J. (2003). Textile dyeing wastewater treatment in a sequencing batch reactor system. *Journal of Environmental Science and Health Part A—Toxic/Hazardous Substances & Environmental Engineering*, **A38**(10), 2089–2099.

Yu, R.-F., Liaw, S.-L., Chang, C.-N., Lu, H.-J. and Cheng, W.-Y. (1997). Monitoring and control using on-line ORP on the continuous-flow activated sludge batch reactor system. *Water Sci. Technol*, **35**(1), 57–66.

Wastewater reclamation through a combination of natural systems (infiltration-percolation and constructed wetlands): a solution for small communities

E. Huertas, M. Folch and M. Salgot

Soil Science Laboratory, Faculty of Pharmacy, University of Barcelona. Avda. Joan XXIII s/n. 08028, Barcelona. Spain (E-mail: *ehuertas@ub.edu; mfolch@ub.edu; salgot@ub.edu)*

Abstract The decision on technologies used for reclaiming wastewater appears as important as a consequence of the implications on the economic, environmental and health conditions of societies. The problem arises when deciding how to deal with wastewater in small communities, because the costs of implementing and operating small, intensive wastewater treatment plants are unacceptable and can lead to financing and operation problems. Extensive treatment systems can be a good solution, where space is available. This paper presents the combination of two soft technologies (infiltration–percolation and constructed wetlands) to treat and reclaim wastewater. The obtained results show that the infiltration–percolation effluent presents a quality enough to be reused for irrigation of industrial crops, nurseries, fodder, cereals and oleaginous seeds, ornamental flower production; industrial cooling; impoundments, water bodies, and streams for recreational use in which the public's contact with the water is not permitted; and irrigation of forested areas, landscape areas and restricted access areas. Moreover, the combination of infiltration–percolation and constructed wetlands increases the list of final reclamation to the following: irrigation of pasture for milk or meat animals, crops for canning industry, crops not raw-consumed, fruit trees except by sprinkling, aquaculture; and aquifer recharge by localised percolation through the soil.

Keywords Constructed wetlands; infiltration-percolation; reclaimed wastewater; reuse; small communities

Introduction

Integrated water management approaches must include wastewater as an important part of the water sub-cycles: anthropic or natural. The interrelationships among the two are clear, from the disposal of wastewater (more or less treated into water bodies) to the use of natural water masses for obtaining resources for the daily life or economic activities of mankind. Those relationships can and must be quantified and qualified, using tools for the definition of the impacts generated, from the negative of untreated wastewater disposed in the environment to the positive ones of reusing reclaimed (waste) water. If that matter is considered in the arid and semiarid zones, like the Mediterranean, other concerns appear as important, like the health aspects of disposing and recycling wastewater or the lack of dilution in natural water bodies. Then, wastewater treatment, reclamation and reuse are paramount to maintain good economic (water availability) and sanitary (health-related illnesses) conditions in a number of areas (Salgot *et al.*, 2006). The decision on the technologies used for treating and reclaiming wastewater appears as important as a consequence of the implications on the economic, environmental and health conditions of societies.

The solution of the wastewater treatment problem has been extremely clear in the last 100 years on the two extreme sides of the management; the big cities and the single houses or extremely reduced populations (Angelakis *et al.*, 2003; Asano, 2005). The problem arises in the decision on how to deal with wastewater in medium sized towns.

Water Science & Technology **Vol 55 No 7 pp 143–148** © IWA Publishing 2007

doi: 10.2166/wst.2007.138

On one side, the economy of scale leads to intensive systems (energy and reduced space); on the other the on-site systems are the evident solution. In the small size collective systems, if the intensive processes are used, the costs of implementing and operating wastewater treatment plants could be unacceptable and can lead to financing and operation problems. The alternative way, the extensive systems can be a good solution, where space is available.

If reuse or disposal into sensitive environments is to be implemented, there are additional advantages in the use of extensive systems, considering the possibility of obtaining an effluent that can be directly reused or disposed without advanced treatments, thus reducing the economic impact of reclamation. Nevertheless, research is needed on the compliance with standards of extensively treated effluents. In order to test the previous statement, an extensive system based wastewater treatment plant was built in Els Hostalets de Pierola (Barcelona, Spain) for 1200 p.e., using modified infiltration–percolation (mIP) and constructed wetlands (WL) – subsurface horizontal flow.

Materials and methods

The study was done in Els Hostalets de Pierola wastewater treatment plant (WWTP). The facility was designed for a population of 1200 p.e. First of all there is a pre-treatment, followed by a modified Imhoff tank working as the primary system. Settled water passes directly to a homogenization tank, then water is pumped to a mIP system (secondary treatment). Part of the water which is treated by IP could be sent to WL, which works as tertiary treatment. Finally, treated wastewater is disposed of into a small river by gravity (see Figure 1). Table 1 presents the most important characteristics of IP and WL.

Physico-chemical parameters (pH; conductivity; suspended solids; COD; NH_4^+-N; NO_3^--N; PO_4^{3-}), as well as Faecal Coliforms, were analyzed according to *Standard Methods* (APHA *et al.*, 1998) while somatic coliphages and F-specific bacteriophages were determined following ISO/DIS, 1999.

Modified infiltration–percolation is a natural, soft, advanced wastewater treatment system consisting in a biological aerobic sequential filtration through a fine sand porous medium (Auset, 2002; Brissaud *et al.*, 2003; Bancolé *et al.*, 2004) which allows fulfilling of the principal objectives in wastewater treatment:

- almost total reduction of suspended solids
- organic matter and nitrogen oxidation
- pathogenic microorganisms reduction

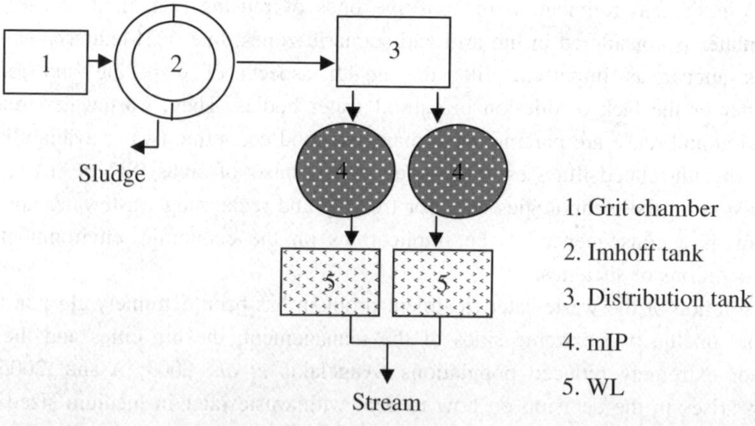

Figure 1 Diagram flow of Els Hostalets de Pierola WWTP

Table 1 IP and WL characteristics in Els Hostalets de Pierola WWTP

Characteristics	Biofilters (IP)	Wetlands (WL)
Total surface	$2 \times 875\,m^2$	$2 \times 200\,m^2$
Surface form	Circular	Rectangular
Impervious material	Yes	Yes
Protection of impervious material	Yes with geotextile 150 cm of sand	Yes with geotextile 60 cm – slope 1%
Depth of filtrating material layer	20 cm of gravel 8–20 cm 30 cm of gravel 20–40 cm	WL1 gravel 2–6 mm WL2 gravel 10–15 mm
Additional aeration (passive)	Yes	No
Water application	Pivot with 22 shovels	Pipe/gabion
Aerosol formation	No	No
Functioning	Discontinue	Continue

Constructed wetlands are also considered soft or extensive technologies. The treatment mechanisms involved are complex and include filtration, sedimentation, chemical precipitation, bacteriological oxidation, and disinfection (Stentström and Carlander, 2001; Tanner and Kadlec, 2003).

Constructed wetlands, with subsurface flux and planted macrophytes, present small aerobic zones around roots and rhizomes, where bacterial colonies are installed. Plants transport the oxygen from leaves and blades to the roots, thus creating a complex aerobic–anoxic–anaerobic system underground (Vymazal, 2001; Cooper *et al.*, 2005).

Results and discussion

Treated flow by IP varies from $43.1\,m^3$/day to $71.5\,m^3$/day; WL treats a flow that fluctuates from $7.3\,m^3$/day to $22.3\,m^3$/day. The IP systems were designed to treat all wastewater generated by the population (real scale); on the contrary, WL were planned as pilots, therefore they cannot treat the same amount of water.

Tables 2 and 3 present the effluent quality after modified infiltration–percolation and the effluent quality after constructed wetlands respectively.

SS results show the correct performance of the filtration mechanism of the both systems (IP and WL). A reduction of almost 99% (SS) when wastewater passes through biofilters, obtaining a maximum concentration of 2.7 mg/L is shown. At the outlet of wetlands, low SS concentrations are maintained. The IP system reaches a COD reduction of 87%, while WL maintains its value at the outlet. IP reduces COD due to filtration

Table 2 Effluent quality after modified infiltration-percolation

Parameters	mIP inlet			mIP outlet			Removal %
	Min	Aver	Max	Min	Aver	Max	
pH	6.96	7.37	8.16	6.71	6.97	7.20	–
Electrical conductivity (μS/cm)	810	1891	2390	1177	1957	2440	–
SS (mg/L)	109	204	587	b.d.l.	2.7	8.7	98.7
COD (mg/L)	345	748	1797	48	98	173	86.9
NH_4^+-N (mg/L)	29.4	54.9	70.2	b.d.l.	9.3	44.2	83.1
NO_3^--N (mg/L)	b.d.l.	0.8	3.7	6.3	30.6	56.7	–
PO_4^- (mg/L)	12.0	23.0	36.2	0.62	9.06	28.2	60.6
FC (μlog/100 mL)	6.64	7.23	7.62	1.00	3.66	5.13	3.57*
Somatic coliphage (μlog/100 mL)	5.94	6.64	7.17	3.24	4.27	7.20	2.37*
Bacteriophage (μlog/100 mL)	5.56	5.83	6.06	1.83	1.99	2.18	3.84*

*Removal in μlog/100 mL. b.d.l. below detection limit

Table 3 Effluent quality after constructed wetlands

Parameters	WL inlet = IP outlet			WL outlet			Removal %
	Min	Aver	Max	Min	Aver	Max	
pH	6.71	6.97	7.20	6.84	7.12	7.62	–
Electrical conductivity (μS/cm)	1177	1957	2440	1560	1939	2557	–
SS (mg/L)	b.d.l.	2.7	8.7	b.d.l.	1.96	1.75	27.4
COD (mg/L)	48	98	173	33	91	143	7.1
NH_4^+-N (mg/L)[1]	b.d.l.	10.45	23.99	9.8	10.82	11.53	–
NO_3^--N (mg/L)[1]	18.84	32.62	56.66	b.d.l.	1.34	3.18	95.9
NH_4^+-N (mg/L)[2]	0.91	5.16	11.48	0.98	8.53	29.01	–
NO_3^--N (mg/L)[2]	22.33	33.11	42.52	9.61	24.60	39.06	25.7
PO_4^- (mg/L)	0.62	9.06	28.2	0.17	4.7	11.5	48.1
FC (μlog/100 mL)	1	3.66	5.13	1.2	2.33	3.11	1.33*
Somatic coliphage (μlog/100 mL)	3.24	4.27	7.20	2.75	3.34	3.81	0.93*
Bacteriophage (μlog/100 mL)	1.83	1.99	2.18	1.62	2.06	2.34	–

*Removal in μlog/100 mL. b.d.l. below detection limit
[1]WL treating < 10 m^3/day
[2]WL treating 20–34 m^3/day

mechanisms and the oxidation of organic matter through the biofilm's aerobic microorganisms. Ammonia nitrogen is transformed into nitrates with total nitrification when the biofilters are established. IP effluent presents a quality suitable to be disposed of in the receiving media (Table 2). In several cases, nitrogen forms can cause problems in sensitive zones, then, an additional treatment is necessary. Constructed wetlands are capable to reduce the concentration of nitrates as shown in Table 3. This removal depends on the management of the systems; two kinds of performances can be obtained: a total denitrification when the system receives low flows (< 10 m^3/day) and a reduction of the denitrification rate when the applied flows increase (> 20 m^3/day).

Fecal coliform reductions in the biofilters reach 3.6 μlog/100 mL. WL improves microbiological characteristics, achieving reductions of 2 μlog/100 mL with an average of 1.3 μlog/100 mL. Inlet water to biofilters has homogeneous viral indicator content. A reduction of 2.4 μlog/100 mL of somatic coliphages and 3.8 μlog/100 mL of bacteriophages was detected in IP systems, while WL only improves microbiological effluent in relation with somatic coliphages (0.9 μlog/100 mL).

Spain is preparing a general recommendation in order to regulate the reuse of reclaimed wastewater. In 1999 the first Spanish guidelines draft appeared which described 14 categories (types) of reclamation. The new draft (2005) groups the 14 categories into 5 general ones (urban uses; irrigation uses; industrial uses; environmental and recreational uses; and aquifer recharge) in order to simplify the use for the recycling managers (see Table 4). 1999 and 2005 drafts include the minimum quality criteria for physical–chemical and biological parameters to be achieved in reclaimed wastewater to be reused for each category. A new aspect has to be taken into account in the 2005 draft: the minimum frequency analysis is included.

Table 5 presents the possibilities of reuse after IP and constructed WL treatment. It is to be noted that 6 sub-groups of reuse can be reached through IP and constructed WL combination (2.2; 2.3; 3; 4.3; 4.4; and 5.1). The IP system presents the capacity to transform ammonia into nitrates and it is capable of reducing microbiological content, besides the reduction of organic matter and suspended solids content. The combination of these processes generates an effluent with enough quality criteria to be reused for, specifically uses 2.3; 3; 4.3; and 4.4. Constructed

Table 4 Categories of reclamation depending on its final use (adapted from AEAS, 2005)

Reclaimed wastewater uses	
1. Urban uses	1.1. Residential uses. Private garden irrigation, toilet flushing, home air conditioning systems, car washing.
	1.2. Urban uses. Irrigation of open access landscape areas (parks, sport fields,...). Street cleaning, fire-fighting, ornamental impoundments and decorative fountains.
2. Irrigation uses	2.1. Greenhouse crops. Irrigation of raw consumed food crops. Fruit trees sprinkler irrigated.
	2.2. Irrigation of pasture for milk or meat animals. Irrigation of crops for canning industry and crops not raw-consumed. Irrigation of fruit trees except by sprinkling. Aquaculture.
	2.3. Irrigation of industrial crops, nurseries, fodder, cereals and oleaginous seeds. Ornamental flower production.
3. Industrial uses	3.1. Industrial cooling, except for the food industry.
4. Environmental and recreational uses	4.1. Golf courses irrigation.
	4.2. Impoundments, water bodies and streams for recreational use in which the public's contact with the water is permitted (except bathing).
	4.3. Impoundments, water bodies, and streams for recreational use in which the public's contact with the water is not permitted.
	4.4. Irrigation of forested areas, landscape areas and restricted access areas. Forestry.
5. Aquifer recharge	5.1. Aquifer recharge by localised percolation through the soil
	5.2. Aquifer recharge by direct injection

Table 5 Possibilities of reclaimed wastewater reuse after IP and WL

Use of the reclaimed wastewater	Quality criteria*	Required treatment
2.2. Irrigation of pasture for milking or meat animals. Irrigation of crops for canning industry and crops not raw-consumed. Irrigation of fruit trees except by sprinkling. Aquaculture.	*E. coli* < 1,000 cfu/100 mL SS < 35 mg/L Nematode eggs <1 egg/10L	IP + WL
2.3. Irrigation of industrial crops, nurseries, fodder, cereals and oleaginous seeds. Ornamental flower production.	*E. coli* < 10,000 cfu/100 mL SS < 35 mg/L Nematode eggs <1 egg/L	IP IP + WL
3. Industrial cooling, except for the food industry	*E. coli* < 10,000 cfu/100 mL SS < 35 mg/L	IP IP + WL
4.3. Impoundments, water bodies, and streams for recreational use in which public contact with the water is not permitted.	*E. coli* < 10,000 cfu/100 mL SS < 35 mg/L	IP IP + WL
4.4. Irrigation of forested areas, landscape areas and restricted access areas. Forestry.	SS < 35 mg/L	IP IP + WL
5.1. Aquifer recharge by localised percolation through the soil.	*E. coli* < 1,000 cfu/100 mL SS < 35 mg/L Total Nitrogen < 35 mg/L	IP + WL

*AEAS (2005)

WL is capable of improving water quality coming from IP, because WL presents disinfection mechanisms and reduces nitrate content due to the denitrification process. Consequently, the combination of both systems (IP + WL) permits us to increase this list in two more uses: 2.2 and 5.1. It is to be noted that intestinal nematode eggs must be controlled in most of the uses.

Conclusions

Modified Infiltration–Percolation is an adequate secondary wastewater treatment generating excellent quality effluent for the disposal or specific types of reuse.

Effluent obtained through IP presents a quality enough to be reused for irrigation of industrial crops, nurseries, fodder, cereals and oleaginous seeds, ornamental flower production; industrial cooling (except for the food industry); impoundments, water bodies, and streams for recreational use in which the public's contact with the water is not permitted; and irrigation of forested areas, landscape areas and restricted access areas.

Constructed wetland as tertiary wastewater treatment is also able to improve the quality of the treated wastewater. Nitrate contents are reduced when low flows are applied to the system. Microbiological quality is also improved using this system.

The combination of IP + WL increases the list of final reclamation uses. Irrigation of pasture for milk or meat animals, irrigation of crops for canning industry and crops not raw-consumed, irrigation of fruit trees except by sprinkling, aquaculture; and aquifer recharge by localised percolation through the soil must be added to the list presented for IP.

Els Hostalets de Pierola wastewater treatment plant, with the combination of two soft technologies (Infiltration–Percolation + constructed Wetlands) has low costs of operation and management, is fully integrated in the environment, reduces generation of by-products and could be considered an ideal treatment and reclamation system for small communities.

References

AEAS (2005). Draft for standards for wastewater reuse in Spain, Madrid. Not published.

Angelakis, A.N., Bontoux, L. and Lazarova, V. (2003). Challenges and prospectives for water recycling and reuse in EU countries. *Water Supply*, **3**(4), 59–68.

APHA, AWWA, WEF (1998). *Standard Methods for the Examination of Water and Wastewater*, 20th edn., Washington D.C.

Asano, T. (2005). Urban water recycling. *Water Science Technology*, **51**(8), 83–89.

Auset, M. (2002). "Mécanismes de la décontamination microbienne en infiltration percolation". PhD Thesis, Montpellier II, France.

Bancolé, A., Brissaud, F. and Gagne, T. (2004). Oxidation processes and clogging in intermittent maturated infiltration. *Water Science and Technology*, **48**(11), 139–146.

Brissaud, F., Xu, P. and Auset, M. (2003). Extensive reclamation technologies, assets for the development of water reuse in the Mediterranean. *Water Supply*, **3**(4), 209–226.

Cooper, D., Griffin, P. and Cooper, P. (2005). Factors affecting the longevity of sub-surface horizontal flow systems operating as tertiary treatment for sewage effluent. *Water Science and Technology*, **51**(9), 127–135.

ISO/DIS (1999). "Water quality – detection and enumeration of bacteriophages", pp. 1–15.

Salgot, M., Huertas, E., Weber, S., Dott, W. and Hollender, J. (2006). Wastewater reuse and risk: definition of key objectives. *Desalination*, **187**, 29–40.

Stentström, T.A. and Carlander, A. (2001). Occurrence and die-off of indicator organisms in the sediment in two constructed wetlands. *Water Science and Technology*, **44**(11–12), 223–230.

Tanner, C.C. and Kadlec, R.H. (2003). Oxygen flux implications of observed nitrogen removal rates in subsurface-flow treatment wetlands. *Water Science and Technology*, **48**(5), 191–198.

Vymazal, J. (2001). Constructed wetlands for wastewater treatment in the Czec Republic. *Water Science and Technology*, **44**(11–12), 369–374.

Wastewater reclamation systems in small communities

L. Alcalde, M. Folch, J.C. Tapias, E. Huertas, A. Torrens and M. Salgot

Soil Science Lab., University of Barcelona, Av.Joan XXIII s/n. 08028 Barcelona, Spain
(E-mail: *lalcalde@ub.edu*)

Water Science & Technology **Vol 55 No 7 pp 149–154** © IWA Publishing 2007

Abstract The demands established in the rules and regulations by the administration in Catalonia seem to exclude small communities from wastewater reclamation and reuse, due to the comparatively high costs associated with the practice at small scale. In the framework of the DRAC project (Demonstration on Wastewater Reclamation and Reuse in Catalonia) two different pre-treatment systems, one extensive (infiltration–percolation) and another intensive (ring filter), each one followed by chlorine dioxide disinfection, were tested in order to be applied for small communities wastewater reclamation and reuse. The results of this study show that infiltration–percolation systems remove very efficiently physico-chemical contaminants and microorganisms. The ring filter system does not show a significant removal rate of contaminants. The use of infiltration–percolation as a pre-treatment for advanced chemical disinfection allows reducing the dose of disinfectant and the contact time needed to achieve a specific water quality, and diminishes disinfection by-products (DBPs) generation. Therefore, this reclamation line is suitable for small communities due to its efficiency and low cost. However, further studies are needed in relation to the removal mechanisms of microorganisms, organic compounds in IP systems and the possible DBPs formation using chlorine dioxide.

Keywords Chlorine dioxide; infiltration-percolation; ring filter; small communities; wastewater reclamation

Introduction

In the Mediterranean region, wastewater reclamation and reuse has to be integrated in the planning and development of water resources in an appropriate and sustainable way. Due to the conditions and demands established in rules and regulations by the administration in Catalonia, small communities seem initially to be excluded from this possibility, due to the comparatively high costs associated with that practice at small scale. Nevertheless, the scarcity and drought conditions generate a need in rural or small urban areas for reusing even small amounts of reclaimed water.

Usually, wastewater reuse requires disinfection of reclaimed water, except for some extensive treatment systems capable of guaranteeing an adequate quality for specific uses. Nowadays, the most common wastewater reclamation systems are based on conventional technologies, including disinfection, based on chemicals' use or alternative physical systems, like UV (Jacangello and Trussell, 2002). Chemical disinfection systems require an appropriate water quality, in order to minimize disinfection by-products (DBPs) generation. DBPs are of great concern because of their potential impact on public health and the environment. Extensive treatment systems are mainly low-cost, easy to operate and maintain, and can be the most appropriate technologies for wastewater reclamation for small communities. Those systems would provide water with a quality suitable for further advanced disinfection technologies, thus being useful as a pre-treatment for disinfection.

The framework of this study was the DRAC project (Demonstration on Wastewater Reclamation and Reuse in Catalonia), a research consortium with the participation of universities (University of Barcelona and University of Montpellier II), private companies (AGBAR, Searsa) and the administration (Government of Catalonia). The aim of one

doi: 10.2166/wst.2007.139

stage of the DRAC project was to evaluate the efficiency/suitability of different filtration systems as pre-treatments for disinfection technologies capable of guaranteeing an effluent with enough quality to be reused. Among others, two different pre-treatment systems, one extensive (infiltration–percolation) and another intensive (ring filter), each one followed by chlorine dioxide disinfection were tested. Chlorine dioxide was chosen because, despite the higher costs, it has proved to be more effective in achieving inactivation of viruses than chlorine, and does not react with ammonia to form the potentially toxic chlorinated DBPs (Metcalf and Eddy, 2003).

Material and methods

The study was carried out in the Vall-Llobrega WWTP, located in Girona, Spain. The WWTP is an activated sludge facility that serves several municipalities in a summer tourist area, with 32,000 inhabitants in winter and up to 60,000 during the peak of the summer season. Part of the secondary effluent from this WWTP was reclaimed by using an extensive pre-treatment (infiltration–percolation) and an intensive pre-treatment (ring filter) working in parallel. The pre-treatment systems were followed by a chlorine dioxide disinfection technology. The main characteristics of the systems are the following:

Extensive filtration system

Infiltration–percolation (IP): circular shape, surface area 554.7 m^2, depth 1.50 m of sand (granulometry: 98% below 1 mm diameter), 20 cm of fine gravel and 30 cm of coarse gravel, hydraulic load 0.43 m/day.

Intensive filtration system

Ring filter (RF): two filtration modules in series: first module with two filters, theoretical pore size 15 $\mu m \pm 10–15\%$; second module with three filters, theoretical pore size 25 $\mu m \pm 10–15\%$. Maximum flow 12–13 m^3/h.

Disinfection system

Chlorine dioxide (ClO_2): the pilot has a 1 m^3 reactor, a homogenization tank and a ClO_2 generator. Preliminary tests were performed to determine the minimum dose required and the minimum residual ClO_2, while trying to maintain the disinfection throughout the time. The ClO_2 dose applied for the IP system was 3 ppm, and for the RF was 8 ppm, with a contact time of 10 minutes and a flow rate of 1 m^3/h.

Physico-chemical and microbiological parameters were determined at the inlet and outlet of each filtration system and the outlet of the disinfection system. The analysis was performed according to the *Standard Methods* (APHA, 2000), with some exceptions; bacteriophages determination and enumeration were performed according to the ISO standards (ISO 1995 and 2000); *Giardia* and *Cryptosporidium* determination were performed according to the EPA standards (EPA, 2001), volatile organic compounds (VOCs) analyses were performed by closed loop stripping analysis (CLSA), a method developed by Grob (1973). The CLSA extracts were further analyzed by GC/MS. The particle count analysis was performed following the indications of the particle counter model manufacturer (Hiac Royco Pacific Scientific 8000, sensor HRLD-150 and sampler 3000 A).

Results and discussion

The main results obtained for the physico-chemical parameters are presented in Table 1, 2 and 3. In relation with organic matter, the IP system shows an average removal percentage 56.7% for the TOC and 35.9% for the COD (Table 1). The BOD_5 of the final

Table 1 Physico-chemical parameters evaluated for the two reclamation trains (mean values)

Physico-chemical parameters	IP + ClO_2			RF + ClO_2		
	InIP	OutIP	OutClO_2	InRF	OutRF	OutClO_2
pH	7.58	7.12	7.55	7.52	7.51	6.84
Temperature (°C)	23.7	23.1	23.3	22.8	22.2	22.5
Conductivity (μS/cm)	2037	2167	2198	2383	2065	2066
Dissolved oxygen (mg/L)	7.6	7.7	10.0	6.3	6.3	6.3
Turbidity (NTU)	3.97	0.16	0.55	1.38	1.13	1.06
TSS (mg/L)	10.80	3.64	3.44	9.02	6.72	6.60
COD (mg/L)	104.2	66.7	76.0	85.7	53.8	85.6
BOD_5 (mg/L)	10.4	< 5	< 5	13.6	12.9	< 5
TOC (mg/L)	12.0	5.2	12.0	15.0	14.5	14.5
P-PO_4 (mg/L)	5.82	5.58	5.64	5.25	5.80	5.45
N-TKN (mg/L)	42.9	20.8	20.4	37.3	34.3	33.4
N-NH_4 (mg/L)	38.5	16.0	15.3	35.4	32.1	31.7
N-NO_3 (mg/L)	0.25	12.5	12.6	0.65	0.85	0.65

Note: b.d.l: below detection limit (1 ulog cfu or pfu/100 mL), cfu: colony forming units, pfu: plaque forming units

effluent was always below the detection limit. The IP system is able to reduce organic matter because it works mainly as an aerobic biological reactor (Brissaud *et al.*, 1991; Auset, 2002).

The removal of N-TKN and N-NH_4 was of an average of 51.42% and 58.44%, respectively, showing the high nitrification rate in the IP system, due to the aerobic biofilm inside the filter. The content of P-PO_3 does not show any important variation.

Table 2 Volatile organic compounds evaluated for the two reclamation trains (mean values)

Volatile organic compounds (VOCs)		VOCs concentration (μg/L)					
		IP + ClO_2			RF + ClO_2		
		InIP	OutIP	OutClO_2	InRF	OutRF	OutClO_2
Antioxidants	2-*t*-butylcyclohexanol	0.06	< 0.01	< 0.01	3.55	3.66	4.01
	2-*t*-butylcyclohexanol (isomer)	0.02	< 0.01	< 0.01	1.73	1.73	2.06
	2-*t*-butylcyclohexanone	0.30	0.06	0.06	1.22	1.14	1.12
	4-*t*-butylcyclohexanone	0.38	0.02	< 0.01	1.31	1.28	1.28
	4-*t*-butylcyclohexanol	0.35	< 0.01	0.06			
	2,6-di-*t*-butyl-*p*-cresol				1.03	1.12	1.35
	Dicyclopentadiene				0.02	0.02	0.01
Musk compounds	Musk ketone	0.02	< 0.01	< 0.01			
	Possible musk n.i.				0.96	1.03	0.93
Miscellaneous	Tonalide	1.17	0.21	< 0.01	0.91	1.02	0.99
	Cedranol				0.17	0.19	0.20
	Galaxolide	0.70	0.1	0.25	1.76	1.95	1.94
	Geranyl acetone				0.04	0.02	0.05
	Geranyl nitrile				0.19	0.17	<0.02
	Unidentified compound (tr 60.7 min)	0.72	< 0.01	< 0.01			
	Unidentified compound ($m/z = 191$) 3 isomers	1.26	0.08	< 0.01	2.73	2.89	1.61
	2-ethylhexanoic	0.08	0.03	0.1			
	t-butylciclohexile acetate				0.45	0.47	0.49
	Dimethyldecadienone	0.18	< 0.01	0.03			
	Possible alcohols (tr 62 min) several isomers	< 0.01	< 0.01	4.27			
	i-propyl myristate	0.34	0.03	0.74			

Note: m/z is the base peak of its mass spectrum, n.i.: non identified

Table 3 Microbiological parameters evaluated for the two reclamation trains (mean values)

Microbiological parameters	IP + ClO_2			RF + ClO_2		
	InIP	OutIP	OutClO_2	InRF	OutRF	OutClO_2
Fecal coliforms (log cfu/100 mL)	5.00	2.36	b.d.l.	4.84	4.84	b.d.l.
Somatic coliphages (log pfu/100 mL)	4.81	2.53	1.7	3.15	3.00	b.d.l.
F-specific bacteriophages (log pfu/100 mL)	3.38	1.00	b.d.l.	3.84	3.74	1.49
Giardia sp. (cysts/L)	20	< 1	< 1	72	57	25
Cryptosporidium sp. (oocysts/L)	49	< 1	< 1	5	5	3

Note: b.d.l: below detection limit (1 (μlog cfu or pfu/100 mL), cfu: colony forming units, pfu: plaque forming units

The IP system has an average percentage of TSS removal of 66.3% and turbidity has diminished 95.6%. The particle count results show an almost total elimination of particles with a diameter from 1 to 130 μm (Figure 1). This result proves the high efficiency of the IP system as a filter.

The IP show a high percentage of removal for different VOCs (Table 2). The system removes efficiently a major number of antioxidants, musk compounds and different miscellaneous compounds present in the secondary effluent. This great efficiency can be associated with the high wastewater retention time in the biofilter, thus improving the adsorption mechanisms. Nevertheless, it is necessary to study in depth the removal mechanisms of the organic compounds.

Regarding microbiological parameters, IP did show a high disinfection efficiency (Table 3). The removal of fecal coliforms was 2.64 μlog, for somatic coliphages 2.28 μlog, and for F-specific bacteriophages 2.38 μlog. *Giardia* sp. and *Cryptosporidium* sp. contents in the final effluent were always below the detection limit. The results obtained in the particle counts compared with the microorganims size, confirm the importance of filtration and adsorption mechanisms in the disinfection efficiency of the IP system (Figure 1). Unsaturated hydraulic conductivity and the air-water interface produce the greatest effect on the fate and transport of viruses in unsaturated media (Faulkner *et al.*, 2003).

The RF system did not show significant removal of TOC and BOD_5, although a COD removal of 37.2% was stated (Table 1). The RF presents a 25.5% reduction of TSS. The RF does not influence ammonia transformation; consequently, there are almost no nitrates in the effluent.

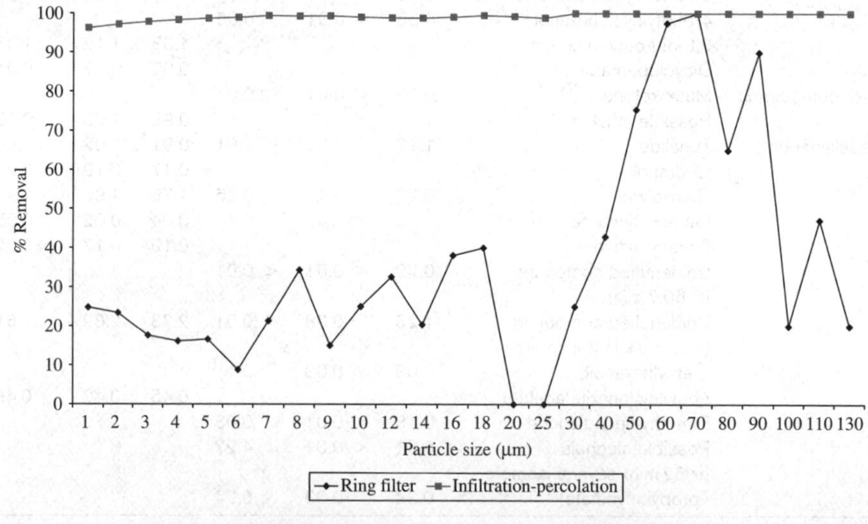

Figure 1 Particle size removal of infiltration–percolation and ring filter systems

The particle count results show an almost total elimination of particles with a diameter of more than 25 μm, as expected from the design characteristics of the RF system (Figure 1). There is also a slight elimination of particles with a diameter of less than 25 μm due to the attachment effect to large particles. The low reduction of turbidity (18.1%) is due to the fact that this parameter is mainly influenced by small particles' content, the ones less removed in the RF system.

Regarding the VOCs, the RF does not eliminate significantly either the antioxidants or the most common polycyclic compounds (galaxolide and tonalide), musk compounds, and miscellaneous compounds (Table 2). The RF system does not remove the microorganisms evaluated at a significant rate (Table 3).

The application of ClO_2 decreased the content of microorganisms from both IP and RF systems, although the dosage and the contact time of ClO_2 needed for achieving this disinfection were lower in the case of the effluent from the IP system. Consequently, the combination of IP + ClO_2 generates less DBPs and reduces reagent costs. However, the environmental impacts associated with the use of chlorine dioxide as a wastewater disinfectant are not well known (Metcalf and Eddy, 2003). There is a slight reduction of VOC content after chlorination. Oxidative processes such as chlorination are effective for reducing the concentration of several classes of microcontaminants. Macrolide and sulfonamide antibiotics as well as estrogens and phenazones are readily oxidized by ClO_2 although removal efficacy is a function of the contaminant structure and oxidation dose. However, many compounds did not react at an appreciable rate with ClO_2 (Pinkston and Sedlak, 2004; Huber *et al.*, 2005).

It has to be noticed that the quality of the final effluent from the IP + ClO_2 reclamation train complies with the draft of the Spanish regulation for wastewater reuse for most of the uses (AEAS, 2005).

Conclusions

Based on the results of this study, the following conclusions can be drawn.

- The infiltration-percolation system removes very efficiently organic matter, nitrogen compounds, TSS, VOCs and microorganisms (bacteria, viruses and protozoa). This great efficiency could be attributable to the filtration and aerobic degradation mechanisms.
- The ring filter system does not show a significant removal rate for physical-chemical contaminants and microorganisms removal.
- The use of infiltration-percolation as a pre-treatment for advanced chemical disinfection allow to reduce the dosage and the contact time needed to achieve a specific water quality, and diminishes the production of DBPs.

The infiltration-percolation system is a low cost and easy to operate and maintain technology that improves the quality of a secondary effluent in order to apply afterwards a disinfection system. This reclamation line is suitable for small communities due to its efficiency and low cost. However, further studies are needed in relation to the removal mechanisms of microorganisms, organic compounds in IP systems, and the possible DBPs formation using chlorine dioxide.

Acknowledgements

The authors would like to thank the collaboration of CIRSEE (ONDEO-Suez/Lyonnaise des Eaux), Kemira, Consorci de la Costa Brava, Agència Catalana de l'Aigua (Water Catalan Agency), SEARSA, AGBAR Foundation, Sanitation Mangement of AGBAR Group; and Catchwater (ENV4-CT98-0790) and Inco-Copernicus (15-CT98-0105) European research projects.

References

AEAS (2005) Draft for standards for wastewater reuse in Spain, Madrid. Not published.

APHA (2000). *Standard Methods for the Examination of Water and Wastewater*, 20th edn, American Public Health Association, Washington, DC.

Auset, M. (2002) Approche des mécanismes de décontamination en Infiltration-percolation. PhD thesis. University of Barcelona.

Brissaud, F., Rodier, C., Lasala, M. and Salgot, M. (1991). Wastewater reuse: a sanitation and management strategy for the Mediterranean coastal areas. *Toxic. Envir. Chemistry*, **31**, 549–557.

EPA (2001) Method 1623: *Cryptosporidium* and *Giardia* in water by filtration/IMS/FA. Environmental Protection Agency, 821-R01-025, U.S.

Faulkner, B.R., Lyon, W.G., Khan, F.A. and Chattopadhyay, S. (2003). Modeling leaching of viruses by the Monte Carlo method. *Water Res.*, **37**(19), 4719–4729.

Grob, K. (1973). Organic substances in potable water and in its precursor. Part I. Methods for their determination by gas liquid chromatography. *J. Chromatogr.*, **84**, 255–273.

Huber, M.M., Korhonen, S., Ternes, T.A. and von Gunten, U. (2005). Oxidation of pharmaceuticals during water treatment with chlorine dioxide. *Water Res.*, **39**(15), 3607–3617.

ISO (1995) *Water quality-detection and enumeration of bacteriophages*. Part 1:Enumeration of F-specific RNA bacteriophages. ISO 10705-1, International Organization for Standardization, Geneve.

ISO (2000) *Water quality-detection and enumeration of bacteriophages*. Part 2: Enumeration of somatic coliphages. ISO 10705-2, International Organization for Standardization, Ginebra.

Jacangelo, J.G. and Trussell, R.R. (2002). International report: water and wastewater disinfection-trends, issues and practices. *Wat. Sci. Tech.: Water Supply*, **2**(3), 147–157.

Metcalf and Eddy, Inc. (2003). *Wastewater Engineering, Treatment and Reuse*, 4th ed., McGraw-Hill Publishing Co., New York.

Pinkston, K.E. and Sedlak, D.L. (2004). Transformation of aromatic ether-and amine-containing pharmaceuticals during chlorine disinfection. *Environ. Sci. Technol.*, **38**, 4019–4025.

Water Science & Technology Vol 55 No 7 pp 155–161 © IWA Publishing 2007

Integrated constructed wetland systems: design, operation, and performance of low-cost decentralized wastewater treatment systems

L.L. Behrends, E. Bailey, P. Jansen, L. Houke and S. Smith

Tennessee Valley Authority, Environmental Technologies, CEB1-C, Muscle Shoals, Alabama USA 35662 (E-mail: *llbehrends@tva.gov*)

Abstract Several different types of constructed wetland systems are being used as decentralized treatment systems including surface-flow, subsurface-flow, vertical-flow, and hybrid systems. Archetypical wetland systems have design strengths and weaknesses, and therefore it should be possible to design combined (integrated) systems to optimize a number of important treatment processes. This study provides comparative efficacy data for two integrated wetland treatment systems (IWTS) designed to enhance treatment of medium strength wastewater generated from a pilot-scale intensive fish farm. Results from the twenty eight months study included consistently high removal of COD (84% +) and ammonia nitrogen (93%) in both systems. Initially, phosphorus removal was also high (>90%) in both systems, but removal efficacy declined significantly over time. Nitrate removal was significantly better in the system that provided sequential aerobic and anoxic environments. Short hydraulic retention times coupled with sustained removal of COD and ammonia indicate that the ReCip components could be a least-cost wastewater treatment technology in the decentralized market sector.

Keywords Constructed wetland; decentralized wastewater treatment; denitrification; nitrification; reciprocating wetland

Introduction

Interest in decentralized wastewater treatment systems has continued to develop because of the need for low-cost modular wastewater treatment systems that are economical, aesthetic and ecologically sustainable (Donnell *et al.*, 2003). Over the past three decades, scientists and engineers have developed a wealth of practical information related to constructed wetlands technology (CWT). CWT has a rich diversity of design and operational features that can be customized for treatment of domestic, agricultural and industrial wastewater (EPA, 2000; IWA, 2000). Several different types of constructed wetland systems are being used throughout the world for decentralized wastewater treatment. These include surface-flow, subsurface-flow, vertical-flow, and hybrid systems (Platzer, 1996). Constructed wetlands have several inherent advantages including low capital costs, less infrastructure, lower operating costs, simplicity of design and ease of operation. The archetype wetland treatment systems, such as surface and subsurface-flow, each have design strengths and weaknesses. However, it should be possible to design a hybrid system that combines the best concepts of existing technologies and thereby improves overall performance and potentially reduces capital and operating costs (Platzer, 1996).

The Tennessee Valley Authority (TVA, www.tva.gov/environment/envservices/alw_wetlands.htm), is a United States government corporation. Recently, TVA scientists designed, operated, and monitored two integrated hybrid wetland wastewater treatment systems to evaluate potential synergies in treatment processes and the associated potential for reductions in capital costs.

doi: 10.2166/wst.2007.140

Methods

Integrated wetland treatment systems

Two mesocosm-scale integrated wetland treatment systems (IWTS), were constructed inside an environmentally controlled greenhouse located at TVA's Water Processes Research Station, Muscle Shoals, Alabama, USA. The systems were put into operation and monitored for 28 months (Feb. 2002 to May 2005). Figure 1 provides a conceptual illustration of the two systems. IWTS I consisted of a surface-flow wetland (free water surface), followed by a 2-cell vertical flow reciprocating wetland, while IWTS II was comprised of a 2-cell vertical-flow reciprocating wetland followed by a subsurface-flow wetland. Table 1 provides design information for components of both systems. Reciprocation (ReCip®) is a patented vertical-flow process (Behrends, 1999, United States Patent 5,863433), in which water is pumped back and forth between contiguous treatment cells to provide controlled aeration of substrate biofilm and wetland plant roots (Behrends *et al.*, 2001). Although the biofilms and plant roots are saturated with atmospheric oxygen on a controlled and recurrent basis, the bulk water is only poorly aerated. This fortuitous combination of recurrent aerobic and anoxic–anaerobic environments is required for removal of many natural and man made chemicals (Zitomer and Speece, 1993).

After the soil, pea-gravel and underdrain plumbing were installed in each of the respective cells, all cells were planted with canary grass (*Phalaris arundinacea* L.), with eight plantings per cell to augment and promote removal of nitrate via plant uptake and denitrification (Zhu and Sikora, 1995). Clarified wastewater from a pilot-scale recirculating aquaculture system (Behrends *et al.*, 1999), was pumped with a two channel peristaltic pump on a continuous basis to the inlet header of each integrated system. The pump's controls were adjusted to maintain an influent flow rate of 80 ml/minute for a hydraulic loading rate of 3.5 cm/day (350 m^3/ha/day). For the duration of the 28 month study ReCip cells were recurrently filled and drained on a two hour cycle (12 cycles/day) using small submersible pumps and two-cycle digital timers. Wastewater influent (Table 2) was of intermediate strength and similar in many respects to septic tank effluent, although the COD concentration was somewhat less.

The respective inlets and outlets of each system were outfitted with platinum tipped redox probes to facilitate monitoring of redox potential. Water quality parameters including temperature, dissolved oxygen and pH were monitored at least weekly with a multi-parameter hand held sonde (Yellow Springs Instruments, www.ysi.com). Whole column

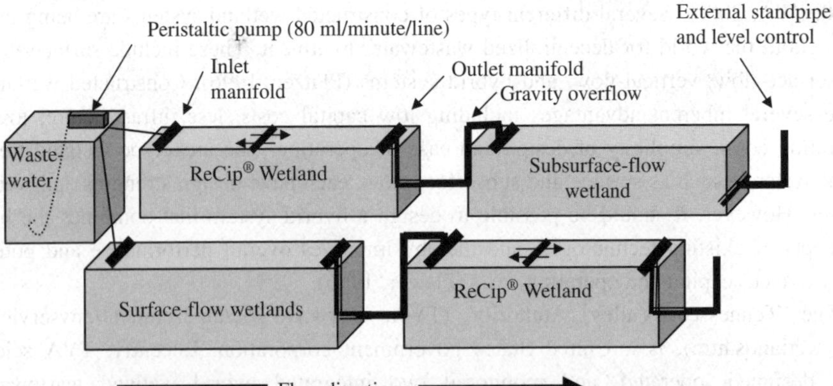

Figure 1 Conceptual illustration of two integrated (hybrid), wetland treatment systems. Both systems received continuous flows of 80 ml/minute (3.5 cm/day), of medium-strength wastewater Total hydraulic retention time (HRT), for each of the two integrated systems was approximately 5.1 days

Table 1 Design criteria for IWTS I and II. See Figure 1 for additional details

	Cells	Substrate	Void space (%)	Surface dimensions (cm)	Water depth (cm) and volume (L)	HRT (days)
IWTS I						
Surface-flow	1	Soil	n.a.	240 × 90	18 (388)	3.4
ReCip I	A&B	Pea gravel	38	120 × 90 ea. cell	45 (388)	1.7*
IWTS II						
ReCip II	A&B	Pea gravel	38	120 × 90 ea. cell	45 (388)	1.7*
Subsurface-flow	1	Pea gravel	38	240 × 90	45 (388)	3.4

*With ReCip systems, HRT is calculated on the basis of void space (38–40%), one cell being full the other cell is near empty

water samples were also collected biweekly or monthly from the respective inlets and outlets of the two systems, iced and/or preserved with acid, and transported to TVA's analytical laboratory for analyses of chemical oxygen demand (COD), total ammonia nitrogen (TAN), total Kjeldahl nitrogen (TKN), nitrate-nitrogen (NO_3), ortho-phosphorus (PO_4), and alkalinity. All analyses were performed according to *Standard Methods* (APHA, 1985) or by methods approved by US EPA. During the course of the study, the canary grass was sampled on several occasions to estimate gross productivity (g/m^2 dry weight), and to analyze proximate composition (Behrends *et al.*, 1994).

At the conclusion of the study water was slowly drained from the experimental units and discrete replicate sludge samples were collected from the surface-flow wetland and discrete gravel samples (20 L /sample) were collected from the top, middle and bottom of each of the other treatment cells. Each of the dewatered sludge and gravel samples, including the interstitial sludge, were weighed to the nearest gram. Subsequently, clean water was added to each gravel sample until the void spaces were filled, and the sample was weighed again. Sludge was removed from the respective gravel samples via systematic agitation and washing. Clean water was then added back to the washed gravel until all void spaces were filled, and the samples were weighed again. Using this process, it was possible to gravimetrically and volumetrically determine the pre-washed and post-washed void space and the amount of sludge (suspended solids, biosolids, and precipitates), which was deposited in the voids during the 28 month study.

Table 2 Influent and effluent water quality values for two Integrated Wetland Treatment Systems. Values for COD and nutrients are averages based on 27 sampling events during the twenty eight month study. Values for other water quality parameters are based on 150 sampling events

Parameters[1]	IWTS I				IWTS II			
	Influent Conc.[1]	SF Wetland Effluent	ReCip II Effluent	Change (%)	Influent Conc.[1]	ReCip I Effluent	SSF Wetland Effluent	Change (%)
COD	172	107	28	−84	182	46	24	−87
TAN	50.9	45.9	3.8	−93	51.4	6.0	3.4	−93
TKN	72.2	58.2	3.9	−95	66.3	10.4	3.4	−95
NO_3-N	0.16	0.65	27.91	173 X	0.16	8.79	3.40	20 X
PO_4-P	25.4	24.1	18.1	−29	25.6	19.3	16.2	−36
Redox[2]	−429	−159	−17	96	−455	−29	−54	88
D.O.	0.7	1.4	1.7	131	1.0	1.4	1.0	−9
Temp °C	25.1	23.2	24.1	−4	25.2	24.1	23.6	−6
pH[3]	6.4	7.0	6.9	7	6.4	6.8	6.8	6

[1]Parameters measured as mg/L unless specified otherwise
[2]Redox potential (mV)
[3]pH (unitless)

Results and discussion

Water quality

Integrated systems I and II performed equally well with respect to removal of COD (84–87%), TAN (93%) and TKN (95%). In both systems, the reciprocating component provided the greatest percentage removal of COD, TKN and TAN (Table 2, Figures 2–4). ReCip® components provided sufficient oxygenation of the fixed biofilm to sustain simultaneous nitrification and COD removal (Figures 2 and 4). Bulk water D.O. concentrations remained low in all systems (<2.0 mg/L), but during the ReCip drain, the extensive biofilm on the rock substrate and roots was exposed to atmospheric oxygen for up to 30 minutes. The surface-flow and subsurface-flow wetland components also provided treatment functions such as removal of COD and denitrification respectively (Table 2, Figures 2 and 5).

This was a relatively long study and despite significant deposition of suspended solids and biosolids in both systems (see later discussion), removal of COD, TAN and TKN were remarkably stable for the duration of the study. Both systems also removed phosphorus, albeit at a relatively low percentage (29 and 36% respectively, Table 2). It is worth noting that both systems initially removed phosphorus at relatively high rates (66–93%), but treatment efficacy declined with time until the rates were diminished to less than 10 percent for both systems. This is consistent with other gravel-based wetland treatment studies, in which gravel-based substrates provide a short-term sink for phosphorus removal (Brix *et al.*, 2001).

With respect to nitrate removal (Table 2, Figure 5,), IWTS II was clearly superior to IWTS I. Within IWTS I, the surface-flow wetland acted as a clarifier, removing much of the influent suspended solids and associated organic carbon, as measured by COD.

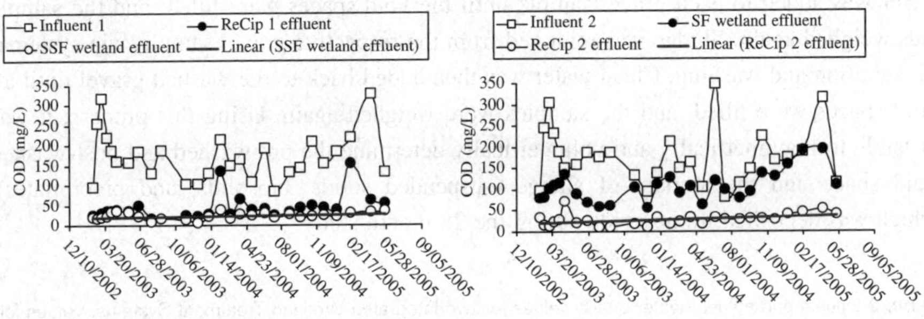

Figure 2 COD removal dynamics as a function of IWTS, sample location and date. The SF component of IWTS 1 provided limited removal of COD, while the ReCip process in both systems provided significant and sustainable removal for the duration of the study

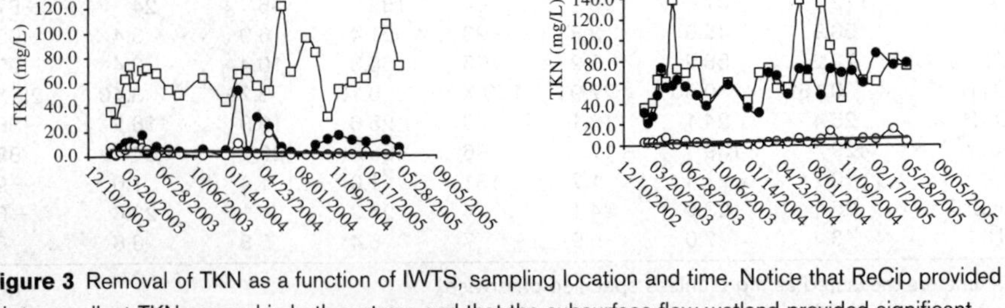

Figure 3 Removal of TKN as a function of IWTS, sampling location and time. Notice that ReCip provided fair to excellent TKN removal in both systems and that the subsurface flow wetland provided significant polishing of effluent from ReCip 1

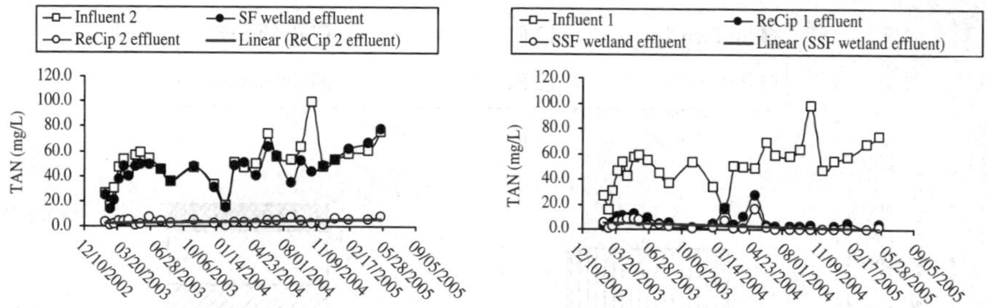

Figure 4 Removal of TAN as a function of IWTS, system component and time. Notice that ReCip provided excellent removal in both systems for the duration of the study. The surface-flow wetland provided very little removal of TAN and the subsurface flow wetland provided marginal polishing of effluent from ReCip 1

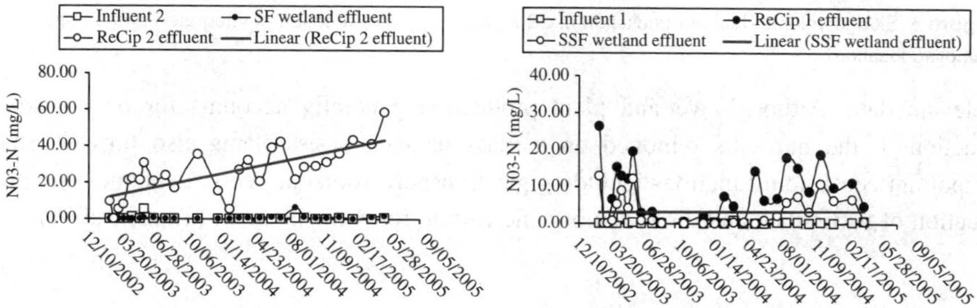

Figure 5 Nitrate-N removal as a function of IWTS, system component and time. Notice the increasing concentration of nitrate-N with time in ReCip 2 effluent as compared to the sustained low nitrate-N concentration in the SSF wetland effluent

However, the surface-flow wetland provided little treatment with respect to removal of nitrogen and phosphorus. Furthermore, the efficient removal of suspended solids and associated organic carbon in the surface flow wetland tended to "starve" the downstream ReCip system of organic carbon. This resulted in reduced denitrification, with effluent concentrations exceeding 27 mg/L. In comparison, IWTS II provided much greater denitrification because of the availability of labile organic carbon (COD) in the anoxic flooded phase of the ReCip process, followed by further nitrate removal in the subsurface flow wetland. Nitrate concentrations were 8.8 and 3.4 in the effluents of the ReCip and subsurface-flow subsystems respectively (Table 2, Figure 5)

Plant production

Canary grass shoots were harvested on Jan 18, 2005, oven dried and processed to determine dry matter production and nitrogen and P content. Table 3 provides a summary of

Table 3 Harvest data and nutrient content for canary grass as a function of treatment system and sampling location. Notice the significantly higher plant production and nutrient content of the surface-flow system which only provided marginal wastewater treatment

	Dry matter (g/m^2)	N content (%)	P content (%)
IWTS I			
Surface-flow	1110	4.5	0.81
ReCip I A	118	3.9	0.73
ReCip I B	105	3.9	1.05
IWTS II			
ReCip II A	272	4.1	0.68
ReCip II B	143	3.8	0.70
Subsurface-flow	664	3.5	0.56

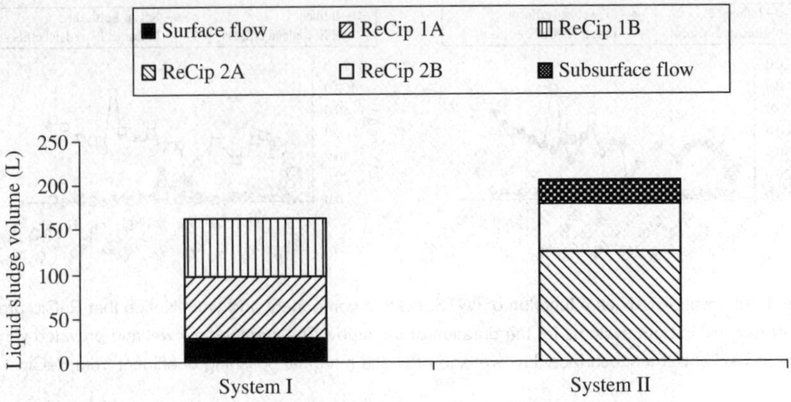

Figure 6 Sludge volume data at conclusion of 28 month study as a function of integrated system and sampling location

relevant data. Although wetland plant production generally accounts for only a small fraction of the nutrients removed on a mass balance basis, plants also provide other important ecosystem functions including gas transport, rootzone redox dynamics and production of particulate and dissolved organic carbon for denitrification (Tanner, 2001).

Sludge and biosolids data

Based on replicated gravimetric and volumetric analyses, total sludge volume for both systems was calculated and is summarized in Figure 6. The largest fraction of sludge and biosolids accumulated in the ReCip components of both systems, which is reasonable since most of the biological treatment (COD and nitrogen removal), took place in these sequential aerobic/anoxic compartments. The sludge itself was very recalcitrant, and had a low oxygen demand (TVA, unpublished data). This is consistent with the findings of Nguyen (2001), who reported that clogging of pore space was due to accumulation of refractory (stable) organic compounds consisting of humic and fulvic acids and humin.

The sludge volume in the surface flow system was approximately 7 per cent of the total water volume, while the sludge volume, as a per cent of the total pore space in the gravel systems, was 37 and 35 per cent respectively for ReCip Cells 1A and 1B; 68 and 29 per cent respectively for ReCip Cells 2A and 2B; and 7% for the subsurface-flow wetland. Even with significant clogging of pore space in the ReCip components of both systems, the integrated systems continued to remove significant amounts of TKN, TAN, and COD. Notwithstanding the sustainability of the two integrated systems over the 28 month trial, it was apparent that sludge management will be required for long term operation of gravel-based treatment systems. TVA has recently concluded a series of studies to evaluate noninvasive methods (chemical, biological and physical) for removing sludge and biosolids from gravel-based treatment systems. Several of the processes evaluated were successful in removing sludge (Behrends *et al.*, 2006 a and b).

Summary and conclusions

Both IWTS I and II provided excellent wastewater treatment with respect to removal of COD, TKN and TAN. Nitrate removal was significantly better in IWTS II which included a two-cell ReCip system followed by a subsurface-flow treatment unit. Phosphorus removal was significant initially in both integrated systems, but removal was dramatically reduced towards the end of the study due to the limited sink provided by gravel-based systems. Both integrated systems produced and retained copious amounts of recalcitrant sludge, especially the ReCip units. Although sludge production was significant, it did not appear to adversely affect wastewater treatment. This is consistent with earlier studies (Behrends *et al.*, 1999).

With respect to system design and economics, it appears that the ReCip components in both systems I and II provided the greatest amount of wastewater treatment, even though the HRT for ReCip was only 1.7 days vs. 3.4 days for the surface- and subsurface-flow components. The surface-flow wetland component of IWTS I was only marginally effective, and appeared to impair down-stream denitrification. Thus, it is concluded that it should be eliminated as a design component of the integrated system. The subsurface-flow treatment component of IWTS II provided additional polishing of wastewater (Table 2). However, given the relatively long retention time, the high cost of gravel substrate, and the marginal removal of nutrients, it too should be eliminated from integrated designs. It is concluded, based on this study and subsequent field trials, that ReCip systems can be designed as stand-alone systems provided that the HRT is increased from 1.7 to 2.5 or 3.0 days. Furthermore, reciprocation cycle times, depth of gravel substrate and depth of reciprocation can be manipulated to further optimize removal of COD, TKN, TAN and nitrate.

References

American Public Health Association (1985). *Standard Methods for the Examination of Water and Wastewater*, 16th edn, Washington DC.

Behrends, L.L., Bailey, E., Bulls, M.J., Coonrod, H.S. and Sikoram, F.J. (1994). Seasonal Trends in Growth and Biomass Accumulation of Selected Nutrients and Metals in Six Species of Emergent Aquatic Macrophytes. In 4th *International Conference on Wetland Systems for Water Pollution Control*, 6–10 November, 1994, Guangzhou, People's Republic of China, pp. 274–289.

Behrends, L. (1999). United States Patent 5,863433. United States Patent Office, Washington DC.

Behrends, L.L. Houke, L., Bailey, E. and Brown, D. (1999). Reciprocating Subsurface-Flow Constructed Wetlands for Treating High-Strength Aquaculture Wastewater. *Wetlands and Remediation Conference*, Salt Lake City, Utah. 16–17 November, 1999.

Behrends, L.L., Houke, L., Bailey, E., Jansen, P. and Brown, D. (2001). Reciprocating Constructed Wetlands for Treating Industrial, Municipal and Agricultural Wastewater. *Wat. Sci. Tech.*, **44**(11–12), 399–405.

Behrends, L.L., Bailey, E., Houke, L., Jansen, P. and Smith, S. (2006a). Non–invasive methods for removing sludge from subsurface-flow constructed wetlands I. Proceedings *NOWRA 2006 Technical Education Conference*. Wastewater Treatment through Integrated Watershed Management, Denver Colorado, August 28–31.

Behrends L.L., Bailey, E., Houke, L., Jansen, P. and Smith, S. (2006b). Non–invasive methods for removing sludge from subsurface-flow constructed wetlands II. *Proceedings 10th International Conference on Wetland Systems for Water Pollution Control*. Lisbon, Portugal, 25–28 September, 2006.

Brix, H., Arias, C.A. and Del Bubba, M. (2001). Media selection for sustainable phosphorus removal in subsurface flow constructed wetlands. *Wat. Sci. Tech.*, **44**(11–12), 47–54.

EPA (2000). Manual Constructed Wetlands Treatment of Municipal Wastewaters (EPA\625\R-99\010), 2000.

Donnell, C., Privett, A. and Behrends, L.L. (2003). Reciprocating constructed wetlands for treating industrial, municipal and agricultural wastewater. *Small Flows Quarterly*, Spring 2003, Vol. 4 No. 2.

IWA (2000). IWA Specialist Group on Use of Macrophytes in Water Pollution Control. In *Constructed Wetlands for Pollution Control; Processes, Performance, Design and Operation*, IWA Publishing, London, UK.

Nguyen, L. (2001). Accumulation of organic matter fractions in a gravel-bed constructed wetland. *Wat. Sci. Tech.*, **44**(11–12), 281–289.

Platzer, C. (1996). Enhanced nitrogen elimination in subsurface flow artificial wetlands – a multistage concept. In *Proceedings of the 5th International Conference on Wetland Systems for Water Pollution Control*. International Water Association, Colchester, UK.

Tanner, C.C. (2001). Plants as ecosystem engineers in subsurface-flow treatment wetlands. *Wat. Sci. Tech.*, **44**(11–12), 9–19.

Zhu, T. and Sikora, F.J. (1995). Ammonium and nitrate removal in vegetated and unvegetated gravel bed microcosm wetlands. *Wat. Sci. Tech.*, **32**(3), 219–228.

Zitomer, D.H. and Speece, R.E. (1993). Sequential environments for enhanced biotransformation of aqueous contaminants. *Environ. Sci. Tech.*, **27**, 227–244.

Design and operation of SBR processes for small plants based on simulations

L. Larrea*, J. Albizuri*, I. Irizar* and J.M. Hernández**

*CEIT and Tecnun (University of Navarra), Manuel de Lardizábal 15, 20018, San Sebastián, Spain
(E-mail: *llarrea@ceit.es*)

**AWA (Alava Water Authority), Spain

Abstract The paper firstly presents an experimental study in a SBR pilot plant operated at 20 °C for nitrogen removal from a very small village wastewater. The plant consisted of only one reactor fed continuously throughout the day and aerated intermittently. Two tests with seven and three intermittences of mixing/aerated phases were conducted and verification of the results by simulations of the activated sludge model (ASM) was also carried out. The experimental results and simulation showed that a wide range of effluent NO_3-N can be obtained using different numbers of intermittences and values of the oxygen transfer coefficient (K_La). At the same time, the paper presents a design procedure for SBR processes based on an iterative process of simulations of the ASM model. After the selection of the cycle time, the mixing/aeration pattern, the initial volume, the solid retention time, and the duration of the phases, the simulation is undertaken, resulting in values for the effluent NH_4-N and NO_3-N, and the suspended solids before settling. Then, the latter parameters are verified to match the effluent and settling requirements. As an application of the design procedure, the effect on design and operation of different SBR configurations and of several operating conditions are analysed in three case studies.

Keywords Design procedure; nitrogen removal; SBR process; simulations

Water Science & Technology **Vol 55 No 7 pp 163–171** © IWA Publishing 2007

Introduction

Applications of SBR in small treatment plants in rural areas (Schleypen *et al.*, 1997; Wilderer *et al.*, 2001; Barjenbruch and Erler, 2005) of many regions throughout the world have been regularly reported in the last years. This is because the SBR provides simplicity and flexibility. In addition, plants can be built compactly and economically. In the province of Alava in Northern Spain, there are more than two hundred small rural villages of less than 5000 inhabitants, in which new biological treatment plants will be built in the next five years. For this purpose, the Alava Water Authority (A.W.A.) has selected the SBR process with one or two reactors as a reference for the biological treatment. In order to achieve an appropriate design and operation of the future facilities, the A.W.A. funded a research project that included both experimental and simulation studies. In an initial phase, the experimental study focused on the use of a single reactor being fed for twenty-four hours. Simulations of the experimental tests were carried out in order to improve the evaluation of the observed behaviour. At the same time, the adoption of a design procedure was another objective. An initial reference was the ATV guideline ATV-M210 (published in 1997) with further contributions being made in several subsequent publications (Artan *et al.*, 2001, 2002; Teichgräber *et al.*, 2001) and culminating recently with the *Technical Report no 19* of the IWA (Artan and Orhon, 2005). At the same time, dynamic simulations based on activated sludge models (ASM) are being increasingly applied in the design of continuous plants. They have been shown to be a user-friendly and efficient method. Consequently, the development of

doi: 10.2166/wst.2007.141

a simulation-based design procedure in which the simulations were nested within the iterative process was considered an interesting objective.

Firstly, this paper presents the evaluation through simulations of the experimental study. Secondly, the new design procedure is described for nitrogen removal and finally, it is applied in order to study the effect of different SBR configurations and operational conditions.

Experimental study

A pilot plant configured with a single SBR reactor was installed in the village of Armiñon, which has about 100 inhabitants. The plant with a maximum volume (V_{max}) of 700 litres and an initial volume (V_o) of 350 litres was equipped with a programmable logic controller (PLC) to regulate the different available operations, i.e.: feeding at a constant flow rate of 10 l/h for 22 hours, aerating intermittently with membrane diffusers, and settling and withdrawing from 3 to 5 a.m. Wasting was carried out before starting settling at 3 a.m. Control of dissolved oxygen was not implemented.

After several measurement programs, the diurnal profiles of the flow rate and concentrations of raw wastewater in dry weather conditions (DWC) were estimated to be approximately as gathered in Figure 1, using normalized values. Estimated average values were: COD total: 550 mg/l, COD filtered: 180 mg/l, NH_4-N: 35 mg/l. Nitrogen removal was only demanded in summer by the Water Authorities, with the aim of achieving NH_4-N values of 0–2 mgN/l and NO_3-N of about 10 mgN/l to meet a total nitrogen requirement of 15 mgN/l. The solids retention time (SRT) was adjusted to about 10 days, and initially the aeration was intermittent with seven repeated cycles of 1 hour anoxic and 2 hours aerated, following typical patterns shown in the literature (Schleypen *et al.*, 1997; Kazmi and Furumai, 2000). Nevertheless, a poor denitrification was observed in these conditions with NO_3-N values of 20–30 mgN/l. The simulation of the experimental test revealed that poor denitrification was due to high oxygen concentrations in unaerated mixing intervals, which in turn was caused by a) the high dissolved oxygen concentration in aerated intervals arising from the use of high values of the oxygen-transfer coefficient (K_La), b) the low oxygen uptake rate (OUR) from using low suspended solids in mixed liquor (X_{ML}) values (1000 mgSST/l), and c) the limited duration of mixing intervals. To overcome the problem, the intermittent aeration was changed to three denitrification–nitrification

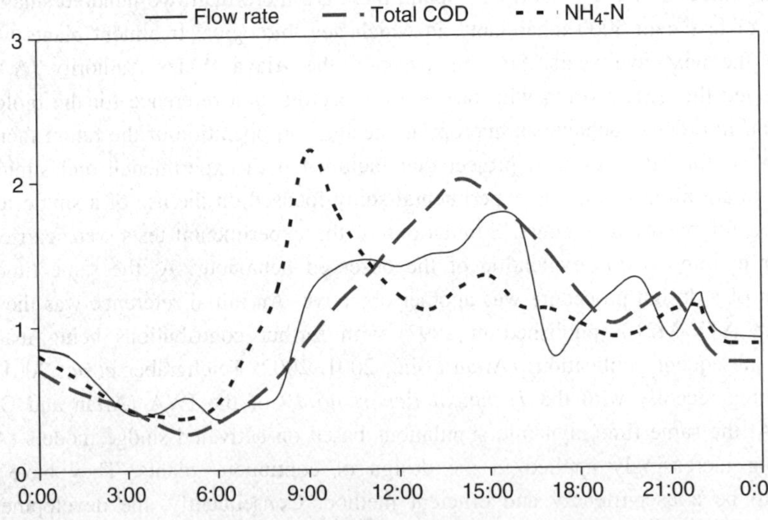

Figure 1 Normalized daily profiles of raw wastewater in dry weather conditions (DWC)

cycles (1 h mixing, 3 h aerated, 4 h mixing, 4 h aerated, 5 h mixing, 5 h aerated). The new operational strategy was successful with effluent NO_3-N values of 5–10 mgN/l as predicted by simulation (Figure 2).

The main reason underlying the successful behaviour in full-scale plants using the intermittent aeration pattern with low durations of the mixing intervals is the general application of low K_La values and high OUR derived from the operation with high SRTs and X_{ML}. Artan and Orhon (2005) indicated that a wide range of effluent NO_3-N can be obtained for different K_La values applied in the intermittent aeration process, concluding that dissolved oxygen control might be important. In this respect, it is thought that a suitable combination of the dissolved oxygen (DO) control pattern and the number of intermittences should be established so that a reasonable value of maximum K_La is obtained. For this purpose, the use of a DO controller which follows a predefined slow DO increase with a ramp pattern at the beginning of the aeration interval is proposed (Larrea *et al.*, 2002). The application of this strategy to the case of Armiñon is described later.

Design procedure for SBR processes

The procedure was developed to address the design and also the operation of generic scenarios where the SBR is being applied for nitrogen removal. It focuses particularly on small plants, although it is also valid for medium and large plants. First of all, several parameters must be defined: i) wastewater characteristics (including the daily variability of the flow rate and concentrations in dry weather (DWC) as well as in wet weather conditions (WWC)); ii) effluent requirements regarding NH_4-N and NO_3-N in composite samples; iii) other requirements such as maximum sludge volume index (SVI), K_La and water height in tank ($H_{w,max}$) as well as temperatures in winter and in summer, minimum SRT or predefined high SRT for stabilization purposes.

Secondly, the configuration of the SBR process should be selected from amongst those generally applied for small plants. Possible configurations include a) continuous filling with intermittent aeration; b) alternated filling of 2 tanks without using a holding tank. (The aeration pattern could be intermittent or continuous using the ramp concept explained above which promotes simultaneous denitrification (SND). Alternated filling of four or six tanks is typical for medium and large plants). c) Use of a holding tank with short filling to two reactors (fill phase duration, $T_f < 1$ h). Step feeding is proposed by Artan and Orhon (2005) because the use of only one pre-denitrification period leads to

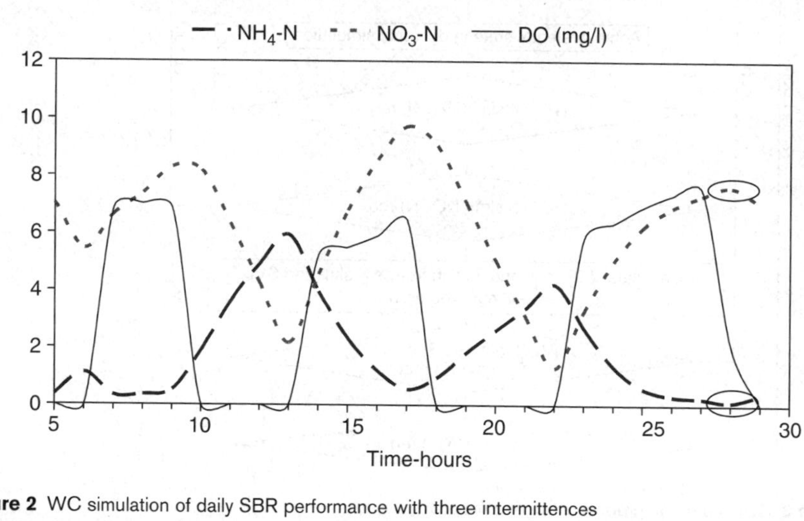

Figure 2 WC simulation of daily SBR performance with three intermittences

the use of a very low cycle time (T_c) and a high K_La, which are not feasible in practice. To overcome these problems, the use of the ramp concept and intermittent aeration might also be applied.

The iterative process of the proposed procedure (Figure 3) then starts with an initial selection of the cycle time (T_c), the filling and mixing/aeration patterns from amongst those described above, and the fill phase duration (T_f) in the case of using a holding tank. It should be noted that only an operational strategy with constant T_c is considered in this paper. The initial volume (V_o) and the duration of the settling phase (T_s) are selected next. As a result, the actual volumetric exchange ratio ($VER_{act} = V_f/V_{max}$) is calculated, where V_f is the fill volume. Then this is followed by the selection of the overall SRT, i.e. it includes the SRT devoted to settling. In the course of the iterative process SRT might either be changed, in order to get the minimum value to meet the effluent requirement, or predefined with a high value for sludge stabilization purposes. The last selection corresponds to the duration of the mixing and aerated phases (T_m and T_a), the latter including either the K_La value or the set points of the dissolved oxygen controller.

The dynamic simulation with an activated sludge model follows until the steady state is achieved. Predictions are provided of the values of relevant parameters in every cycle of the SBR process, such as: effluent NH_4-N and NO_3-N, X_{ML} before settling, and maximum K_La. From this point the procedure continues with the verification of predefined requirements. Firstly, the biological nitrogen removal is verified by means of two loops. In loop 1, the aim is to suitably select the duration of the mixing and aeration phases in order to achieve a simulated value of about 2 mg/l of effluent NH_4-N in composite sample (rather than lower values) since it leads to optimum SRT values. Loop 1 is nested within loop 2. In loop 2 the minimum SRT should be selected to match both NH_4-N and NO_3-N requirements since SRT is correlated with the total nitrogen removal capacity. For high SRTs with sludge stabilization purposes, the criteria might be to match a given

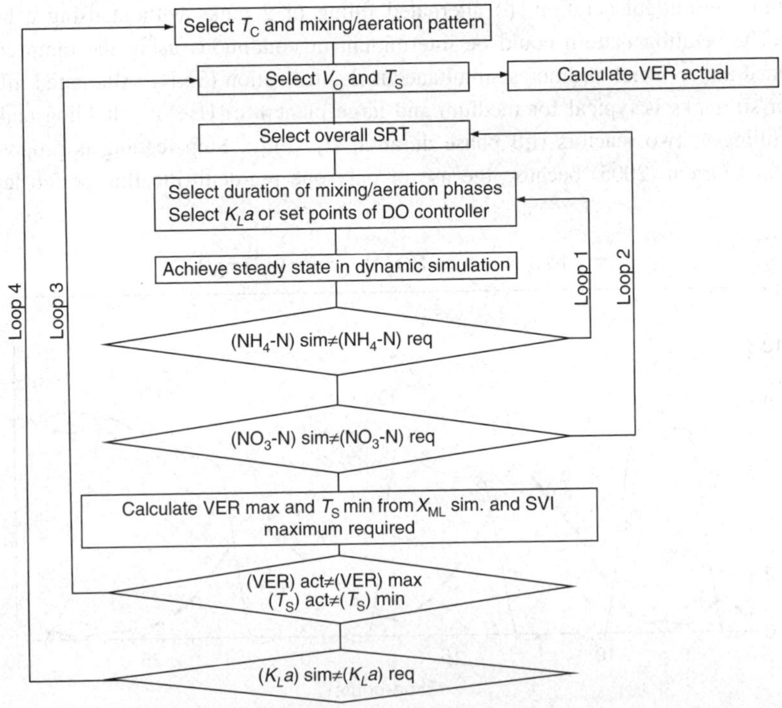

Figure 3 Scheme of the proposed design procedure

requirement for a sludge oxygen uptake rate (OUR). Loop 3 refers to the verification of the settling process. For this purpose, the maximum VER and minimum settling phase (T_s) are calculated from the following equations derived from the ATV guideline ATV-M210, with X_{ML} being taken from the simulation.

$$\mathrm{VER}_{max} = (1 - X_{ML} \cdot (\mathrm{SVI}/1000)) - 0.1$$

where, $H_{w,max}$: predefined maximum water height

$$T_{s,min} = H_{w,max} \cdot (\mathrm{VER}_{act} + 0.1)/V_s$$

$$V_s(\text{settling velocity}) = 650/(X_{ML} \cdot \mathrm{SVI})$$

Then, the actual VER and T_s are checked, allowing V_o to be optimized to match maximum VER and minimum T_s. Loops 1 and 2 are nested within loop 3. Finally, loop 4 aims at the verification of maximum K_La, where either a DO controller has been applied or a constant K_La has been predefined. If maximum K_La is not met, changes in the cycle time or in the filling/aeration pattern, or in V_o should be applied.

As a general rule, different filling and aeration patterns and different cycle times can be tested over several runs where dry weather conditions are applied as the starting step in order to determine the initial volume (V_o) and the remaining operational conditions (SRT, etc.). Then, to determine V_{max}, it is proposed that the obtained V_o for DWC is maintained, if possible, and the WWC applied subsequently. The effluent requirements for NH_4-N and NO_3-N will be met comfortably, and it is also proposed that SRT and V_o are selected to match VER and T_s limits and at the same time to maintain the mass of suspended solids in the reactor nearly constant. Once the V_{max} design value has been selected, DWC operation might use higher V_o or different T_c values than those established in the design process in order to benefit from the V_{max} available.

Application of the design procedure

In this section the effect on design and operation of different SBR configurations and of several operating conditions are analysed in different case studies. A tailored program developed within the WEST platform which makes the changes in the operating conditions of the SBR process easier has been used in the simulations of the ASM2d model. Bioreactions also take place during settling.

Case study 1: Continuous filling and intermittent aeration

The case of the village of Armiñon has been analysed for this study with the dry weather conditions (DWC) presented in Figure 1. The wet weather conditions (WWC) consist of the dry concentrations diluted with a rainfall of four times the average dry weather flow, and the maximum flow accepted in the SBR bioreactor, which corresponds to four times the average dry weather flow. The effluent requirements are $NH_4\text{-N} = 2$ mgN/l and $NO_3\text{-N} = 10$ mgN/l at 20 °C. Other requirements are: maximum SVI of 180 ml/g, minimum SRT, and maximum water height of 4 m ($H_{w,max}$).

Effect of number of intermittences in dry weather conditions. As described above, the aeration pattern is based on a DO controller using the ramp concept. The application of 3, 4 and 5 intermittences of mixing/aerated intervals has been tested using the same V_o and SRT. It is found that the higher the number of intermittences, the higher is the K_La value, even after applying a slower ramp in the DO controller. However, the application of three intermittences requires control of DO at too low levels. Four intermittences might provide an intermediate solution (Table 1 and Figure 4).

Table 1 Simulation results using four intermittences at 20 °C

Argument	Unit	Design	
		DWC	WWC
Simul no		1	2
V_o	m^3	14	18
V_{max}	m^3	35	97
SRT	d	10	14
SST	mg/l	1600	550
Mass SST	kg	60	60
VER act	–	0.60	0.80
VER max	–	0.61	0.80
T_S act	h	2	2
T_S min	h	1.2	0.6
K_L max	1/d	140	50

Wet weather conditions (Sim. no 2). Following the rule proposed above, the procedure results in slightly higher V_o (18 m^3) than in DWC and in V_{max} of 97 m^3. Obviously, K_La decreases considerably in WWC (Table 1).

Case study 2: two reactors with holding tank

The case of a village with 1000 p.e. and a flow rate ratio of 220 l/d p.e., including infiltration is considered. The influent flow rate to the SBR process is assumed constant, with a value of 220 m^3/d. The concentrations are also constant and taken from the average values gathered in Figure 1. The requirements are the same as in case study 1, with the exception of the temperature and SVI, which are 12 °C and 140 ml/g respectively.

Effect of Tc (Sim. no 4–6). The feeding pattern consists of only one fill phase of 0.5 hours at the beginning of the cycle. The aeration pattern consists of two successive intermittences of mixing-aerated intervals using a DO controller in which the initial ramp concept may be applied if appropriate. The optimum simulation results gathered in Table 2 denote that the minimum V_{max} is obtained with a T_c of 6 hours by virtue of the fact that V_f reaches its lowest value and that V_o is almost constant. Nevertheless, this low T_c leads to the highest X_{ML}, the lowest VER, and, significantly, to the highest K_La at the beginning of the first aeration interval. The latter is caused by the impossibility of using the ramp concept. If this is applied, it is not possible to meet the effluent requirements.

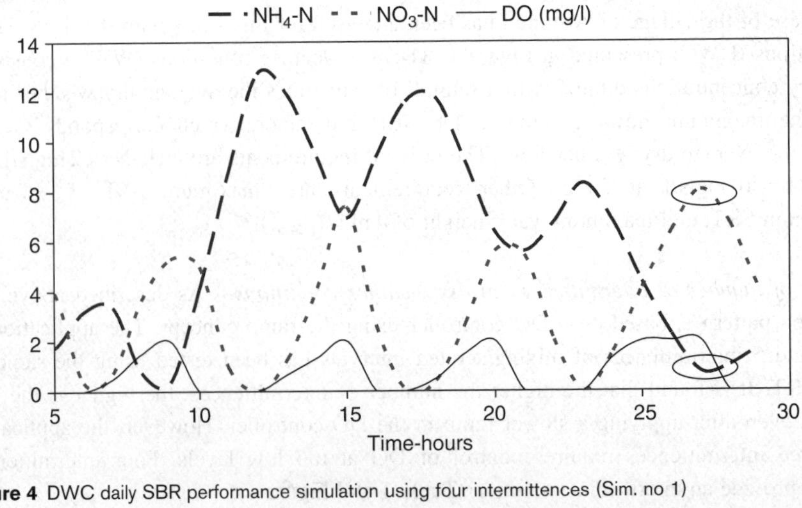

Figure 4 DWC daily SBR performance simulation using four intermittences (Sim. no 1)

Table 2 Simulation results for runs conducted in case study 2

Parameter	Unit	Design			Operation		
		$T = 12\,°C$ SVI = 140 ml/g			$T_c = 8$ h		$T_c = 6$ h
		$T_c = 6$ h	$T_c = 8$ h	$T_c = 12$ h	Step feeding	$T = 20\,°C$	SVI = 180 ml/g
Simul no		4	5	6	7	8	9
V_o	m^3	52	55	57	55	55	64
V_{max}	m^3	80	92	112	92	92	92
SRT	d	10	10	10	10	10	10
SST	mg/l	3850	3450	2900	3300	2900	3300
Mass SST	kg	299	296	308	291	255	295
VER act	–	0.34	0.40	0.49	0.40	0.40	0.30
VER max	–	0.36	0.42	0.49	0.44	0.49	0.31
$T_S$act	H	1.5	1.5	1.5	1.5	1.5	1.5
$T_S$min	H	1.5	1.5	1.5	1.4	1.3	1.5
$K_L$amax	1/d	170	100	75	120	150	130

For a T_c of 8 hours, a ramp of 1 hour is feasible and K_La decreases to $100\,d^{-1}$. For a T_c of 12 hours, all the above observations are intensified.

Effect of fill and aeration pattern (Sim. no 5 and 7) (Figure 5). Taking as initial conditions the design parameter values of simulation no 5 with a T_c of 8 hours, a dual step feeding of 0.5 hours each was introduced, combined with 2 successive mixing/aerated intervals (Figure 5b). It was not found feasible to apply the ramp concept when trying to meet effluent requirements. Consequently, this gave rise to higher K_La than in simulation 5 (Table 2).

Effect of temperature (Sim. no 8). Taking as initial conditions the design parameters of simulation number 5 with a T_c of 8 hours, the temperature was raised to 20 °C in simulation no 8 with the resulting increase in OUR and K_La. To counteract this effect, a slower ramp was feasibly applied because nitrogen removal also increases with temperature. Using the same 10-day SRT as at 12 °C, the effluent total nitrogen is much lower than required. The X_{ML} is also lower, which would allow an operation with higher SVI (170 ml/g). The lower the SRT, the higher the SVI allowed for operation.

Effect of SVI (Sim. no 9). If SVI increases above 140 ml/g, a T_c of 6 hours can be applied, V_o then increases from 55 to 64 m^3, meeting in turn the effluent requirements.

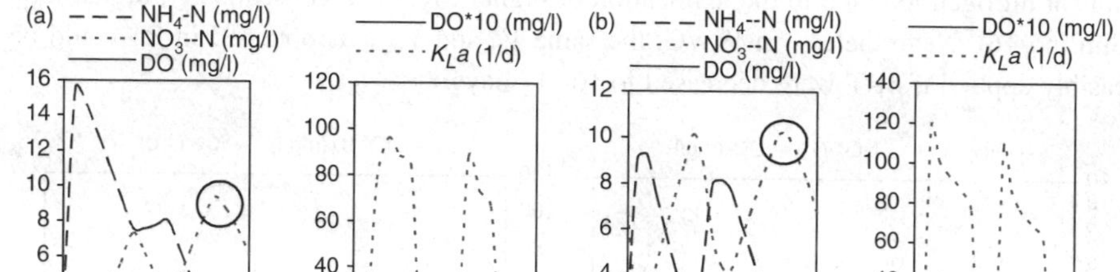

Figure 5 Simulation of a single cycle SBR performance for a T_c of 8 hours, using the ramp concept in the aerated phases (Sim. no 5) (Figure 5a) and dual step feeding (Sim. no 7) (Figure 5b)

Table 3 Simulation results for runs conducted in case study 3

Parameter	Unit	Design at 12 °C						
		R1						R2
		DWC			WWC			DWC
		$T_c = 6$ h	$T_c = 8$ h	$T_c = 12$ h	$T_c = 6$ h	$T_c = 8$ h	$T_c = 12$ h	$T_c = 6$ h
Simul no		10	11	12	13	14	15	16
V_o	m^3	90	100	85	90	100	85	130
V_{max}	m^3	130	143	156	200	245	305	169
SRT	d	20	24	23	20	24	21	26
SST	mg/l	4200	4250	3100	2400	2000	1300	4700
Mass SST	kg	534	624	490	471	473	405	787
VER act	–	0.31	0.30	0.46	0.55	0.60	0.70	0.23
VER max	–	0.31	0.31	0.47	0.56	0.62	0.72	0.24
$T_S$act	h	1.5	1.6	1.5	1.5	1.6	1.5	1.5
$T_S$min	h	1.5	1.5	1.5	1.3	1.2	0.9	1.3
$K_La$max	1/d	110	70	65	55	44	34	85

Case study 3: two reactors without holding tank

The same case of 1000 p.e. is considered. The dry and wet weather variability for flow rate and concentrations are those used in case study 1 (Figure 1). The effluent and other requirements are those applied in case study 2.

Design at 12°C (Sim. no 10–16). As COD and nitrogen loads to the two reactors are different, simulations have been carried out for both with cycle times of 6, 8 and 12 hours where $T_f = T_c/2$. In this case the aeration pattern of only one set of mixing-aerated intervals using a DO controller with the ramp concept was applied, instead of using intermittences in the course of the filling phase. The optimum results shown in Table 3 reveal that the minimum V_{max} is obtained with a T_c of 6 hours, as in case study 2, but that the maximum K_La only amounts to 110 d^{-1} in reactor R1. This is lower than that found in case study 2 and is due to the ramps used during aeration, something which was not feasible in case study 2. Figure 6 indicates when effluent requirements are met in composite samples of the four discharges carried out in the course of the day. In wet weather conditions, V_o for reactor 1 is kept at 90 m^3 and V_{max} rises to 200 m^3. In addition, a SRT of 20 days is sustained, which produces a X_{ML} of 2400 mg/l, that matches the maximum VER. Logically, the V_{max} with a T_c of 6 hours without holding tank doubles the V_{max} with holding tank and a T_c of 8 hours.

For the other reactor, R2, the resulting values of V_o and V_{max} in DWC needed to meet the requirements were higher (130 m^3) than for reactor R1 (90 m^3), since the higher influent nitrogen load led to the application of higher SRT and, consequently, higher X_{ML} (Sim. no 16). Nevertheless, in WWC, the same V_o and V_{max} as obtained in R1 could be feasibly applied if SRT were decreased to 16–17 days.

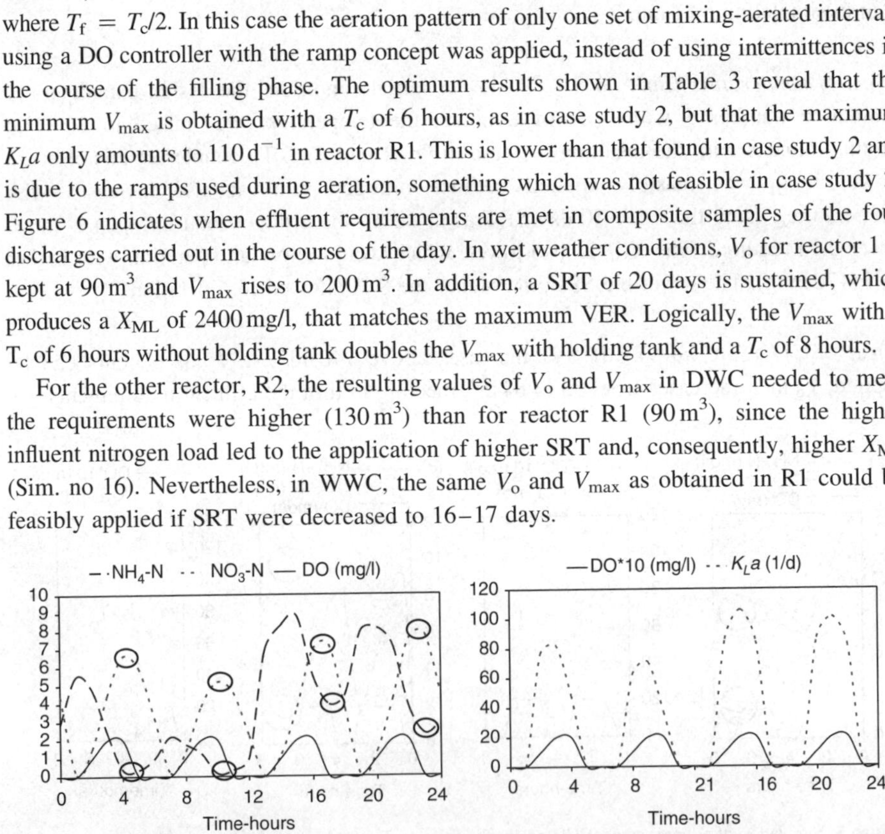

Figure 6 DWC daily reactor R1 performance simulation for a T_c of 6 hours (Sim. no 10)

Conclusions

From the experimental study in a SBR pilot plant continuously fed and aerated intermittently, it was concluded that a suitable combination of a dissolved oxygen control pattern and the number of intermittences could be established to achieve optimum nitrification and denitrification with reasonable values of maximum K_La.

With regard to the design procedure, it is thought that the simple proposed structure embracing selection of operating conditions, simulation and verification in four nested loops makes the iterative process a user-friendly and efficient means of achieving the final results. The application to a case study with a SBR process using only one reactor, fed continuously and aerated intermittently, achieved optimum nitrogen removal when applying four intermittences and a dissolved oxygen controller. In the case of a SBR process using two reactors after a holding tank, it was found that a cycle time of eight hours achieved optimum results without applying step feeding. Finally, in the case of not using a holding tank, a cycle time of six hours was feasible, thereby minimizing the hydraulic retention time. In all cases, the application of a dissolved oxygen controller with a ramp at the beginning of the aeration phase has been proved to be very powerful.

References

Artan, N. and Orhon, D. (2005). *Mechanism and Design of Sequencing Batch Reactors for Nutrient Removal*, Scientific and Technical Report No. 19, IWA Publishing, London, UK.

Artan, N., Wilderer, P.A., Orhon, D., Morgenroth, E. and Ozgur, N. (2001). The mechanism and design of sequencing batch reactor systems for nutrient removal – the state of the art. *Wat. Sci. Technol.*, **43**(3), 53–60.

Artan, N., Wilderer, P.A., Orhon, D., Tasli, R. and Morgenroth, E. (2002). Model evaluation and optimization of nutrient removal potential for sequencing batch reactors. *Water SA*, **28**(4), 423–432.

ATV (1997). Bemessung von Belebungsanlagen im Aufstaubetrieb M210, September 1997.

Barjenbruch, M. and Erler, C. (2005). A performance review of small German WSPs identifying improvement options. *Wat. Sci. Technol.*, **51**(12), 43–49.

Kazmi, A.A. and Furumai, H. (2000). Field investigations on reactive settling in an intermittent aeration sequencing batch reactor activated sludge process. *Wat. Sci. Technol.*, **41**(1), 127–135.

Larrea, L., Irizar, I. and Hidalgo, M.E. (2002). Improving the predictions of ASM2d trough modelling in practice. *Wat. Sci. Technol.*, **45**(6), 199–209.

Schleypen, P., Michel, I. and Siewert, H.E. (1997). Sequencing batch reactors with continuous inflow for small communities in rural areas in Bavaria. *Wat. Sci. Technol.*, **35**(1), 269–276.

Teichgräber, B., Schreff, D., Ekkerlein, C. and Wilderer, P.A. (2001). SBR technology in Germany – an overview. *Wat. Sci. Technol.*, **43**(3), 323–330.

Wilderer, P.A., Irvine, R.L. and Goronszy, M.C. (2001). *Sequencing Batch Reactor Technology*, Scientific and Technical Report No. 10, IWA Publishing, London, UK.

Grey water treatment in UASB reactor at ambient temperature

T.A. Elmitwalli*, M. Shalabi, C. Wendland** and R. Otterpohl****

*Department of Civil Engineering, Benha High Institute of Technology, P.O. Box 13512, Benha El-Gedida, Benha, Egypt (E-mail: *t_elmitwalli@yahoo.com*)

**Institute of Wastewater Management and Water Protection, Hamburg University of Technology, Eissendorfer Strasse 42, D-21073 Hamburg, Germany (E-mail: *moataz.shalabi@tuhh.de*; *c.wendland@tuhh.de*; *otterpohl@tuhh.de*)

Water Science & Technology **Vol 55 No 7 pp 173–180** © IWA Publishing 2007

Abstract In this paper, the feasibility of grey water treatment in a UASB reactor was investigated. The batch recirculation experiments showed that a maximum total-COD removal of 79% can be obtained in grey-water treatment in the UASB reactor. The continuous operational results of a UASB reactor treating grey water at different hydraulic retention time (HRT) of 20, 12 and 8 hours at ambient temperature (14–24 °C) showed that 31–41% of total COD was removed. These results were significantly higher than that achieved by a septic tank (11–14%), the most common system for grey water pre-treatment, at HRT of 2–3 days. The relatively lower removal of total COD in the UASB reactor was mainly due to a higher amount of colloidal COD in the grey water, as compared to that reported in domestic wastewater. The grey water had a limited amount of nitrogen, which was mainly in particulate form (80–90%). The UASB reactor removed 24–36% and 10–24% of total nitrogen and total phosphorus, respectively, in the grey water, due to particulate nutrients removal by physical entrapment and sedimentation. The sludge characteristics of the UASB reactor showed that the system had stable performance and the recommended HRT for the reactor is 12 hours.

Keywords Anaerobic digestion; domestic wastewater; ecological sanitation; grey water; UASB reactor

Introduction

In ecological sanitation, the wastewater is considered not only as a pollutant, but also as a resource for fertiliser, water and energy and for closing water and nutrients cycles (Otterpohl *et al.*, 1999; Otterpohl *et al.*, 2003; Elmitwalli *et al.*, 2006). The ecological sanitation based on separation between grey and black water (and even between faeces and urine), is considered a visible future solution for wastewater collection and treatment. Grey water, which symbolises the wastewater generated in the household excluding toilet wastewater (black water), represents the major volume of the domestic wastewater (60–75%) with low content of nutrients and pathogens (Jefferson *et al.*, 1999; Otterpohl *et al.*, 1999; Eriksson *et al.*, 2002).

Most grey-water treatment plants include a one or two-step septic-tank for pre-treatment (Otterpohl *et al.*, 2003). The grey-water treatment needs both physical and biological processes for removal of particles, dissolved organic-matter and pathogens (Jefferson *et al.*, 1999). Recently, many researchers have studied the grey-water treatment either by application of high-rate aerobic systems, such as the rotating biological contactor (Nolde, 1999), fluidised bed (Nolde, 1999), aerobic filter (Jefferson *et al.*, 2000), membrane bioreactor (Jefferson *et al.*, 2000), or by application of low-rate systems, like slow sand filter (Jefferson *et al.*, 1999), vertical flow wetlands (Otterpohl *et al.*, 2003). Although high-rate anaerobic systems, which are low-cost systems, have both physical and biological removal, no research has been done until now on grey water in these systems. The grey water contains a significant amount (41%) of chemical oxygen demand (COD) in

doi: 10.2166/wst.2007.142

the domestic wastewater (Otterpohl *et al.*, 2003) and this amount can be removed by the high-rate anaerobic systems.

Although high-rate anaerobic systems have been successfully operated in tropical regions for domestic wastewater treatment, the process up till now has not been applied in low-temperature regions. The COD removal is limited for domestic wastewater treatment in high-rate anaerobic systems at low temperatures and, therefore, a long HRT is needed for providing sufficient hydrolysis of particulate organics (Zeeman and Lettinga, 1999; Elmitwalli *et al.*, 2002). The grey water has a relatively higher temperature (18–38 °C), as compared to the domestic wastewater (Eriksson *et al.*, 2002), because the grey water originates from hot water sources, like shower (29 °C), kitchen (27–38 °C) and laundry (28–32 °C). Therefore, high-rate anaerobic systems might run efficiently for on-site grey water treatment, even in low-temperature regions. The upflow anaerobic sludge blanket (UASB) reactor is the most applied system for anaerobic domestic wastewater treatment. Accordingly, the aim of this research is to study the feasibility of a UASB reactor for grey water treatment at ambient temperature.

Materials and methods

Experimental set-up

A UASB reactor was installed in the Institute of Wastewater Management and Water Protection, Hamburg University of Technology, Germany. Figure 1 shows a schematic diagram of the system. The grey water was collected from 'Flintenbreite' settlement in Luebeck city, Germany. A 1 m^3 container was installed in the settlement to collect the grey water, which was transported to the Institute. The collected grey water represented a wastewater of one day and was collected from the inlet sewer-pipe in the first septic-tank in the settlement. The UASB reactor started operation after adding a seed-sludge from an

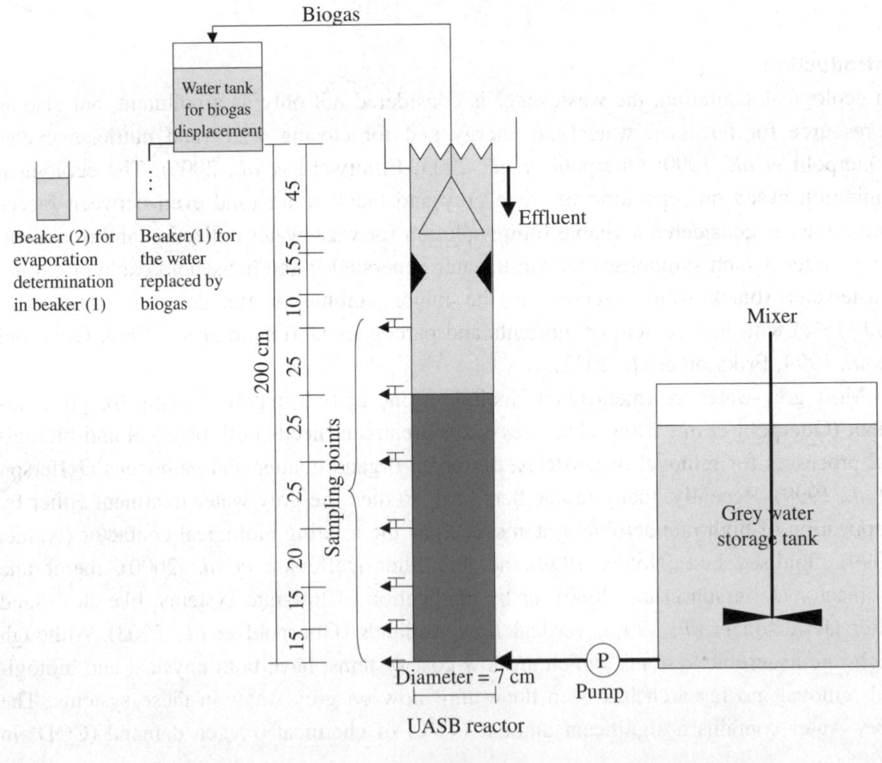

Figure 1 Schematic diagram of the UASB reactor treating grey-water at ambient temperature

anaerobic digester treating primary and secondary sludge. The UASB reactor was operated at an HRT of 20, 12 and 8 hours for 140, 105 and 93 days, respectively.

The septic tank at an HRT of 2–3 days is the most applied system for the pre-treatment of grey water and the main aim of this research is to replace the septic tank by a UASB reactor. Therefore, a column of 7 cm diameter and 2.0 metres height was used for measuring the settling of the grey water. The settling of the grey-water suspended solids (SS) in the septic tank can be represented by the settling of the SS in the column.

At the end of continuous operation of the UASB reactor, batch recirculation experiments were carried out two times to determine the maximum removal of COD fractions (suspended, colloidal and dissolved) in grey water. In each batch recirculation experiment, the storage tank was filled with 28 litres of grey water (4 times of the UASB reactor volume) and after 3 HRT of the UASB reactor, the remaining grey water in the storage tank was recirculated in the UASB reactor for a period of 5 days. The COD fractions of the recirculated wastewater in the storage tank were measured in time for determination of the COD-removal course.

Analysis

COD, total Kjeldahl nitrogen (TKj-N), NH_4-N, ortho and total P, total solids (TS) and volatile solids (VS) were determined as described by *Standard Methods* (APHA, 1998). Raw samples were used for total COD, 4.4 μm folded paper-filtered samples for COD_f and 0.45 μm membrane-filtered samples for dissolved COD. The suspended COD and colloidal COD were calculated by the differences between total COD and COD_f, COD_f and dissolved COD, respectively.

Anaerobic digestibility and maximum specific methanogenic acitivity (SMA_{max}) tests at a temperature of 30 °C were performed for determination of, respectively, the remaining biodegradable fraction and maximum methanogenic activity of the sludge in the UASB reactor at the end of the each operational phase. The digestibility and the SMA_{max} were carried out for a mixture of the sludge from each tap in the reactor. The digestibility tests were carried out for a digestion period of 105 and 96 days for phase 1 and 2 respectively. The SMA_{max} was measured according to Elmitwalli *et al.* (2002), but by measuring COD depletion (instead of acetate depletion) to simplify the test. COD depletion was measured in the second feed with an initial acetate concentration of 1.5 gCOD/l for each feed and a sludge concentration of 2 gVS/l. The SMA_{max} test was done in duplicate for each sample.

Results and discussion

Batch experiments

Settling experiments. Figure 2 presents the course of COD fractions in the supernatant (40–50 cm from the top) of the column, which represent the septic-tank. The results showed that after settling of the grey water for a long period of 2–3 days (similar to the HRT of the septic tank), a limited removal of total COD was obtained (11–14%). Moreover, a release of colloidal COD (negative removal) was achieved. This might be due to the presence of lipids and surfactants in the wastewater, which are mainly coming from kitchen and laundry. Therefore, the septic tank is not an effective system for the pre-treatment of the grey water. The septic tank is mainly suitable for domestic wastewater pre-treatment (not grey water), where it can remove 30–40% of total COD (Metcalf and Eddy, 1999).

Batch recirculation experiments. Figure 3 shows the results of the batch recirculation experiments. The results demonstrated that the UASB reactor is an efficient system for

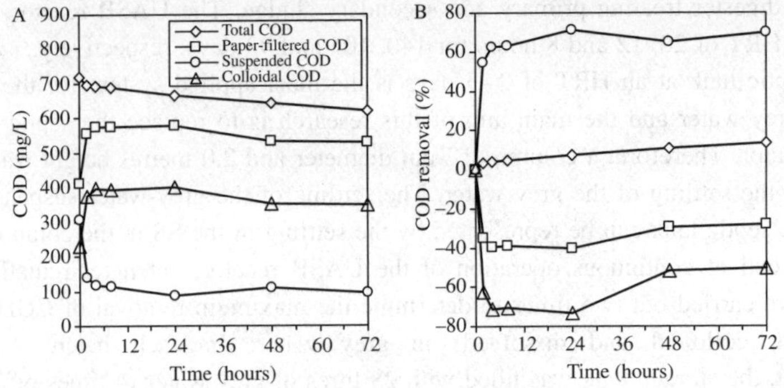

Figure 2 Course of COD-fractions concentration and removal efficiency in the supernatant (40–50 cm from the top) of a column representing a septic-tank (wastewater temperature = 17.5 °C)

grey-water treatment, as the maximum removal efficiencies for total, suspended, colloidal and dissolved COD were 79, 99, 72 and 66%, respectively. The maximum removal efficiency of grey-water COD fractions is relatively higher than those reported by Last and Lettinga (1992), Wang (1994) and Elmitwalli *et al.* (2000) for presettled, pretreated and raw domestic wastewater, respectively.

Performance of the UASB reactor

COD-fractions removal. Table 1 shows the concentration and removal efficiency of COD fractions in grey water treatment in the UASB reactor. The results demonstrated that the suspended COD followed by colloidal COD represented the major part of the grey-water total-COD. In domestic wastewater and black water, the major part of total COD is suspended COD followed by dissolved COD (Levine *et al.*, 1985; Elmitwalli *et al.*, 2002; Gaillard, 2002). The particulate (suspended + colloidal) COD represented about 80% of grey-water total-COD, while it corresponded to 65–75% of domestic wastewater total-COD (Wang, 1994 and Elmitwalli *et al.*, 2002). The higher content of particulate in grey water is due to the increase of the colloidal COD percentage (32–43% of total COD), while it is 25–30% of domestic wastewater total-COD (Elmitwalli *et al.*, 2002). The relatively high percentage of the colloidal COD in the grey water might be due to the presence of surfactants and lipids in the wastewater. The percentage of suspended COD was almost similar in both grey water and domestic wastewater (40–48% of total COD), while the percentage of dissolved COD was higher in domestic wastewater (25–35%), as compared to grey water (18–20%).

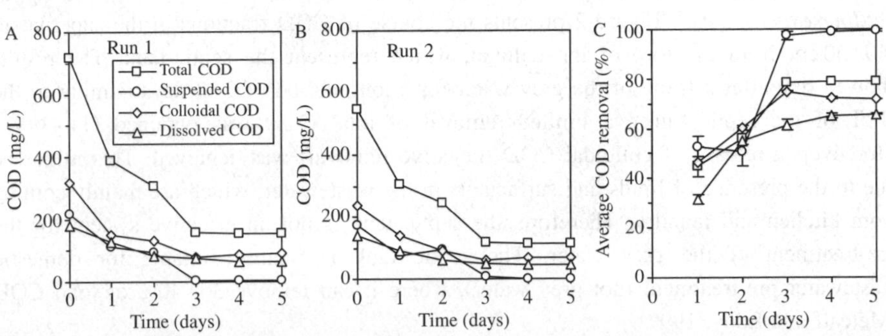

Figure 3 Course of COD-fractions concentration and removal efficiency in the batch recirculation experiments of grey water in the UASB reactor (wastewater upflow velocity and temperature were 0.25 m/h and 20 °C, respectively)

Table 1 Average concentration and removal efficiency of COD fractions and nutrients in grey-water treatment in the UASB reactor at different HRTs. Standard deviations are in brackets

Parameter	Phase 1: HRT = 20 h, temp. = 18 (14–21) °C		Phase 2: HRT = 12 h, temp. = 23 (21–24.5) °C		Phase 3: HRT = 8 h, temp. = 20 (19–21) °C	
	Influent	% Removal	Influent	% Removal	Influent	% Removal
Total COD	681 (124)	31 (5)	647 (137)	41 (13)	682 (106)	33 (7)
Suspended COD	268 (115)	61 (12)	298 (101)	61 (20)	327 (64)	42 (6)
Colloidal COD	291 (122)	18 (13)	231 (86)	28 (22)	219 (71)	28 (27)
Dissolved COD	134 (27)	23 (12)	117 (40)	16 (16)	136 (33)	14 (33)

Although the UASB reactor had the highest HRT (20 hours) in the first phase of operation, the reactor had the lowest total COD removal, mainly because the UASB reactor in phase 1 was operated at a low temperature (average = 18 °C, ranged between 14 and 21 °C). However, the UASB reactor had a better performance than the septic tank at HRT of 2–3 days (11–14% total-COD removal). In phase 2, although the HRT of the UASB reactor was reduced to 12 hours, the UASB reactor had the highest total-COD removal of 41%, because the reactor was operated in the summer period at an average wastewater temperature of 23 °C. In phase 3, the total COD removal decreased to 31%, because the HRT of the UASB reactor was decreased to 8 hours at 20 °C. Accordingly, the UASB reactor treating grey water had a better total-COD removal, as compared to the septic tank, even at a short HRT of 8 hours. From the obtained results, it seemed that the removal of colloidal COD depended on the wastewater temperature, while the removal of suspended and dissolved COD depended on the wastewater temperature and the HRT of the UASB reactor. The results demonstrate that decreasing the HRT of the UASB reactor mainly reduced the total COD removal.

Nutrients removal. Table 2 illustrates the average concentration and removal efficiency of nutrients (N and P) in grey-water treatment in the UASB reactor. The results indicated that the grey water had a limited amount of nitrogen, which was mainly in particulate form (80–90%), while in the black water and domestic wastewater, most of the nitrogen is soluble as NH_4. The concentration of phosphorus in grey water was almost similar to that in domestic wastewater (6.8–8.6 mg total-PO_4-P/L for Dutch municipal wastewater, Elmitwalli *et al.*, 2002). The UASB reactor removed only the particulate nutrients by physical entrapment and sedimentation and, therefore, it had relatively low removal of nutrients. Although the UASB reactor was efficient in the removal of the suspended COD, the reactor had a limited removal of particulate nutrients, especially particulate nitrogen. Therefore, it seemed that most of the particulate nutrients in the grey water were in colloidal form, as the UASB reactor had a limited removal of colloidal COD.

Table 2 Average concentration and removal efficiency of nutrients (N and P) in grey-water treatment in the UASB reactor. Standard deviations are in brackets

Parameter	Phase 1: HRT = 20 h temp. = 18 (14–21) °C		Phase 2: HRT = 12 h temp. = 23 (21–24.5) °C		Phase 3: HRT = 8 h temp. = 20 (19–21) °C	
	Influent	% Removal	Influent	% Removal	Influent	% Removal
Total PO_4-P	9.9 (0.3)	24 (8)	9.7 (0.7)	21.6 (9.1)	9.9 (0.8)	10.1 (2.9)
Ortho PO_4-P	6.6 (1)	16 (5)	8.7 (1.2)	20.1 (9.9)	8.4 (0.1)	15.6 (2.8)
Particulate PO_4-P	3.3 (0.7)	39 (8)	1 (0.5)	53 (32.2)	4.3 (4.3)	33.3 (62.8)
TKj-N	27.1 (3.5)	24 (4)	27.3 (4.5)	35.6 (2.3)	–	–
NH_4-N	5.5 (0.8)	4 (28)	3.9 (1.0)	2.3 (26.1)	3.5 (1.6)	8.6 (72.6)
Particulate N-N	21.6 (3.3)	29 (10)	24.4 (12)	40.2 (6.2)	–	–

Sludge characteristics and biogas production. Figure 4 explains the sludge profile in the UASB reactor at the end of each phase and Table 3 presents the characteristics of the sludge. The results indicate that the sludge concentration in the reactor increased on increasing the HRT and decreasing wastewater temperatures. The average sludge concentration in the UASB reactor (7–11 g VS/L) is lower than that reported in the treatment of municipal wastewater. The relatively low sludge concentration in the UASB reactor treating grey water is mainly due to the lower VS/TS ratio, as compared to that in the treatment of municipal wastewater (60–85%, Elmitwalli *et al.*, 2002; Mahmoud, 2002). The UASB reactor treating grey water had a relatively low VS/TS ratio, because the reactor was treating raw grey water, without grits removal. Therefore, the grits in the raw grey water, like fine sand and inorganic material, precipitate in the reactor, resulting in low VS/TS ratio. The sludge profile, visual observations and COD/VS ratio confirm the previous assumption. The COD/VS ratio is similar to that reported for the treatment of municipal wastewater in the UASB reactor (Mahmoud, 2002). The sludge digestibility results demonstrate that the operation of the reactor in summer (phase 2) stabilised the accumulated sludge in the reactor. Moreover, the SMA_{max} was higher in phase 2, as compared to that of phase 1 and 3, due to relatively higher temperature in phase 2. The long sludge residence time (SRT) and low sludge-digestibility illustrated that the sludge in the reactor had a sufficient stability.

The results of biogas production showed that about 48, 63 and 56% of the removed total COD were converted to methane in, respectively, phase 1, 2 and 3. The previous values were calculated based on the measured biogas production and assumption of 75% methane in the biogas (based on the biogas composition in previous studies in the treatment of municipal wastewater in the UASB reactor, Elmitwalli *et al.*, 2002; Halalshah, 2002; Mahmoud, 2002). The highest conversion of the wastewater to biogas was in phase 2, due to relatively higher temperature. This led to improvement of particulate hydrolysis and methanogenesis in the reactor, as shown by higher SMA_{max} in phase 2.

Final discussion

The batch recirculation experiments showed the application potential of the UASB reactor in grey water treatment, as a high value (79%) of maximum total COD removal was achieved. However, the results of continuous treatment of the grey water in the UASB reactor showed that a relatively low removal of total-COD (31–41%) was obtained at HRT between 8 and 20 hours and wastewater temperature between 14–24 °C. However, this value is higher by 2–3 times than that can be achieved by septic tank (the common system for grey water pre-treatment). Moreover, the real temperature of grey water ranges

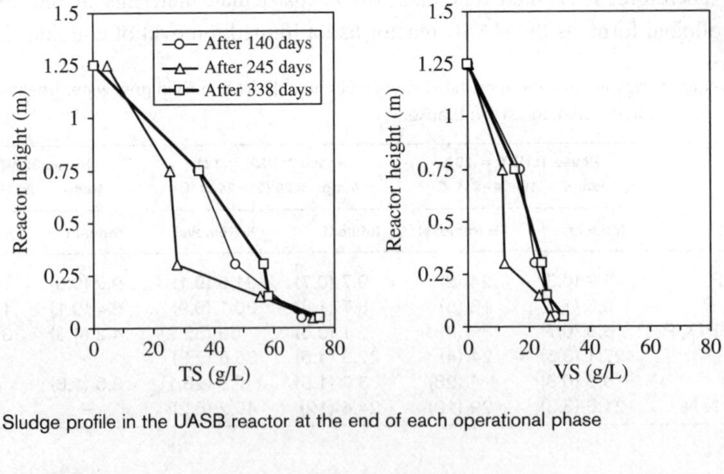

Figure 4 Sludge profile in the UASB reactor at the end of each operational phase

Table 3 Characteristics of the sludge taken from the UASB reactor at the end of each phase. Standard deviations are in brackets

Parameter	Influent Phase 1: HRT = 20 h, temp. = 18 (14–21) °C	Influent Phase 2: HRT = 12 h, temp. = 23 (21–24.5) °C	Influent Phase 3: HRT = 8 h, temp. = 20 (19–21) °C
Average VS (g/L)	11	7	8.7
Average VS/TS (%)	59	41	43
COD/VS (mg/mg)	2.1	1.90	1.86
SRT* (days)	432	312	330
Digestibility (% kg COD/kg COD)	25 (3)	16 (2)	–
SMA_{max} (kg COD/(kg VS.day))	0.05 (0.01)	0.08 (0.015)	0.07 (0.01)

*Calculated based on the excess sludge at the end of each phase

between 18 and 38 °C (Eriksson *et al.*, 2002), which is higher than that applied in this research. Therefore, the on-site treatment of grey water in a UASB reactor can result in a higher total COD removal, as compared to that achieved in this research. The UASB reactor treating grey water should have a longer HRT, as compared to that treating domestic wastewater, because the grey water has a high peak flow factor of 6.2 (Imura *et al.*, 1995). Grey water is produced over short periods, whereas toilet flushing takes place more consistently through the day. Therefore, a HRT of 12 hours can be considered the suitable HRT for grey water treatment in a UASB reactor at ambient temperature. The UASB reactor can be applied for on-site grey water treatment in urban and rural areas. In urban areas, the effluent of the UASB reactor, which is almost free of SS, can be transported by a small bore sewer system (low cost system) to the nearby available areas, where it can be post treated and reused in urban agriculture.

The application of the UASB reactor for grey water treatment will increase the application potential of ecological sanitation based on separation between grey and black water. It is assumed within the ecological sanitation that the produced biogas will be utilised for electricity production and/or heating. For example, in the settlement of 'Flintenbreite' in Luebeck city, Germany, the produced biogas from anaerobic digester of black water and kitchen organic-wastes is used for heating and electricity production. In the settlement, the grey water is pre-treated in a two-step septic-tank followed by vertical wetlands. Such a system needs a long HRT, while at application of a UASB reactor for the pre-treatment of the grey water, a short HRT can be applied. Moreover, the produced biogas can be utilised for heating and production of electricity, if it is combined with the produced biogas from the anaerobic digester of the black water and kitchen organic-wastes. Also, due to higher removal efficiency of total COD and partial removal of nutrients in the UASB reactor, the hydraulic loading of the post-treatment system (like vertical wetlands in the settlement) can be increased and, accordingly, a small land area will be required.

Conclusions

The grey water contained a higher amount of colloidal COD, as compared to that reported in domestic wastewater. The grey water had a limited amount of nitrogen, which was mainly in particulate form, while the concentration of phosphorus in grey water was almost similar to that in domestic wastewater. The batch recirculation experiments showed that a maximum total COD removal of 79% can be obtained in treatment of the grey water in a UASB reactor. The continuous operational results of a UASB reactor treating grey water at different HRTs of 20, 12 and 8 hours at ambient temperature (14–24 °C), showed that 31–41% of the total COD can be removed. These results are significantly higher than that achieved by a septic tank (11–14%), at HRT of 2–3 days. Moreover, the UASB reactor removed 24–36% and 10–24% of, respectively, total nitrogen and total phosphorus in the grey water.

Acknowledgements

The first author acknowledges the Alexander von Humboldt foundation for giving him a fellowship to carry out this research. The authors are grateful to Birgit Buest and Stefan Deegener for technical support.

References

APHA, AWWA and WEF (1998). *Standard Methods for the Examination of Water and Wastewater*, 20th edn., Washington, DC, USA.

Elmitwalli, T., van Dun, M., Bruning, H., Zeeman, G. and Lettinga, G. (2000). 'The role of filter media in removing suspended and colloidal particles in anaerobic reactor treating domestic sewage'. *Biores. Tech.*, **72**(3), 235–242.

Elmitwalli, T.A., Zeeman, G., Oahn, K.L.T. and Lettinga, G. (2002). Treatment of domestic sewage in a two-step system anaerobic filter/anaerobic hybrid reactor at low temperature. *Wat. Res.*, **36**(9), 2225–2232.

Elmitwalli, T.A., Feng, Y., Behrendt, J. and Otterpohl, R. (2006). Anaerobic digestion potential for ecological and decentralised sanitation in urban areas. *Wat. Sci. Tech.*, **53**(9), 45–54.

Eriksson, E., Auffarth, K., Henze, M. and Ledin, A. (2002). Characteristics of grey wastewater. *Urban Wat.*, **4**(1), 85–104.

Gaillard, A. (2002). Waste(water) characterisation and estimation of digestion kinetics. M.Sc. Thesis, Wageningen University The Netherlands.

Halalsheh, M. (2002). Anaerobic Pre-treatment of Strong Sewage: A Proper Solution for Jordan. Ph.D. Thesis, Wageningen University The Netherlands.

Imura, M., Sato, Y., Inamori, Y. and Sudo, R. (1995). Development of a high-efficiency household biofilm reactor. *Wat. Sci. Tech.*, **31**(9), 163–171.

Jefferson, B., Laine, A., Parsons, S., Stephenson, T. and Judd, S. (1999). Technologies for domestic wastewater recycling. *Urban Wat.*, **1**, 285–292.

Jefferson, B., Laine, A.L., Judd, S.J. and Stephenson, T. (2000). Membrane bioreactors and their role in wastewater reuse. *Wat. Sci. Tech.*, **41**(1), 197–204.

Last, A.R.M.van der and Lettinga, G. (1992). Anaerobic treatment domestic sewage under moderate climatic (Dutch) conditions using upflow reactors at increased superficial velocities. *Wat. Sci. Tech.*, **25**(7), 167–178.

Levine, D., Tchobanaglous, G. and Asano, T. (1985). 'Characterization of the size distribution of contaminants in wastewater: treatment and reuse implications'. *Journal of Water Pollution Control Federation*, **57**(7), 805–816.

Mahmoud, N. (2002). Anaerobic pre-treatment of sewage under low temperature (15 °C) conditions in an integrated UASB-digester system. Ph.D. thesis, Wageningen University The Netherlands.

Methcalf & Eddy Inc. (1999). *Wastewater Engineering, Treatment, Disposal and Reuse*, 3th edition, McGraw Hill Inc., New York, USA.

Nolde, E. (1999). Greywater reuse systems for toilet flushing in multi-storey building-over ten years experience in Berlin. *Urban Wat.*, **1**, 275–284.

Otterpohl, R., Albold, A. and Oldenburg, M. (1999). Source control in urban sanitation and waste management: ten systems with reuse of resources. *Wat. Sci. Tech.*, **39**(5), 153–160.

Otterpohl, R., Braun, U. and Oldenburg, M. (2003). Innovative technologies for decentralised water, wastewater and biowaste management in urban and peri-urban areas. *Wat. Sci. Tech.*, **48**(11/12), 23–32.

Wang, K. (1994). Integrated anaerobic and aerobic treatment of sewage. Ph.D. thesis, Wageningen University The Netherlands.

Zeeman, G. and Lettinga, G. (1999). The role of anaerobic digestion of domestic sewage in closing the water and nutrients cycle at community level. *Wat. Sci. Tech.*, **39**(5), 187–194.

Experimental study on nitrogen components during composting process of feces

S. Hotta, T. Noguchi and N. Funamizu

Department of Environmental Engineering, Graduate School of Engineering, Hokkaido University, Kita-13, Nishi-8, Kita-ku, Sapporo 060-8628, Japan (E-mail: *zhenye@eng.hokudai.ac.jp*; *tomopiro@eng.hokudai.ac.jp*; *funamizu@eng.hokudai.ac.jp*)

Abstract We measured nitrogen components during composting process of feces in a batch test in which sawdust was used as a matrix. Further decomposition rates of fecal nitrogen and carbon were obtained in the batch tests of different feces loading. In composting material that was a mixture of sawdust and fresh feces, fecal organic matter decomposed to CO_2 and fecal nitrogen mineralized to ammonia during the composting process. The biological response of organic matter and nitrogen in the composting material was evaluated by oxygen consumption (OUR) and ammonia production that was a sum of volatilized ammonia gas and ammonia remaining in the composting material. Since composting material contains two different sources of organic matter from feces and sawdust, the OUR by using the sawdust matrix only was evaluated in preliminary tests. The fecal contribution to the OUR in the composting material was therefore calculated by subtraction of the result in the preliminary tests from the one in the composting material. The ammonia production from the fecal nitrogen was obtained by the same procedure. The decomposition rates of input organic matter in feces were approximately 83 and 70% respectively, whereas ammonia production rates were approximately 73 and 58% of input fecal nitrogen. There was an interesting time lag of the peak time between volatilisation rates of ammonia and CO_2 during the composting process while fecal carbon and nitrogen simultaneously decomposed to ammonia and CO_2 in the composting material.

Keywords Ammonia production; biodegradation of organic matter in feces; composting process of feces; fecal nitrogen

Water Science & Technology **Vol 55 No 7 pp 181–186** © IWA Publishing 2007

Introduction

Human waste (feces and urine) is treated in a composting toilet by using sawdust as a matrix in an onsite wastewater differentiable treatment system (OWDTS) (Lopez *et al.*, 2002). The composting toilet is a kind of dry closet that requires no water for flushing. The toilet waste goes through a composting process and finally stabilizes in the sawdust matrix. The sawdust matrix contains useful characters, for instance, it provides a suitable balance of water and air retention for aerobic biodegradation and is tolerant to physical friction and biochemical reaction. Thus sawdust can work as the matrix in the bioreactor for more than a year in normal conditions. From a laboratory scale experiment, it had been reported that fecal organic matter and nitrogen immediately decomposes CO_2 and ammonia in the sawdust matrix (Hotta *et al.*, 2005a). Since sawdust matrix used for a long time should contain nutrients for plant growth, it can be used as compost or soil conditioner in farmland (Kitsui and Terazawa, 1999). In contrast, nutrient losses, especially nitrogen loss, have been well reported in the conventional composting process of organic waste. Witter and Lopez-Real, (1987) have shown nitrogen loss of more than 50% of initial input in the composting process of sewage sludge of and manure. More than 70% of nitrogen loss was reported in the composting process of household waste without matrix (Eklind and Kirchmann, 2000). Sanchez-Monedero *et al.* (2001) have provided further knowledge for nitrogen transformation and loss during the composting process of organic waste.

doi: 10.2166/wst.2007.143

Compared to the conventional composting process the composting toilet is a unique system in the following aspects, (1) input of toilet waste continues for a long time, (2) input is much smaller, (3) temperature and mixing frequency of the sawdust matrix are well controlled. Therefore it might be hard to apply the previous knowledge obtained in the area of conventional composting process to the composting toilet system.

Previous researches in the composting toilet have shown that a great amount of biodegradable organic matter in feces decomposed to CO_2 and residual organic matter was stabilized in approximately 48 hours in the sawdust matrix (Lopez *et al.*, 2004a, b). Since most of nitrogen in the toilet waste is from urea (Fittschen and Hahn, 1998), they are unstable in the sawdust matrix especially when feces is mixed with urine (Hotta *et al.*, 2005b). Hotta *et al.* (2005c) have showed that a large amount of the input nitrogen was lost as ammonia gas during the practical operation. Source separation of urine from feces might improve the nitrogen loss. Then CN balance in the sawdust matrix would change significantly. Since biodegradability of fecal nitrogen is also unknown in the sawdust matrix, still more knowledge for transformation of fecal nitrogen with attendance of carbonic evolution is therefore required. The aim of this study is therefore to propose a procedure to evaluate transformation of fecal nitrogen in the composting process by using sawdust as a matrix.

Materials and methods

Experimental device for batch tests

Transformation of the nitrogen components during the composting process of feces by using sawdust as a matrix was measured in a series of batch tests (RUN1 to RUN3). Figure 1 shows an experimental device in which it was possible to control several important conditions for biological activities, for instance, reaction temperature (50°C), air supply (100 ml/min) and moisture content (55 to 65%). The air compressor was placed in front of the first O_2 sensor with stopcock 1 to provide air supply to the reactor.

By using two O_2 sensors as shown in Figure 1, OUR (Oxygen Utilization Rate) was obtained to evaluate aerobic biodegradation of organic matter during the composting process of feces (Lopez *et al.*, 2004a, b). To avoid drying of composting material in the reactor, a water bottle was placed in front of the reactor. To capture the volatilized ammonia gas, the bottle of sulfuric acid ($0.6\,\mathrm{mol\,l^{-1}}$) was put just behind the reactor. To calculate ammonium nitrogen volatilization rate (ANVR), bottles of sulfuric acid were replaced at suitable intervals. Since the higher pressure was maintained in a system of the experimental device rather than outside, airflow must be suspended by turning off three stopcocks during exchanging the bottle.

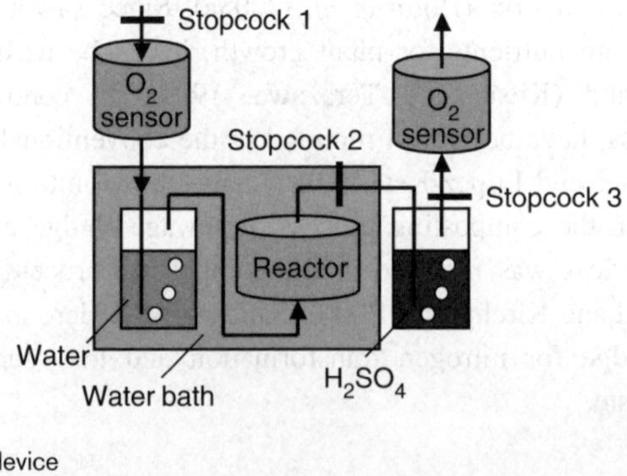

Figure 1 Experimental device

Preparation of materials

The composting material was prepared by mixing of sawdust, fresh feces and distilled water. The load of feces in the composting material was prepared to 5% (RUN1, RUN3) and 20% (RUN2) of F/S ratio, which indicated percentage of the fecal matter on the sawdust. The fresh feces had been stored a couple of days in a refrigerator (5°C) until it was used. To obtain the sawdust that had never touched urine in its history, the sawdust had been prepared in the bioreactor in which only feces had been fed for approximately 6 months. Initial moisture content of the composting material was prepared to 60%, which was supposed to be suitable in the composting process (Horisawa *et al.*, 2000).

Analysis of nitrogen components

To observe the nitrogen transformation in the early stage of the composting process, two lines of batch tests were prepared (RUN1-1, RUN1-2). Accurate OUR and ANVR were obtained in RUN1-1. Intermittent sampling and analysis of the composting material were conducted in RUN1-2. Then total nitrogen in liquid phase (TNL) and total ammonia (TA) were measured by following the semi-Kjeldahl method and the Indophenol one (Weatherburn, 1967; Kempers and Zweers, 1986; Dora, 1975). Total nitrogen in solid phase (TNS) was measured by NC analyser (Sumigraph NC-1000: Shimadzu Corporation). The lost components of fecal nitrogen and carbon in the composting process were calculated in two more batch tests (RUN2, RUN3). The 20% of FS ratio was prepared in RUN2 to minimize an experimental error. Since a practical bioreactor contains much more sawdust than feces, the FS ratio might be much smaller than 20% in a real situation. The RUN3 with FS ratio of 5% was therefore prepared. To complete the biological reaction in RUN2 and RUN3 batch tests continued much longer than in RUN1-1 and RUN1-2 (longer than 350 h). The net loss of fecal nitrogen and carbon was obtained by subtraction of the sawdust contribution from the data obtained in the composting material. The sawdust contribution had been obtained in the preliminary test.

Results and discussion

Composting process of feces

Figure 2 shows the time course of the OUR profile in the two lines of the batch test in RUN1. As seen, the profile was stable and level at the end of the test. It was therefore assumed that most of the biodegradable organic matter aerobically decomposed and residual organic matter stabilized to the biologically inert type at the end. Since there was no significant difference between the two OUR profiles, it was assumed that almost the same biological reactions were obtained in the composting material in the two reactors of RUN1-1 and RUN1-2.

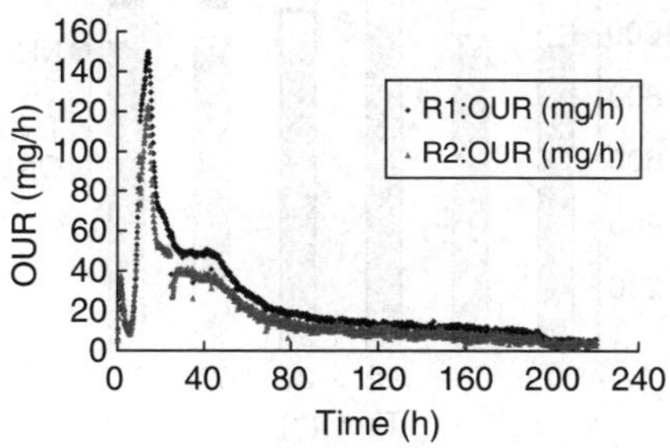

Figure 2 Time course of OUR and CO_2PR in RUN1-1

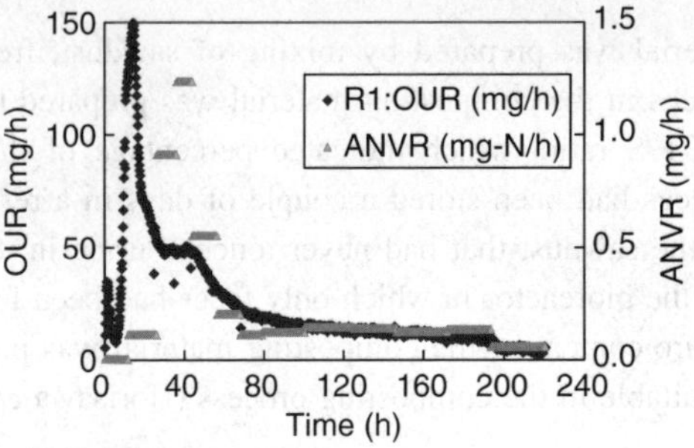

Figure 3 Comparison of OUR and ANVR in RUN1-1

Nitrogen transformation during the composting process

Figure 3 shows the time course of the OUR and the ANVR. In this study, O_2 consumption indicates CO_2 production in the reactor because it was already confirmed that OUR profiles followed the profiles of CO_2 production rate before. As seen in Figure 3, there was an obvious time lag between the volatilization rate of carbon and nitrogen especially in the first 70 hours while decomposition of carbon and nitrogen in feces simultaneously occurred. In first 14 hours until which the OUR begun increasing, the ANVR kept being on quite low levels. The results showed that time lag of two peak times for the OUR and the ANVR was approximately 24 hours. The profile of ANVR was also on the stable level at the end of the experiment while both profiles of the OUR and ANVR were detectable.

Figure 4 shows the time course of nitrogen component in the composting material during the test. From Figure 4, it was found that most nitrogen had been in the organic type in the composting material through the composting process. Ammonia nitrogen ranged less than 15% of the total nitrogen until 222 hours. Since the sawdust matrix used in this study had received and treated feces in preliminary preparation, stable organic nitrogen could remain in the sawdust matrix.

Calculation of nitrogen component after the composting process

Table 1 shows the component data of carbon and nitrogen during the composting process in RUN2 and RUN3. To complete decomposition of biodegradable organic matter, batch tests in RUN2 and RUN3 continued for more than 350 hours until the OUR reached the undetectable stage. Decomposed carbon during the composting process was calculated by an integration of the OUR profile: they were 8536 mg and 3601 mg as listed in Table1-1. Whereas ammonified nitrogen (volatilised nitrogen + ammonia nitrogen remaining in composting material) from the composting material were 881 mg and 205 mg. In the case

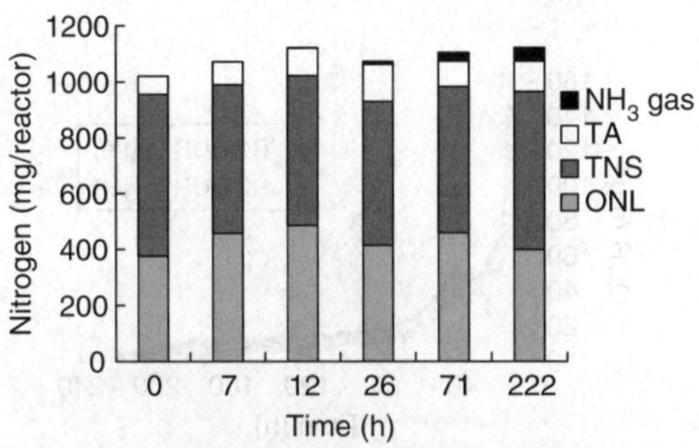

Figure 4 The ratio of nitrogen component in composting process in RUN1-2

Table 1 Calculation of lost component from feces during composting process

	Carbon	Nitrogen
TABLE 1-1 In case of RUN2		
Loss from composting material	8536	881
Loss from sawdust	1547	6
Loss from feces	6989	875
Input in feces	8427	1202
Lost component	83%	73%
Residual component	17%	27%
Table1-2 In case of RUN3		
Loss from composting material	3601	205
Loss from sawdust	1917	5
Loss from feces	1685	200
Input in feces	2402	343
Lost component	70%	58%
Residual component	30%	42%

unit: mg / reactor

of RUN2 input fecal carbon and nitrogen were 8427 mg and 1202 mg respectively. Inputs in RUN3 were 2402 mg and 343 mg respectively. The preliminary batch test by using sawdust matrix without feces had already shown the loss from sawdust as listed in Table 1. Simple subtraction of the preliminary results from the data in RUN2 and RUN3 finally provided loss from feces in the composting material. Sanchez-Monedero *et al.* (2001) have reported no nitrification in the composting process at temperatures higher than 40°C; we therefore eliminated nitrification in the nitrogen transformation in this study.

As listed in Table 1, 83% (RUN2) and 70% (RUN3) of input fecal carbon decomposed to CO_2 during the test. In the case of nitrogen, 73% (RUN2) and 58% (RUN3) of input fecal nitrogen were ammonified. The biologically inert type of fecal nitrogen could be calculated by the subtraction of ammonified nitrogen from input fecal nitrogen. Calculation results showed 27% (RUN2) and 42%(RUN3) of the input fecal nitrogen.

Conclusion

The net loss of fecal nitrogen as ammonia gas during the composting process was calculated at different feces loading in the sawdust matrix. They were approximately 73% and 58% of input fecal nitrogen. The biologically inert type of nitrogen was estimated to be 27% and 42% of input fecal nitrogen. There was an obvious lag of peak times between the volatilization rate of carbon (the OUR) and nitrogen (the ANVR) in the composting process of feces.

Acknowledgements

The authors would like to thank Prof. Minoru Terazawa (Hokkaido University) for his help in sawdust preparation. We thank Prof. Miguel Angel Lopez Zavala for his great help in our experiment. This work has been supported by CREST (Japan Science and Technology Corporation).

References

Dora, S. (1975). Determination of ammonia and Kjeldahl nitrogen by indophenol method. *Water Research*, **10**(1), 31–36.

Eklind, Y. and Kirchmann, H. (2000). Composting and storage of organic household waste with different litter amendments. II: nitrogen turnover and losses. *Bioresource Technology*, **74**(2), 125–133.

Fittschen, I. and Hahn, H.H. (1998). Characterization of the municipal wastewaterpart human urine and a preliminary comparison with liquid cattle excretion. *Water Science and Technology*, **38**(6), 9–16.

Kempers, A.J. and Zweers, A. (1986). Ammonium determination in soil extracts by the salicylate method. *Communications in Soil Science Plant Analysis*, **17**(7), 715–723.

Kitsui, T. and Terazawa, M. (1999). Environmentally friendly toilets for the 21st century, Bio-toilet. *Proceedings of the 10th ISWPC. Yokohama*, **3**, 120–121.

Horisawa, S., Tamai, Y., Sakuma, Y., Doi, S. and Terazawa, M. (2000). Effect of moisture content of a wood matrix on a small-scale biodegradation system for organic solid waste. *Journal of Wood Science*, **46**, 317–321.

Hotta, S., Noguchi, T. and Funamizu, N. (2005*a*). Nitrogen transformation during the aerobic biodegradation of feces. *Environmental Engineering Research*, **42**, 325–333.

Hotta, S., Hiroki, N. and Funamizu, N. (2005*b*). The basic study of the effect of feces contamination on urea hydrolysis. *Proceedings of the International Water Association (IWA) Conference*, 2005 Xi'an 63–70.

Hotta, S., Lopez Zavala, A.M. and Funamizu, N. (2005*c*). Blackwater treatment in the composting toilet by using sawdust as a matrix. *Proceedings of the ECOSAN 9th International Conference Mumbai*, 80–91.

Lopez Zavala, A.M., Funamizu, N. and Takakuwa, T. (2002). Onsite wastewater differential treatment system: modeling approach. *Water Science and Technology*, **46**(6–7), 317–324.

Lopez Zavala, A.M., Funamizu, N. and Takakuwa, T. (2004*a*). Temperature effect on aerobic biodegradation of feces using sawdust as a matrix. *Water Research*, **38**, 2406–2416.

Lopez Zavala, A.M., Funamizu, N. and Takakuwa, T. (2004*b*). Modeling of aerobic biodegradation of feces using sawdust as a matrix. *Water Research*, **38**, 1327–1339.

Sanchez-Monedero, M.A., Roig, A., Paredes, C. and Bernal, M.P. (2001). Nitrogen transformation during organic waste composting by the Rutgers system and its effect on pH, EC and maturity of the composting mixtures. *Bioresource Technology*, **78**, 301–308.

Weatherburn, M.W. (1967). Phenol-Hypochlorite Reaction for Determination of Ammonia. *Analytical Chemistry*, **39**(8), 971–974.

Witter, E. and Lopez-Real, J. (1987). The potential of sewage-sludge and composting in a nitrogen recycling strategy for agriculture. *Biological Agriculture and Horticulture*, **5**(1), 1–23.

Anaerobic digestion of blackwater from vacuum toilets and kitchen refuse in a continuous stirred tank reactor (CSTR)

C. Wendland, S. Deegener, J. Behrendt, P. Toshev and R. Otterpohl

Institute of Wastewater Management and Water Protection, Hamburg University of Technology, Germany (TUHH) (E-mail: *c.wendland@tu-harburg.de*)

Abstract The objective of this research was mesophilic anaerobic digestion of blackwater from vacuum toilets (BW) and kitchen refuse (KR) in a CSTR within an ecological sanitation system. A detailed investigation of the BW characteristics was carried out. Research on anaerobic digestion was performed with CSTR of 10 l volume at HRT of 10, 15 and 20 days. The digestion of BW at 20 days HRT showed stable performance without inhibition effects, in spite of relatively high ammonium concentrations. The removal of total and particulate COD was 61% and 53%, respectively, and the methane yield 10 l CH_4/cap/day. The addition of kitchen refuse (KR) improved the performance of the CSTR in terms of COD removal efficiency and methane yield. At 20 days HRT the removal of total and particulate COD increased up to 71% and 67%, respectively, and the methane yield to 27 l CH_4/cap/day. The results at 15 days HRT showed similar performance. At HRT of 10 days, the anaerobic treatment was limited but reached steady state conditions at higher VFA concentrations in the effluent, with a decrease of COD removal of 30 to 33% and of methane yields of 19 to 21%.

Keywords Anaerobic digestion; blackwater; CSTR; ecological sanitation; kitchen refuse; vacuum toilets

Water Science & Technology **Vol 55 No 7 pp 187–194** © IWA Publishing 2007

Introduction

In recent years, a new approach to wastewater management has been developed and implemented in many projects worldwide, which is referred to as ecological sanitation (ecosan) or decentralised sanitation and reuse (desar) (Otterpohl and Lange, 2000; Lens *et al.*, 2001). As an answer to the tremendous sanitation problems worldwide as well as problems realised with centralised conventional wastewater treatment, ecological sanitation focuses on source separation of organic waste and wastewater on a household level, followed by an appropriate treatment of each stream in decentralised or semi-centralised systems and consequent reuse of water and nutrients. Anaerobic digestion plays a key role as an adequate treatment technology for concentrated wastewaters such as BW, as it requires less energy than aerobic treatment. Anaerobic digestion even produces energy in the form of biogas, and does not generate sewage sludge but an organic fertiliser with highly concentrated nutrients like nitrogen and phosphorus compounds.

In the pilot project Flintenbreite in Luebeck, Germany, source separation was realised in the year 2000 within a housing estate for 350 to 400 inhabitants (Otterpohl *et al.*, 1997; Wendland and Oldenburg, 2003). The concept comprises vacuum toilets with subsequent pasteurisation and anaerobic digestion of BW in a semi-centralised biogas plant and finally recycling of the digested anaerobic effluent in agriculture (Figure 1). The vacuum toilets need 0.7 to 1.0 l water per flush and hence produce a small but high strength flow of about 5 l/cap/day BW. The second highly concentrated stream from households, the kitchen refuse (KR) is milled by a grinder and added to the BW before the pasteurisation step. The pasteurisation of 24 hours at 55°C or 1 hour at 75°C is regulated by German legislation for organic waste (BioAbfV, 1998). The biogas resulting

doi: 10.2166/wst.2007.144

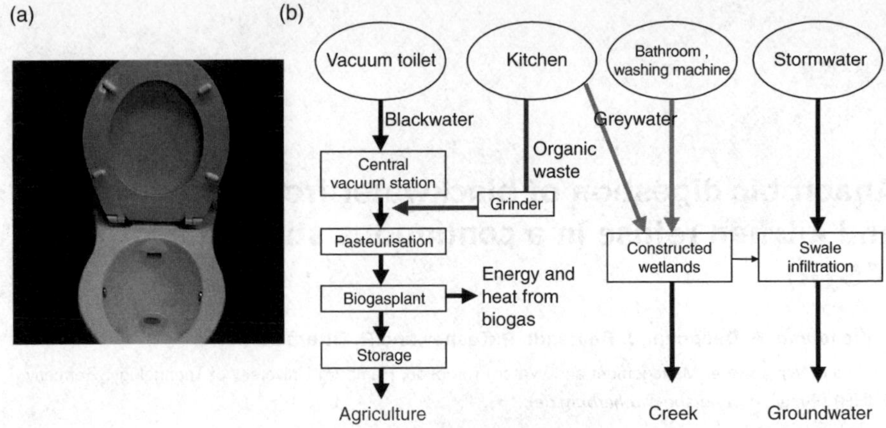

Figure 1 (a) Vacuum toilet installed in Luebeck-Flintenbreite; (b) Scheme of the pilot project Flintenbreite, Luebeck, Germany

from this process is to be utilised on-site in the combined power and heat generator (heating for houses/digester and production of electricity).

The preliminary studies for the biogas plant design in Luebeck-Flintenbreite presented here aimed at investigating the performance of anaerobic digestion of blackwater from vacuum toilets and kitchen refuse (KR) in a CSTR, in particular

- characterisation of BW from vacuum toilets,
- COD removal efficiency and methane yield of BW digestion at 20 days hydraulic retention time (HRT),
- COD removal efficiency and methane yield of BW + KR digestion at 10, 15 and 20 days HRT.

Methods and materials

Experimental set-up and analysis

The CSTR of 10 l volume (Figure 2) was kept at mesophilic temperature in a heated water bath. It was continuously mixed by an overhead stirrer. The headspace (about 1.5 l) was connected to a gas meter continuously recording the produced quantity of biogas by displacement. By means of online measurements of air temperature and barometric pressure, the standard volume of biogas was calculated continuously in an automated way. Gas samples were taken from time to time and analysed in a gas chromatograph. The pH was measured three times a week in influent and effluent samples. Total and dissolved ($<0.45\ \mu m$) COD, VFA (volatile fatty acids), VSS (volatile suspended solids), TOC (total organic carbon), NH_4-N (nitrogen as ammonium) and TN (total nitrogen) were analysed weekly in influent and effluent samples. Total P and SO_4-S were analysed only in two to four grab influent samples during the duration of one experiment. All parameters were analysed according to APHA (1995), COD and NH_4-N by cuvette tests.

Reactor feeding

BW was collected from the storage tank fed by the vacuum pumps at the housing estate Luebeck-Flintenbreite. KR was collected from the student restaurant at TUHH once in a sufficient quantity. It represented typical fresh KR with about 30% raw and 70% cooked material. It was a mixture of potatoes, pasta, bread, vegetables, salad and meat. The KR was milled to particles with less than 2 mm diameter and subsequently stored at 6°C.

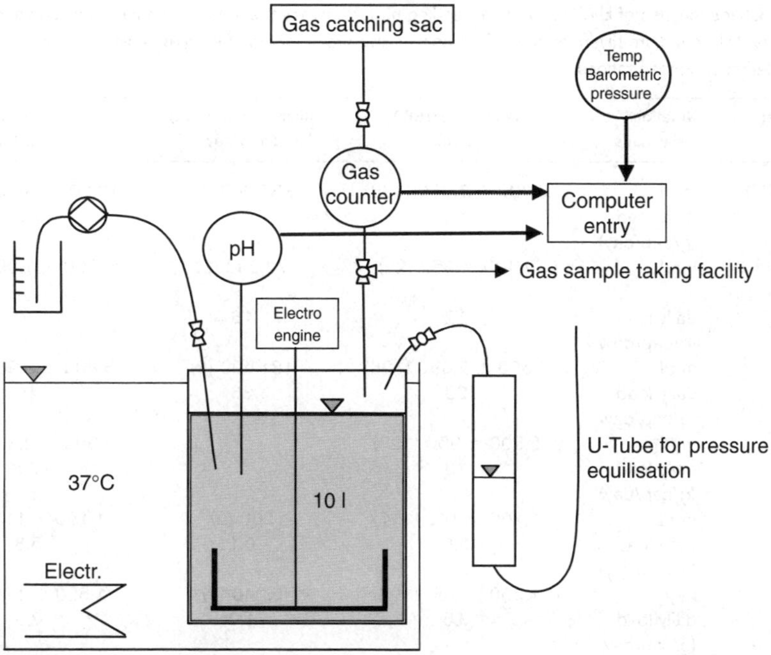

Figure 2 Scheme of the experimental set-up

The feeding mixture of BW and KR was prepared weekly and subsequently fed to the reactors or stored at 6°C until needed for feeding. The reactor was fed discontinuously three times a week in case of HRT of 20 and 15 days and six times a week in case of 10 days HRT.

The addition of KR to the BW was done according to statistical German values of 0.2 kg/cap/day. This amounts to 40 g KR per litre BW. The CSTR was started with adapted sludge (during preliminary experiments) and the feeding was as shown in Table 1.

Results and discussion

Characteristics of BW, KR and the mixture

Most of the samples analysed here were BW. Table 2 shows the concentrations in measured BW, KR and the mixture of both and the load per capita and day calculated for each parameter. Compared to typical raw domestic wastewater, BW from vacuum toilets is very homogeneous due to the vacuum system with a pressure of 0.8 bar. Also the KR was well homogenised by the grinder. However, concentrations are varying highly, especially for the organic parameters such as COD, VSS and TOC. High standard deviations are mainly caused by the high solid content of BW and even more pronounced in the KR. The particulate COD in BW represents 72% of the total COD which is in the typical range of domestic and municipal wastewater, which shows particulate organic fractions varying from 65% to 80% as reported by Wendland *et al.* (2006), Elmitwalli *et al.* (2002) and Wang (1994). When adding KR to the BW, the concentration of

Table 1 The four phases of the investigation program

Phase	Feeding	HRT	Duration
Phase 1	Raw BW	20 days	413 days
Phase 2	Raw BW + KR	20 days	70 days
Phase 3	Raw BW + KR	15 days	36 days
Phase 4	Raw BW + KR	10 days	54 days

Table 2 Characteristics of BW, KR and the feeding mixture; concentrations ± standard deviation and respective daily loads are given; numbers in brackets give the numbers of samples collected for determination of concentrations

Parameter	Notation Daily loads	Blackwater (BW) 5.0 l/day	Kitchen refuse (KR) 200 g/day	BW + KR 5.2 l/day
Total COD	mg/l	8,700 ± 3,980 (126)	335,000 (3)	19,200 ± 6,550 (25)
	daily load (g/cap/day)	43	67	96
Dissolved COD	mg/l	2,400 ± 650 (22)	79,000 (3)	6,800 ± 1,050 (25)
	daily load (g/cap/day)	12	16	34
VSS	mg/l	4,500 ± 2,680 (126)	181,500 (3)	8,800 ± 3,900 (25)
	daily load (g/cap/day)	23	36	44
TOC	mg/l	2,500 ± 950 (126)		5,900 ± 2,300 (13)
	daily load (g/cap/day)	13		29
NH_4-N	mg/l	1,100 ± 140 (116)	280 (2)	1,150 ± 110 (25)
	daily load (g/cap/day)	5.6	0.1	5.8
Total N	mg/l	1,500 ± 250 (126)	5,140 (2)	1,520 ± 170 (25)
	daily load (g/cap/day)	7.5	1.0	7.9
Total P	mg/l	202 (3)	577 (2)	171 ± 51 (4)
	daily load (g/cap/day)	1.0	0.1	0.9
SO_4-S	mg/l	43 (3)		
	daily load (g/cap/day)	0.2		

particulate COD decreased to 65% although it was expected to increase due to the high particulate COD fraction in the KR of 77%. This difference is assumed to be caused by the fact that there was an immediate start of hydrolysis after mixing of BW and KR and that there were up to 24 hours time difference between mixing, sampling and analysing.

Considering the daily loads of BW, the values were 30 to 45% lower than the loads reported for human excreta by Larsen and Gujer (1996), Fittschen and Hahn (1998), and Otterpohl *et al.* (2003). These differences are caused mainly by two facts: first, the housing estate Luebeck-Flintenbreite is a residential area, so employed persons are not present most of the day; second, about 40% of the inhabitants are children who produce less excreta and are also not at home during half of the day. In extensive investigations in Swedish housing estates, Jönsson *et al.* (1999) found that only 65% of the human excreta are excreted at home.

The daily loads of KR given here are based on average German values but they might be different according to the people's motivation to collect kitchen refuse separately. As statistical data show, the variation of amounts of collected KR is naturally higher than of BW.

Performance of CSTR at varying HRT

COD mass balance. Because of the varying concentrations of organic matter in BW and KR, it is important to calculate mass balances to find out the level of experimental error. The calculated COD mass balances for the different conditions are given in Figure 3. They are calculated based on the COD loads in each phase after at least one HRT. The total COD masses of input and output show relatively small differences with a maximum of 7% for phase 1 (BW at 20 days HRT) which might be caused by sampling or analytical errors.

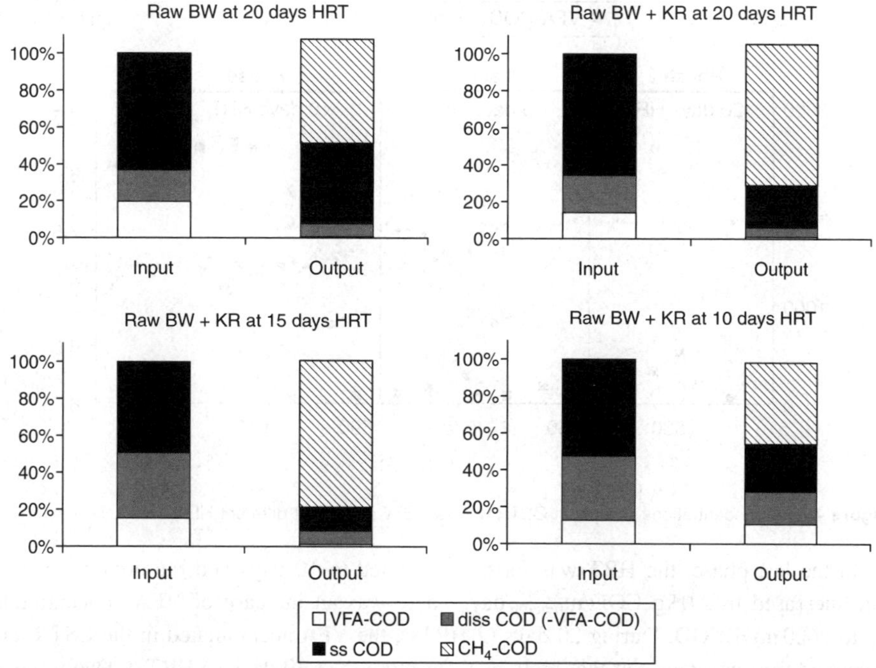

Figure 3 COD mass balances for all phases of experiments

Removal of COD fractions. The removal of total and particulate COD is shown in Table 3. For BW digestion at 20 days HRT, the removal of total and particulate COD was 61% and 53% respectively. Low VFA concentrations in the effluent indicate that the process was stable and uninhibited in spite of high nitrogen concentration. When adding KR, the organic loading rate increased from 0.5 to 1.0 kg COD/$m^3_{reactor}$/day at 20 days HRT. The removal ratio was elevated to 71% and 67% for total and particulate COD because of the higher anaerobic biodegradability of KR. With 15 days HRT, the total COD removal efficiency of 75% was even higher whereas the removal of particulate COD was at the same level. The increase of total COD removal performance with decreased HRT can be explained by varying particulate COD loads in phase 2 and 3 (Figure 3). The good and stable performance of the CSTR at 15 and 20 days HRT was further corroborated by the low concentrations of volatile fatty acids (Figure 4). The VFA concentrations (as criteria for the stability of the anaerobic process) were varying, but were typically below 150 mg/l as COD. In BW digestion, pH was stable around 7.5, as well as in phase 2 with KR addition. In phase 3, the pH slightly decreased to 7.3 but was increasing to 7.5 after a few days.

Table 3 Removal efficiency of organic matter and total N at different HRT in % with standard deviation

Parameter	Phase 1 BW (20 days HRT)	Phase 2 BW + KR (20 days HRT)	Phase 3 BW + KR (15 days HRT)	Phase 4 BW + KR (10 days HRT)
Total COD	61 ± 12	71 ± 13	75 ± 7	50 ± 15
Particulate COD	53 ± 24	67 ± 15	67 ± 13	53 ± 22
VSS	51 ± 19	65 ± 20	69 ± 12	51 ± 21
TOC	55 ± 12	72 ± 12	71 ± 29	52 ± 15
NH_4-N	−11 ± 14	−14 ± 8	−15 ± 6	−5 ± 10
Total N		Below 2		

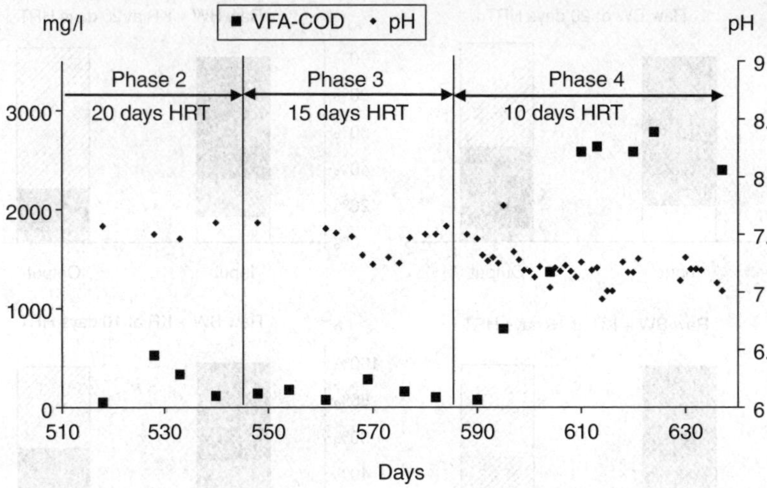

Figure 4 VFA concentrations and pH in CSTR with raw BW and KR at different HRT

In the last phase, the HRT was further decreased to 10 days and the organic loading rate increased to 2.0 kg $COD/m^3_{reactor}$/day. There was an increase of VFA concentration up to 2600 mg/l COD. During 20 days (2 HRTs), the VFA accumulated in the CSTR and remained then constantly at that high level for more than 30 days (3 HRTs). Due to good buffering capacity of BW, the pH did not fall below 7.0 and a steady state process was achieved. The methanogenic activity and thus the COD removal was limited during the low HRT experimental phase. The COD removal was stable at 50% which is 30 to 33% below the removal at uninhibited conditions (15 to 20 days HRT). Particulate COD removal was about 53%, which is 20% lower than the removal at 15 and 20 days HRT. Thus, also the hydrolysis was partly inhibited at 10 days HRT. However, these results show that a stable digestion with BW and KR can be realised with major COD removal even at HRT of 10 days.

For further evaluation of the performance of the CSTR and to deepen the knowledge under operation conditions with varying influent quantities and qualities, the application of the ADM1 model has recently been started by Feng *et al.* (2006).

Removal of VSS and TOC. The parameter VSS and TOC represent the organic matter similar to COD. The VSS is a value for organic SS like particulate COD and the TOC is similar to total COD. The removal efficiencies (Table 3) indicate these parallels.

Removal of total nitrogen and ammonium. As expected, the analyses of total nitrogen in influent and effluent show no significant nitrogen removal. There was an increase of ammonium by 5 to 15% in the effluent caused by the conversion of organic nitrogen to ammonium due to protein hydrolysis. Fertilising experiments with the effluent of the CSTR showed very good results as quick acting fertiliser in field and greenhouse tests (Simons and Clemens, 2004).

Methane yield. In Table 4, methane yields related to total COD_{added} as well as related to capita and day for the different experimental phases are shown. For BW digestion, 240 l CH_4 were formed per kg COD_{added} equivalent to 10 l CH_4/cap/day (Table 2). If BW is treated together with KR, the methane yield is 270/280/205 l CH_4/kg COD_{added} and 27/28/21 l CH_4/cap/day at 20/15/10 days HRT respectively. If it is assumed that only 65% of the produced BW is collected in the housing estate, the expected methane yield

Table 4 Methane yield based on the loads of each phase and fraction with standard deviation

	Phase 1 BW (20 days HRT)	Phase 2 BW + KR (20 days HRT)	Phase 3 BW + KR (15 days HRT)	Phase 4 BW + KR (10 days HRT)
Methane yield				
(l CH_4/kg COD_{added})	240	270	280	205
(l CH_4/cap/day)	10	27	28	21
Expected methane yield based on 100% BW collection and treatment*				
(l CH_4/cap/day)	16	32	33	26
Methane-fraction in the biogas (% by weight)	0.76%	0.65%	0.65%	0.65%

*It is assumed that only 65% of the COD BW load of the inhabitants is excreted in Luebeck-Flintenbreite

for 100% BW collection and treatment can be calculated accordingly to 16 l CH_4/cap/day. In case of BW + KR digestion at 15 to 20 days HRT, the expected methane yield would be 32 to 33 l CH_4/cap/day corresponding to 35.6 l CH_4/cap/day reported by Kujawa-Roeleveld *et al.* (2003) for an accumulation digester at longer HRT for a similar BW + KR load. The 10% higher methane yield might be due to the longer HRT or indicates that even less than 65% of the daily excreted COD load is collected in Luebeck-Flintenbreite.

Conclusions

Mesophilic anaerobic digestion in CSTR is an appropriate process to treat high strength BW as well as mixtures of BW and KR. Although the composition of BW was not ideal for anerobic digestion due to high ammonium concentrations, the process at 20 days HRT was very stable. The feeding of the CSTR can be discontinuous (3 to 6 times per week) but should be regular. By adding KR, the performance of anaerobic treatment is highly improved in terms of COD removal and methane yield. At HRT of 15 and 20 days, a stable and uninhibited process was achieved when BW and KR mixture was fed. Even at HRT of 10 days, a steady state process with a decreased COD removal of 30 to 33% and decreased methane yield by 19 to 21% could be achieved, however with increased VFA concentrations in the effluent. Thus, for designing a CSTR for BW + KR, an HRT of 15 days is recommended, if the influent feeding is regular and the influent load is in a common range. In the case of more fluctuating influent loads, e.g. caused by varying kitchen refuse quantity and quality, an HRT of 20 days is recommended. As expected, nitrogen removal does not take place, but this is not required as the effluent is supposed to be used in agriculture and the contained nutrients can partly replace chemical fertiliser.

Acknowledgements

The authors would like to thank Otterwasser GmbH, Luebeck, and Oswald-Schulze-Foundation for their support.

References

APHA (1995). *Standard Methods for the Examination of Water and Wastewater*, 19th edn, American Public Health Association, Washington DC, USA.

BioAbfV (1998). German ordinance on application of organic wastes in agriculture, in German. Verordnung über die Verwertung von Bioabfällen auf landwirtschaftlich, forstwirtschaftlich und gärtnerisch genutzten Böden (Bioabfallverordnung) 21.09.1998 (BGBl. I pp. 2955).

Elmitwalli, T.A., Zeeman, G., Oahn, K.L.T. and Lettinga, G. (2002). Treatment of domestic sewage in two step system anaerobic filter/anaerobic hybrid system at low temperature. *Wat. Res.*, **36**(9), 2225–2232.

Feng, Y., Behrendt, J., Wendland, C. and Otterpohl, R. (2006). Parameter analysis of the IWA Anaerobic Digestion Model No 1 for the anaerobic digestion of blackwater with kitchen refuse. *Wat. Sci. Tech.*, **54**(4), 139–147.

Fittschen, I. and Hahn, H.H. (1998). Characterisation of municipal wastewater part of human urine and a preliminary comparision with liquid cattle excretion. *Wat. Sci. Tech.*, **38**(6), 9–16.

Jönsson, H., Vinneras, B., Höglund, C. and Stenström, T.-A. (1999). Source separation of urine. *Wasser und Boden*, **51**(11), 21–25.

Kujawa-Roeleveld, K., Elmitwalli, T., Gaillard, A., van Leeuwen, M. and Zeeman, G. (2003). Co-digestion of concentrated blackwater and kitchen refuse in an accumulation system within the DESAR (decentralised sanitation and reuse) concept. *Wat. Sci. Tech.*, **48**(4), 121–128.

Larsen, T.A. and Gujer, W. (1996). Separate management of anthropogenic nutrient solution (Human urine). *Wat. Sci. Tech.*, **34**(3–4), 87–94.

Lens, P., Zeeman, G. and Lettinga, G. (2001). *Decentralised Sanitation and Reuse. Concepts, Systems and Implementation*, IWA Publishing, London, UK.

Otterpohl, R. and Lange, J. (2000). *Wastewater – Handbook towards a sustainable water management,* in German, Mall Beton Verlag, Donaueschingen-Pfohren, Germany.

Otterpohl, R., Braun, U. and Oldenburg, M. (2003). Innorvative technologies for decentralised water, wastewater and biowaste management in urban and peri-urban areas. *Wat. Sci. Tech.*, **48**(11/12), 23–32.

Otterpohl, R., Grottker, M. and Lange, J. (1997). Sustainable water and waste management in urban areas. *Wat. Sci. Tech.*, **35**(9), 121–133.

Simons, J. and Clemens, J. (2004). Urine fertilisation and ammonia emissions on arable land. *Proceedings of the IWA Congress in Marrakech*, Morocco, September 2004.

Wang, J.-Y. (1994). Integrated anaerobic and aerobic treatment of sewage. Ph-D thesis, Wageningen University, The Netherlands.

Wendland, C., Oldenburg, M. (2003). Operation experiences with a source separating project, In: *Proceedings of the 2nd International Symposium on Ecological Sanitation*, "ecosan – closing the loop", April 2003, Luebeck, Germany.

Wendland, C. Behrendt, J., Elmitwalli, T.A., Al Baz, I., Akcin, G., Alp, Ö. and Otterpohl, R. (2006). UASB reactor followed by constructed wetland and UV radiation as an appropriate technology for municipal wastewater treatment in Mediterranean countries. *Preprints of the 7th Specialised Conference on Small Water and Wastewater Systems*, March 2006 in Mexico-City.

Wastewater treatment with activated pre-clarifier and planted soil filters

C. Lindenblatt, M. Wichern and H. Horn
Institute of Water Quality Control, Technical University Munich, Am Coulombwall, 85748 Garching, Germany (E-mail: *c.lindenblatt@bv.tum.de*)

Abstract With a series of buffer tank, activated pre-clarifier (SBR) and planted soil filters, it is possible to get a stable degradation and good effluent values in the case of unsteady inflow and changing concentrations. In the process presented here the activated pre-clarifier is working as a denitrification tank and the soil filter as a nitrification reactor. An automatic control manages the storm-water runoff, the water-recirculation returns the nitrate and provides a minimal feed. Experiences with this plant from 1999 to 2005 are given in this paper, research has been done in the project "Planted soil filters as a Biotechnological Process", founded by the German Federal Environment Foundation (DBU). A full scale pilot plant was built to treat wastewater from composting sides, with a 42 m^3 SBR and a bed area of 2 × 550 m^2 of the planted soil filters. Now, after six years of operation, the results are still satisfactory. Besides this pilot plant, landfill leakage and municipal wastewater have been treated in a technical scale plant with the same process in an 80 L SBR and 0.75 m^2 vertical flow soil filter with good results.
Keywords Compost site; constructed wetlands; landfill leakage; planted soil filter; SBR; wastewater treatment

Water Science & Technology **Vol 55 No 7 pp 195–202**

Introduction

The paper presented summarizes the experiences obtained at the Institute of Water Quality Control, Technical University of Munich, using a special plant combination to treat different types of waste water. It starts with a pilot plant at a composting site followed by some experiments in half technical scale with landfill leakage and municipal waste water. The objective of this research was to find out whether a planted soil filter connected with a specific pre-clarifier would be able to treat different kinds of waste-water. Furthermore it will be shown if the combined system is in a long-term an efficient and cost effective method for treatment. Clogging, which is reported as one main factor for the treatment efficiency of filters (Platzer and Mauch, 1997; Winter and Goetz, 2003; Kayser and Kunst, 2005) was also investigated during the long term operation of the plants. Biomass growth and inflow particles are reported to be the main reason for clogging (Siegrist, 1987; Spychala and Blazejewski, 2003).

Example 1 - wastewater from a composting site

The WGV Recycling Company, in Eurasburg/Germany, operates a plant for composting organic residual substances in Quarzbichl, Oberbayern. The site is located near the rural district of Bad Tölz-Wolfratshausen and the district of Weiheim with more than 210,000 habitants. The factory produces over 24,000 t/a of high quality compost. Wastewater at this site results from rain fall on the 8500 m^2 traffic area, from cleaning procedures on the camp surfaces and from process water.

In the context of the research project "Planted Soil Filters as a Biotechnological Process" (www.bodenfilter.de; Fehr *et al.*, 2002), which was promoted by the German Federal Environment Foundation (DBU), a pilot plant was developed in 1998. The operation

doi: 10.2166/wst.2007.145

began with the wastewater treatment at the WGV in June 1999. Up till then, tank trucks transported more than 12,000 m^3/a wastewater to the surrounding wastewater treatment plants annually.

First, the new established Pilot Plant consisted of an aggregate for mechanical solid separation, a reservoir acting as a sediment basin (42 m^3), and two vertical flow sand filters with a surface area of 2 × 550 m^2 (0.8 m depth). Laboratory and essentially, half-technical experiments, for pre-treatment in a sequencing batch reactor (SBR) with mixer and aeration system, showed the suitability of an activated preclarifier for solid separation, denitrification and also for biological phosphorus removal. Two years after start-up, the settling tank was converted to activated pre-treatment (05/2001) and run up to the integration of the re-circulation into the SBR control (04/2003), without denitrification. In the final report of the research project (Geller *et al.*, 2002) the results are documented during the time of July 1999 until November 2001.

For the further scientific monitoring of the sewage system, the WGV Recycling Company assigned the Institute of Water Quality Control with further investigations and the documentation of the purification plant. In the meantime, the water recirculation was integrated in the sequencing batch process and the automatic process control was optimised (2005).

The wastewater of the compost area flows into an old buffer tank with a volume of 300 m^3. To avoid unwanted deposits, a pump stirs the settled sludge at intervals, so that an effective movement from the sump pump is made possible. From the beginning, the wastewater treatment plant was not designed to treat 100% of the resulting compost wastewater. A part of the water is still taken by trucks to a purification plant in the neighbourhood.

After a raw water settling time of more than 2 hours, the settled raw water was pumped out of the buffer tank into the activated pre-treatment tank (PT). The filling takes place after a preset program that controls the volume exchange of the reactor, the mixing time, aeration and settling time. The activated sludge volume in the pre-clarifier is adjusted to 250 mL/L and the surplus sludge is automatically pumped back to the buffer tank. The effluent of the activated pre-treatment is lifted and pumped to the soil filters.

Four times per day, for three minutes, ~6.9 m^3 are pumped to one bed surface using a distribution system. Every 14 days (autumn, winter, spring) and 7 days in the summer, the two filters are alternately fed. The soil filter outlet flows to the return pit, from which a subset was re-circulated to the pre-clarifier. After the control pit, the effluent (E) flows

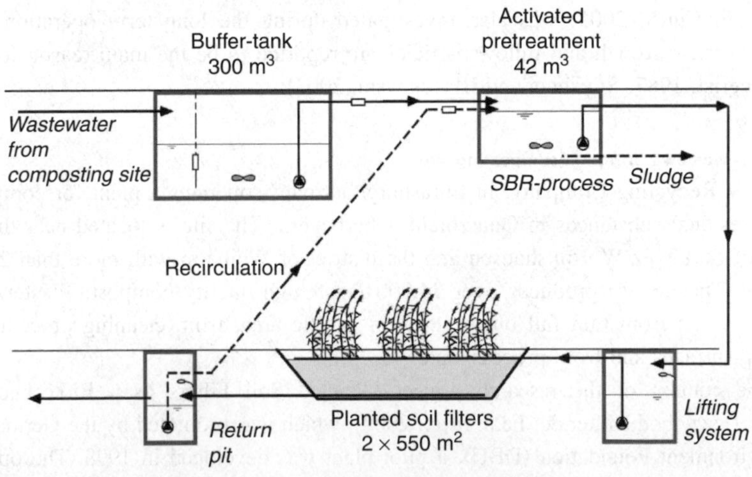

Figure 1 Scheme of treatment system (taken from Geller *et al.*, 2002)

to the river Loisach. Figure 1 shows the flow chart of the treatment plant. Inductive flow measuring instruments measured the water quantities at the pre-clarifier and the return flow. For operation control, the staff measures twice per week the conductivity in the buffer-tank and the sludge volume in the SBR.

Example 2 - treatment of landfill leakage

The same tests with landfill leakage were made in the institute from September 2002 to February 2003 with a half technical plant. The leakage flows from a landfill site with an area of 6 hectares in which 480,000 m^3 solid wastes have been deposed for 42 years. Five years after operation, sealing and re-cultivation, the wastewater flows out of the landfill with 50 to 120 m^3/d depending on weather conditions. Inside the landfill, stable conditions were assumed and the COD concentration decreases slowly while ammonia increases. A necessary treatment is forecasted for the next 50 years. The average value of COD was 275 mg/L and the total nitrogen 132 mg/L.

In different conditions the flexibility of the process was tested and optimised to the most effective nitrogen reduction. The idea was to use the available carbon at the earliest possible moment for denitrification and conduct a stable operation with an external carbon (methanol) source. Therefore a sequencing batch reactor with 80 L volume was used as a first treatment step with continuous inflow. In five cycles per day the reaction time starts with a discontinuous mixing time of 3 to 4 hours and ends with a short aeration time of 10 minutes to avoid denitrification during sludge settling. The effluent was pumped directly to the soil filter and the next cycle started again with the recirculation of some effluent to the pre-clarifier. The recycling ratio was 1. As a secondary treatment step a soil filter was built with 0.75 m^2 and 0.7 m height and filled with washed sand consisting of 0 to 3 mm grain size. A mathematical model was developed for this application.

Example 3 - municipal wastewater treatment

In 2003 and 2004 some tests were made at the Institute of Water Quality Control to treat municipal wastewater with the same equipment. The aim was to produce a high quality of treated water for reuse with a small wastewater treatment plant. Wastewater was pumped from the Garching wastewater treatment plant to the laboratory. During an 8 week measuring period the efficiency of coliform bacteria reduction was analysed and over 3 months the efficiency at different hydraulic loads and recycling water rates with the activated pre-clarifier (80 L) and a soil filter (0.75 m^2) were measured.

Analytical methods

The analytical methods that were used for pH, conductivity and TKN are described in the "German Methods for Water-, Wastewater- and Sludge-Analyses" (DEV, 2004). NH_4-N, NO_3-N were measured with a flow injection analyser (FIA), COD and phosphorus with a test kit from Hach/Dr.Lange. The biological parameters were measured with the MPN method with a lauryl sulphate medium (Fluorocult/Merck). Information about the modelling of the systems can be found in Wichern *et al.* (2006).

Results

Treatment of municipal wastewater (Example 3)

A 99.9% reduction of coliforms was measured under operating conditions with a recirculation ratio of 2 and a hydraulic bed-load of 180 L/(m^2 d). The COD concentration was below 20 mg/L in the outlet and ammonia was less than 1.0 mg/L. 45% of COD was reduced in the pre-clarifier with a short aeration time. With a recirculation ratio of 0.5 to 1.0 and a hydraulic load of 200 L/(m^2 d), the COD efficiency was 92% in the filter bed. So more than 50% of the total COD degradation took place in the second step. Figure 2 shows the disposition of nitrogen under the conditions of a recirculation ratio of 1 and ~ 200 mm hydraulic bed load. The total nitrogen reduction was more than 70%. It decreased with lower recirculation ratio.

Treating of landfill leakage (Example 2)

In the pre-clarifier a nearly complete reduction of the recirculation nitrate occurs if there was added enough organic carbon. To assure an adequate process stability, average effluent nitrate concentrations 50% below the standards should be reached. The use of methanol as a carbon source adds 0.2 €/m^3 to the operational costs. An almost total nitrification took place at room temperature in the soil filter even at ammonia loads of more then 10 g_{NH_4}-N/(m^2 d) and 200 L/(m^2 d) hydraulic load. The COD concentration was up to 60 mg/L in the outlet.

Composting site wastewater treatment (Example 1)

During the intensive measurement period from 1998 to 2001, samples were taken two times per week, and were then analysed for the relevant wastewater parameters. Heavy metals and other substances, like absorbable organic halogenide (AOX), have not been a problem on the composting site (Geller *et al.*, 2002). One of the research points of interest was the separation of solids out of the raw-water by using special mechanical equipment, like hydro-cyclones and different filter methods. Since 50% of the solids were finer than 63 μm, settling was the best available technology for solid separation.

Several tracers were used to measure the retention time of water flow through the sand filters. We found out, that bromide, deuterium and a special fluorescence tracer were the best indicators of the water flow in this three-phase-system containing biofilm, fluid and air. In both quartzous sand substrates, which were used in the beds, the retention time varied from ~3.0 days in bed 1 (d_{60} = 0.4 mm) to ~2.5 days in bed 2 (d_{60} = 0.6 mm).

In the year 2004, it was possible to treat 5210 m^3 of settled raw water from the storage tank with this two step treatment plant. This volume is 86.3% of the average value of the last three years. The smaller input took place as a result of the pond strategy which was made in the time of March to May to avoid weeds. A failure of the control system in July and the change in the installation for company-owned equipment, led to downtimes.

Raw-water from composting site

The raw wastewater in the storage tank corresponded in 2004 to the characteristically measured values of the intensive measurements from 1998 to 2001 concerning the middle

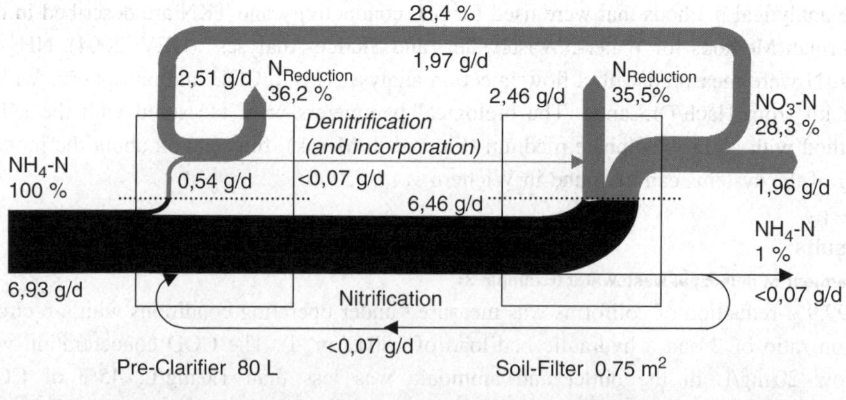

Figure 2 Nitrogen-flow-diagram for the model plant, operation phase 1, July 2004 (Feed 72.2 L/d, hydr. filter load ~200 mm, return ratio 1)

COD concentration of 1700 mg/L of COD in the settled sample (613–3.025 mg/L). Phosphorus concentration was 13.3 mg/L. The nitrogen concentration reached 120 mg/l (TKN) in the settled sample. The ratio of 1.82 between TKN and NH_4-N shows that nearly half of the nitrogen in the settled sample is organically bound. This fact is particularly remarkable, because under anaerobic conditions an ammonification can occur in the storage tank. These conditions increased the ammonium concentration rapidly, especially in summer at higher temperatures. The average values of the measured parameters in the settled raw-water from the storage tank in 2004 are given in Table 1.

With the values of storage volume and raw-water conductivity, the flow-rate was set up between 10 and 25 m^3/d. The ammonium concentrations measured by WGV with a test kit are corresponding to the conductivity with reasonable accuracy.

Pre-treatment and sludge balance

In the activated pre-treatment tank (PT) a volume exchange of ~17% takes place four times per day. The tank is filled with return water depending on the quantity of the raw water and the operating water level. Before the two-hour setting off period, the tank is aerated. The daily batch effluent quantity amounts to 27.5 m^3/d. For one bed this is a hydraulic load of 50 mm/d.

The activated sludge is measured regularly by operational staff. The pump intervals for the surplus sludge withdrawal can be adjusted according to the determined sludge volume of 250 mL/L. Surplus sludge was pumped back into the storage tank and from there it is taken out together with the other removable materials. The sludge removal of 400 L three times per week was taken from the SBR. With a sludge concentration of 20 g/L it gives 1250 kg dry-substance per year. This sludge quantity agrees with the sludge increase, which can be calculated by the COD reduction.

In the storage tank, a mean sludge content of 1.1 kg/m^3 was measured from 1999 to 2001. By considering the total WGV wastewater quantity of 10,370 m^3 in 2004, the yearly sludge from the compost-place in the storage tank is 11 to 13 tons dry weight, which needs to be disposed. The portion of SBR sludge was less then 12%. During the transition period, October 2003 to March 2004, 90 kg/a of a precipitant (iron III chloride, 40%) was used for testing. It contained approximately 32 kg/a (2.5 kg/kg_{Fe}) precipitation sludge and corresponded to a portion of 0.3% of the total sludge volume. The sludge age of the activated pre-clarifier was more than 26 days.

Water outlet quality

The annual average effluent concentrations and standard deviations from pre-treatment and outlet are given in Table 2.

With respect to the mean value the COD from the effluent of the soil filter is 77% of the limit value (AbwV, 2002) and nitrogen is 72.7%. This proves a good process stability

Table 1 Average concentrations in the settled raw-water (SB) from the compost-site 2004

Parameter	Unit	Value SB	Standard deviation	Number of samples
pH	–	7.1	0.3	12
Conductivity	mS/cm	3.13	0.96	12
COD	mg/l	1713	739	11
TKN	mg/l	120	32	11
NH_4-N	mg/l	76	31	12
P	mg/l	13.3	3.4	8
PO_4-P	mg/l	8.4	4.1	7

Table 2 Average concentrations in the outlet of pre-treatment (PT) and soil-filter (E)

Parameter	Unit	Value pre-treatment	Standard deviation	Number of samples	Value soil-filter	Standard deviation	Number of samples
pH	–	7.6	0.2	13	7.4	0.3	16
Conductivity	mS/cm	2.56	0.67	13	2.17	0.51	15
COD	mg/l	731	474	14	108	46	17
TKN	mg/l	61	20	15	–	–	–
NH_4-N	mg/l	36	14	15	1.3	1.7	16
NO_x-N	mg/l	1.6	2.6	15	29.3	18.6	17
P	mg/l	5.3	3.7	13	–	–	–
PO_4-P	mg/l	2.8	2.8	10	1.8	1.2	19

for the COD and N removal. The limit value is sometimes reached for phosphorus. The average of the phosphorus concentration in the effluent is about 60% of the limit. During the intensive measurements (1999–2001) seasonal influences were found, which led to the assumption that at the end of the vegetation period a certain phosphorus release occurred. So a precipitation of phosphorus in the effluent of the soil filters has to be installed in the future. The precipitated sludge is returned to the pre-treatment with the back flow stream.

With the recirculation of treated water in the SBR, it was possible to keep the nitrate effluent concentrations for the entire year 2004. The nitrification in the soil filter can reach an average of up to 96.4% per year. In summer, the TKN reduction at temperatures over 12 C was found to be 99%. The summer temperatures and the increased nitrogen load led to an undesired nitrification in the pre-treatment (SBR). Enough time must be given for denitrification of the nitrate from the recirculation. A larger nitrification in the SBR is not possible, as there should be enough time for enhanced biological phosphorus removal. Otherwise, the aeration time should be adjustable. Due to the easy control of the pre-treatment, only an unstable P-removal was possible. Precipitation experiments with $FeCl_3$ resulted in a 20% phosphorus reduction.

Efficiency

The annual average values of the loads and reductions are summarized in Table 3 and are given as specific values of the treatment plant. The soil-filter data is calculated with two beds in alternating operation.

The elimination capacity of the plant in the single treatment steps is given in Figure 3 related to the raw water load. The recirculation ratio was 0.6 in the 2004 annual average. Mass balance differences between +4.5% and −3.1% are within measuring tolerance.

The COD and the nitrogen loads are reduced in the pre-treatment by 40%. The nitrification and hydrolysis of the organic nitrogen takes place mainly (~70%) in the second process step (97.9% inside the soil filter). Because of recirculation denitrification inside

Table 3 Specific load and reduction inside the treatment steps

	Pre-treatment load g/(m^{3}d)	(PT) Volume reduction g/(m^{3}d)	42 m^3 $\eta_{PT\ intern}$ %	Soil-filter (E) load g/(m^{2}d)	2 × 550 m^2 reduction g/(m^{2}d)	$\eta_{SFintern}$%
COD	730	290	39.7	18.1	15.3	84.5
TKN	52.5	13.8	26.3	1.44	1.41	97.9
NH_4-N	31.2	9.40	30.1	0.84	0.81	96.4
NO_x-N	6.49	5.87	90.4	< 0.05	–	–
P_T	6.18	2.82	45.6	0.115	0.059	51.3

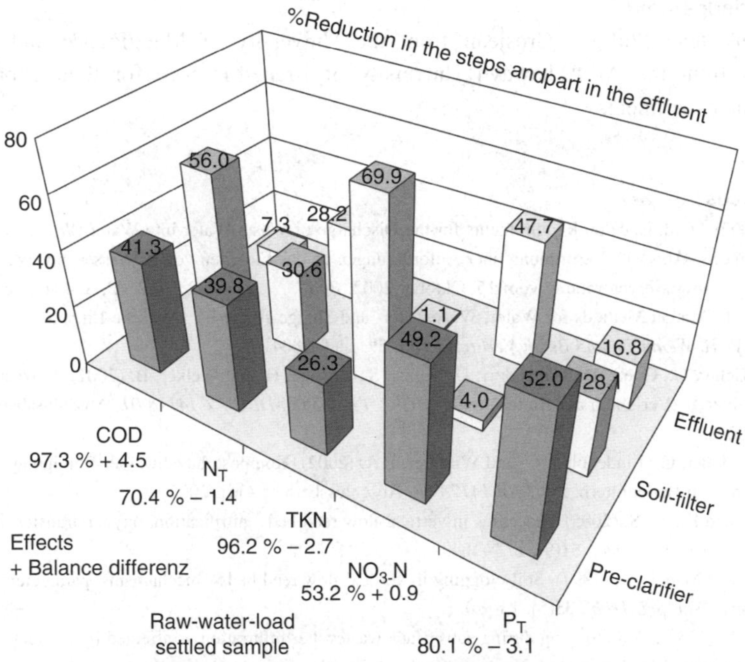

Figure 3 Elimination capacity in the single treatment steps and in the effluent as percentage of the raw influent wastewater

the pre-clarifier of nearly 50% is reached. Also, the phosphorus is being reduced by a half in the pre-treatment and in the soil filter by 28.1%.

There is no indication of a change of the permeability of the soil filter after 6 years of operation. The costs of the compost-place wastewater management at the location of WGV have decreased since 1998 down to 25%.

Conclusion

From a research project (1999–2005) resulted a pilot plant for compost-place wastewater treatment with SBR as an activated pre-treatment and two soil filters with water recirculation. This process shows also the ability for treatment of municipal wastewater and for landfill leakage.

In 2004, the treatment plant of WGV Recycling Company, in Quarzbichl, cleaned 5210 m^3 of compost-place wastewater with a modified back flow integration in the SBR pre-treatment. With this activated pre-clarifier, more operating stability is achieved for the soil passage. In the annual average, the COD is reduced to 97.3%, and nitrogen and phosphorus are reduced to 70.4% and 80.1%.

With an additional aeration inside the storage tank and a regular sludge withdrawal, anaerobic conditions can be avoided. This allows, together with an improved automated control, an increase of the inflow particularly after heavy rain, which happens especially near the Alps from June to August. Clogging, which is reported as one main factor for the treatment efficiency of filters (Platzer and Mauch, (1997), Winter and Goetz, (2003), Kayser and Kunst, (2005) didn't occur during the long term operation of the plants. The results show that the inflow particles sedimented and degradated well in the activated pre-clarifier.

Acknowledgements

We thank Jean Philippe Grosjean from the University of Metz/France and Ahmed Legrouri from the Al Akhawayn University of Ifrane/Morocco for their cooperation during the experiments.

References

AbwV (2002). "Ordinance on Requirements for the Discharge of Waste Water into Water (Waste Water Ordinance – AbwV)". Verordnung über Anforderungen an das Einleiten von Abwasser in Gewässer, AbwV – Abwasserverordnung vom 15. Oktober 2002. *BGBl. Nr. 1 vom 23.10.2002*, S. 4047, ber. S.4550.

DEV (2004). German Methods for Water, Wastewater- and Sludge-Analyses. Deutsche Einheitsverfahren. *Wiley-VCH, Weinheim and Beuth Verlag, Berlin*, **59**. Lief. 2004.

Fehr, G., Geller, G., Goetz, D., Hagendorf, U., Kunst, S., Rustige, H. and Welker, B. (2002). Bewachsene Bodenfilter als Verfahren der Biotechnologie. *UBA Texte 5/03. DBU AZ 14178-01*, Abschlussbericht April 2002.

Geller, G., Höner, G., Lindenblatt, C. and Wilderer, P.A. (2002). Kompostplatzabwasser-Reinigung mit Bewachsenen Bodenfiltern. *DBU AZ 14178-04*, Abschlussbericht April 2002.

Kayser, K. and Kunst, S. (2005). Processes in vertical-flow reed beds: nitrification, oxygen transfer and soil clogging. *Wat. Sci. Tech.*, **51**(9), 177–184.

Platzer, C. and Mauch, K. (1997). Soil-clogging in vertical flow reed beds - mechanisms, parameters and solutions? *Wat. Sci. Tech.*, **35**(5), 13–20.

Siegrist, R.L. (1987). Soil clogging during subsurface wastewater inflitration as affected by effluent composition and loading rate. *J. Environ. Qual.*, **16**(2), 181–187.

Spychala, M. and Blazejewski, R. (2003). Sand filter clogging by septic tank effluent. *Wat. Sci. Tech.*, **48**(11–12), 153–159.

Wichern, M., Lindenblatt, C., Lübken, M., Sezerino, P.H., Kehl, O., Wilderer, P.A. and Horn, H. (2006). Modelling Carbon and Nitrogen Removal in a biofilm wastewater treatment system. *Proceedings of 7th Specialized Conference on Small Water and Wastewater Treatment System*, Mexico-City, 7.-10.3. 2006.

Winter, K.J. and Goetz, D. (2003). The impact of sewage composition on the soil clogging phenomena of vertical flow constructed wetlands. *Wat. Sci. Tech.*, **48**(5), 9–14.

Water Science & Technology **Vol 55 No 7 pp 203–210** © IWA Publishing 2007

Use of real-time PCR to examine the relationship between ammonia oxidizing bacterial populations and nitrogen removal efficiency in a small decentralized treatment system 'Johkasou'

G. Nakagawa*, Y. Ebie, S. Tsuneda***, M. Matsumura* and Y. Inamori****

*Graduate school of Life and environmental science, University of Tsukuba, 1-1-1 Tennodai, Tsukuba, Ibaraki 305-0006, Japan (E-mail: *nakagawa.gou@nies.go.jp*)

**National Institute for Environmental Studies, 16-2 Onogawa, Tsukuba, Ibaraki 305-8506, Japan

***Department of Chemical Engineering, Waseda University, 3-4-1 Ohkubo, Shinjuku-ku, Tokyo 169-8555, Japan

Abstract The aim of this study was to examine the relationship between ammonia oxidizing bacterial populations and biological nitrogen removal in a small on-site domestic wastewater treatment system "Johkasou". The population dynamics of ammonia oxidizing bacteria (AOB) in six full-scale advanced Johkasous was surveyed using real-time PCR assay over a period of one year. These Johkasous were selected to compare the AOB populations in different treatment performance. When the effluent NH_4-N concentration was higher than 2 mg L^{-1}, it was difficult to meet the effluent standard of advanced Johkasous (T-N 10 mg L^{-1}). In contrast, the nitrogen removal efficiency was hardly affected by nitrite oxidation and denitrification in these systems. In other words, ammonia oxidation was a rate-limiting step. Furthermore, we focused on the relationship between NH_4-N loading per AOB cell and nitrogen removal. Real time PCR monitoring results demonstrated that it is important to regulate NH_4-N loading per AOB cell below 210 pg $cell^{-1}$ day^{-1} to meet the effluent standard of advanced Johkasou. It is considered that NH_4-N loading per AOB cell is a useful parameter for determining suitable nitrogen loading and small decentralized system design.

Keywords Ammonia oxidizing bacteria; biological nitrogen removal; domestic wastewater; Johkasou; real-time PCR; small decentralized system

Introduction

Recently, an advanced Johkasou ("Johka" and "Sou" mean purification and tank respectively in Japanese) as a small onsite treatment system for household domestic wastewater has become a center of attraction due to its high ability to mitigate the nitrogen impact on the water environment. Since Johkasou is inexpensive to set up and its running cost is economical, it is very useful for nitrogen removal on a small scale.

A key process in the biological nitrogen removal during wastewater treatment is through the 2-step oxidation of ammonium (NH_4^+) to nitrite (NO_2^-) and nitrite to nitrate (NO_3^-). This process required the coordination of two distinct chemolithoautotrophic groups: ammonia-oxidizing bacteria (AOB) and nitrite-oxidizing bacteria (NOB). The failure of biological nitrogen removal systems to establish stable nitrification is generally attributed to the slow growth of nitrifying bacteria and their sensitivity to environmental factors (Okabe *et al.*, 1999). Consequently, it is very important to keep nitrifying bacterial populations for maintaining stable nitrification in the small onsite treatment system.

The recent development of molecular biological techniques such as real-time PCR has enabled us to monitor nitrifying bacteria in activated sludge (Harms *et al.*, 2003). Reliable monitoring of nitrifying bacteria in a biological nitrogen removal system utilizing

doi: 10.2166/wst.2007.146

molecular techniques can improve nitrification efficiency and stability by increasing our understanding of microbiology and ecology of the nitrogen removal system.

This study attempted to utilize real-time PCR to monitor the number of AOB cells adhering to porous carriers in six full-scale advanced Johkasous. We focused on the NH_4-N loading per AOB cell as an important parameter. The objectives of this study are to clarify the relationship between AOB populations and the nitrogen removal efficiency of advanced Johkasous, and to increase our understating of key factors for regulating the on-site biological nitrogen removal process.

Materials and methods

Full-scale advanced Johkasou

Six advanced Johkasous (A, B, C, D, E, F), as on-site household domestic wastewater treatment facilities, were surveyed. These Johkasous were selected to compare the AOB populations in different treatment performance. The water quality of domestic wastewater flowing into Johkasous has been estimated as follows; BOD is $200\,mg\,L^{-1}$, T-N is $50\,mg\,L^{-1}$ and T-P is $5\,mg\,L^{-1}$. These systems consisted of five compartments: primary anaerobic tank, secondary anaerobic tank, aeration tank, treated water storage tank, and disinfection tank (Figure 1, Table 1). The biofiltration process filled with a porous ceramics carrier was applied in the aeration tank.

The advantages of the biofiltration process compared to the conventional activated sludge processes are high volumetric reaction rates and enrichment of slow growing microorganisms (Rother and Cornel, 2004). The porous ceramics carrier was spherical,

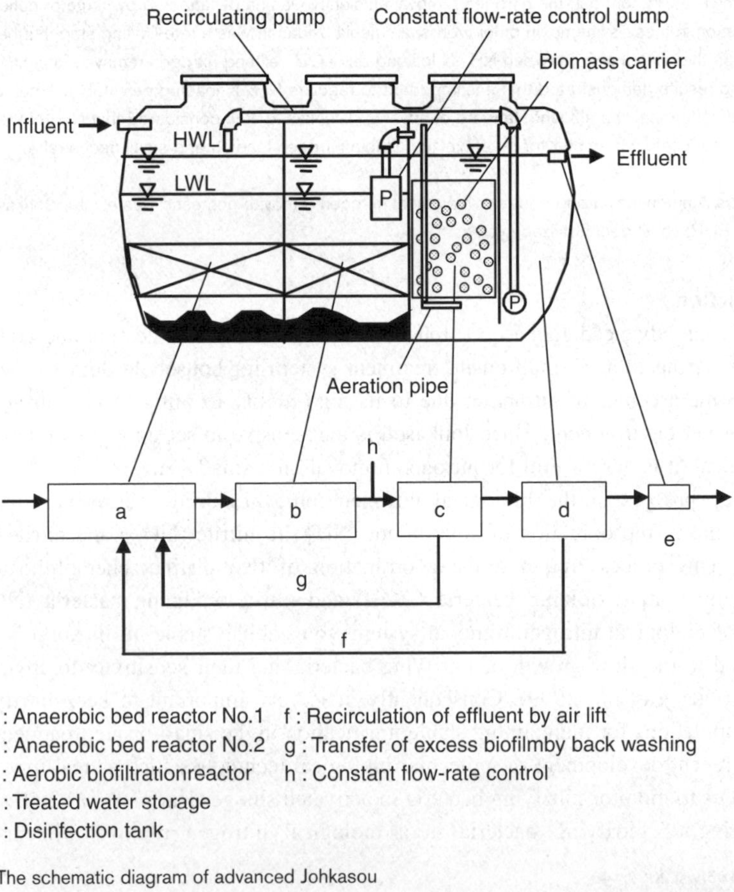

Figure 1 The schematic diagram of advanced Johkasou

Table 1 The capacity of each tank of advanced Johkasou (Unit: m)

Johkasou name	Tanks				
	a	b	c	d	e
A, C, D, E	1.216	0.660	0.279	0.276	0.020
B, F	1.815	0.944	0.391	0.391	0.023

and the average diameter was 7 mm. Recirculation ratios, which mean the volume of circulating water to that of influent, were set at 4.

Sampling

Effluent water samples were collected nearly every week over a period of one year. Influent water samples of the aeration tank and biofilm-attached carriers in the aeration tank were collected several times every two months over a period of one year. Water temperature, pH, and dissolved oxygen (DO) concentration in the aeration tank were monitored every two months over a period of one year.

DNA extraction

The biofilm was detached from biofilm-attached carriers with a sonicator and a mixer, and were transferred into a sterile tube with sterilized water. DNA was extracted from the biofilm with an Isoplant DNA extraction kit according to manufacture's instructions (Nippon Gene) with minor modification. A biofilm (approximately 1.0 mg-VSS) in 2.0-ml tubes was supplemented with 300 μL of sodium dodecyl sulfate solution, 150 μL of benzyl chloride solution, and 1 mg of zirconium and silica. The solution was homogenized with MINI-beadbeater for 2 min, and incubated for 15 min at 50°C and then 150 μL of sodium acetate was added. After setting on ice for 15 min, the solution was centrifuged (12,000 × g, 15 min). The supernatant sample was removed and transferred to a tube containing about 1 mL of 100% methanol. After the sample was centrifuged, the supernatant was removed and transferred to a tube containing 300 μL of 70% methanol. After the sample was centrifuged, the supernatant was discarded, and the total precipitated DNA was dissolved in 50 μL of TE buffer (pH 8.0). The genomic DNA concentration was measured with a spectrometer (BioRad) at 260 nm. The purity of genomic DNA was determined by calculating the ratio of the absorbance measured at 260 nm to that measured at 280 nm.

Real-time PCR assay

Real time PCR assays for ammonia monooxygenase encoding gene (*amoA*) were run on a LightCycler (Roche Diagnostics GmbH, Germany). PCR primers used in this study are summarized in Table 2. The biofilm was detached from the biomass carrier, and DNA was extracted from the biofilm. PCR products amplified from *Nitrosomonas europaea* (IFO 14298) with AMO-F (Sinigalliano *et al.*, 1995) and *amoA* 2R primer set were used to establish a standard curve that was included in each run. The standard curve ranging from 1.25×10^2 to 2×10^7 *amoA* gene copies per PCR setup was constructed.

Table 2 PCR primers for *amoA* gene

Primer	Nucleotide sequence (5′-3′)	Sequence position	Reference
AMO-F	AGAAATCCTGAAAGCGC	21–38	Sinigalliano *et al.* (1995)
amoA-1F	GGGGTTTCTACTGGTGGT	332–349	Rotthauwe *et al.* (1997)
amoA-2R	CCCCTCKGSAAAGCCTTCTTC	802–822	Rotthauwe *et al.* (1977)

The primers *amoA* 1F and *amoA* 2R (Rotthauwe *et al.*, 1997) were used for the SYBR Green I real-time PCR.

Quantification of *amoA* gene

Real-time PCR was performed in 20 μL reaction mixtures by using buffer supplied with LightCycler-FastStart DNA Master SYBR Green I Kit (Roche Diagnostics GmbH, Germany). PCR amplification of a 491-bp fragment of the *amoA* gene was carried out by using the *amoA*-1F and *amoA*-2R primer set specific for AOB belonging to the ß subclass of *Proteobacteria.* The PCR mix with a total volume of 20 μL contained 1 × LightCycler-FastStart DNA Master SYBR Green I; 4 m $mol\,l^{-1}$ of $MgCl_2$; 250 n $mol\,L^{-1}$ of each primer; 5 mg of bovine serum albmin (BSA) and 5 μL of extracted DNA. The thermal profiles included an initial denaturing step consisting of 95°C for 10 min followed by 38 cycles of denaturation at 95°C for 20 s, annealing at 60°C for 8 s with the *amoA* primers, and elongation at 72°C for 20 s.

Results and discussion

The nitrification performance and total nitrogen removal

The effluent and influent water samples of the aeration tank of six on-site advanced Johkasous were analyzed over a period of one year. The relationship between the nitrification (ammonia-oxidizing and nitrite-oxidizing reactions) performance and total nitrogen (T-N) removal is shown in Figure 2. The relationship between the denitrification performance and T-N removal is shown in Figure 3. In Japan, the most stringent effluent standard of the advanced Johkasou is set at BOD 10 $mg\,L^{-1}$, T-N 10 $mg\,L^{-1}$ and T-P 1 $mg\,L^{-1}$. In the present results, when the effluent NH_4-N concentration was higher than 2 $mg\,L^{-1}$, almost all of the effluent T-N concentrations were not able to meet effluent standard (T-N 10 $mg\,L^{-1}$). In contrast, NO_2-N concentrations of no less than 94% of the effluent water samples were kept below 1 $mg\,L^{-1}$, and in fact untreated NO_2-N was very low over a period of one year. During the experiment, NO_x-N concentrations of no less than 89% of the effluent water samples were kept below 1 $mg\,L^{-1}$. No noticeable accumulation of NO_x-N in the influent water was detected (Figure 3). These results demonstrated that nitrogen removal efficiency was hardly affected by nitrite oxidation and denitrification in these systems. In other words, ammonia oxidation was a rate-limiting step.

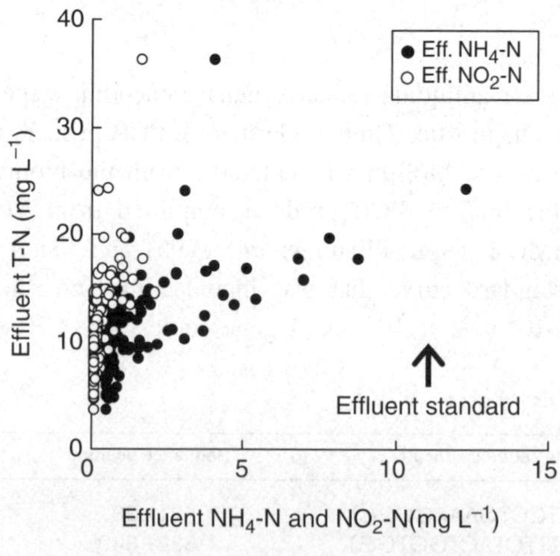

Figure 2 The relationship between effluent NH_4-N, NO_2-N and T-N concentrations

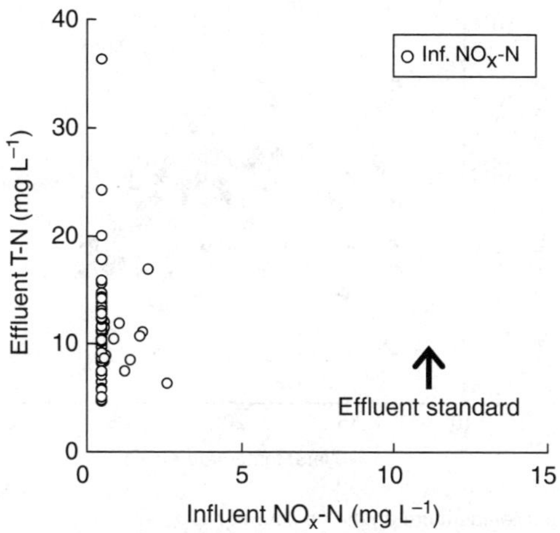

Figure 3 The relationship between influent NO_x-N of the aeration tank and effluent T-N Concentrations

AOB populations and operational characteristics

The number of AOB cells in the aeration tanks of the on-site six advanced Johkasous was monitored by real-time PCR several times every two months over a period of one year. The quantitative results of six Johkasous are shown in Table 3. The real-time PCR data were triplicate and standard deviation was evaluated for each biofilm sample. Assuming that one cell of AOB contains 2 copies *amoA* gene based on the copies reported for *Nitrosomonas europaea* (McTavish *et al.*, 1993), we calculated AOB cell number per volume of the aeration tank. Comparison with each system revealed that the difference in the mean number of AOB cells was approximately two orders of magnitude.

Operational characteristics of the aeration tank, the influent of the aeration tank, and the effluent were monitored. The annual characteristics of six Johkasous are shown in Table 4. During the experiment, no significant trends associated with water temperature

Table 3 Annual mean values of *amoA* gene number and AOB cell number

Johkasou name	*amoA* gene number (copies m^{-3})	AOB cell number (cells m^{-3})
A	$0.9 \pm 1.0 \times 10^{12}$	$4.7 \pm 5.1 \times 10^{11}$
B	$2.1 \pm 2.4 \times 10^{12}$	$1.0 \pm 1.2 \times 10^{12}$
C	$1.4 \pm 0.6 \times 10^{11}$	$7.2 \pm 3.0 \times 10^{10}$
D	$1.3 \pm 1.1 \times 10^{12}$	$6.4 \pm 5.3 \times 10^{11}$
E	$4.9 \pm 7.5 \times 10^{11}$	$2.5 \pm 3.8 \times 10^{11}$
F	$5.2 \pm 9.7 \times 10^{12}$	$2.6 \pm 4.9 \times 10^{12}$

Table 4 Annual characteristics of the aeration tank. Data are presented as means (± standard deviation). All values are in mg L^{-1} unless otherwise stated

Johkasou name	Water temp.(°C)			pH(−) Mean (Std dev.)
	Mean	Max.	Min.	
A	18.6	23.2	13.0	7.82 (0.20)
B	20.4	24.6	13.6	7.53 (0.19)
C	22.2	26.7	17.6	7.33 (0.16)
D	19.2	22.6	13.0	7.58 (0.11)
E	22.2	26.2	17.2	7.36 (0.21)
F	19.8	24.2	13.0	7.61 (0.11)

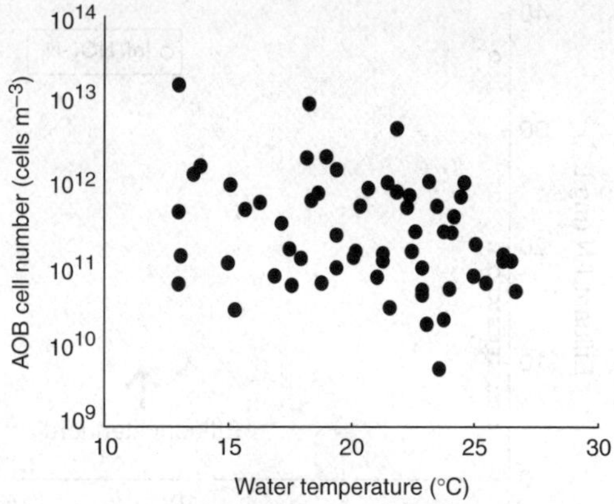

Figure 4 The effect of water temperature on the number of AOB cells

were detected in nitrification (data not shown), or AOB population dynamics (Figure 4), although water temperature was varied from 26.7 to 13.0. A biomass carrier in the aeration tank might offer an option for mitigating the effect of water temperature. Welander *et al.* (1997) reported that low temperature had very little effect on nitrification activity in a suspended carrier biofilm process. In all Johkasous, a stable pH (approximately 7.5) was obtained with no maintenance (Table 4), and pH seemed to have little impact on the difference in nitrification and the number of AOB cells in these Johkasous.

Relationship between NH_4-N loading per AOB cell and nitrogen removal

Figure 5 shows that no correlation was confirmed between the number of AOB cells and effluent T-N concentrations. However, Figure 6 indicates that there was a significant relationship between NH_4-N loading per AOB cell and T-N removal. When mean NH_4-N loading per AOB cell was lower than 210 pg cell^{-1} day^{-1}, the mean T-N concentration was kept below the effluent standard of advanced Johkasous (T-N 10 mg L^{-1}). When mean NH_4-N loading per AOB cell was higher than the 210 pg cell^{-1} day^{-1}, no Johkasou met the effluent standard.

Furthermore, the mean number of AOB cells of Johkasou E was the second lowest value, although NH_4-N loading per AOB cell of the Johkasou was approximately 8 times higher than other Johkasous. AOB in Johkasou E did not proliferate in response to the abundance of NH_4-N. This might be due to trouble in BOD removal, because the mean

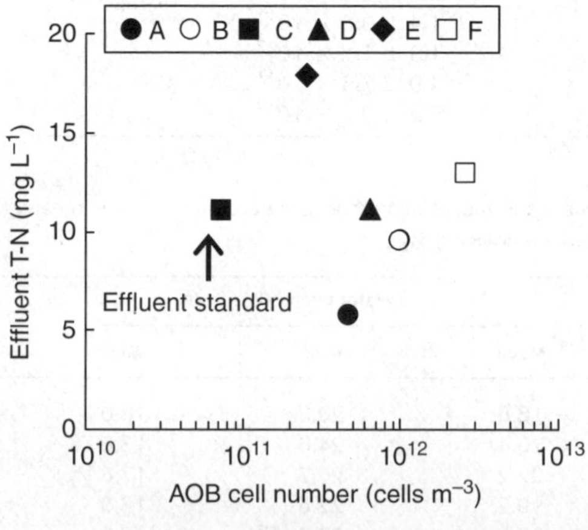

Figure 5 The relationship between the number of AOB cells (annual mean values) and T-N removal (annual mean values)

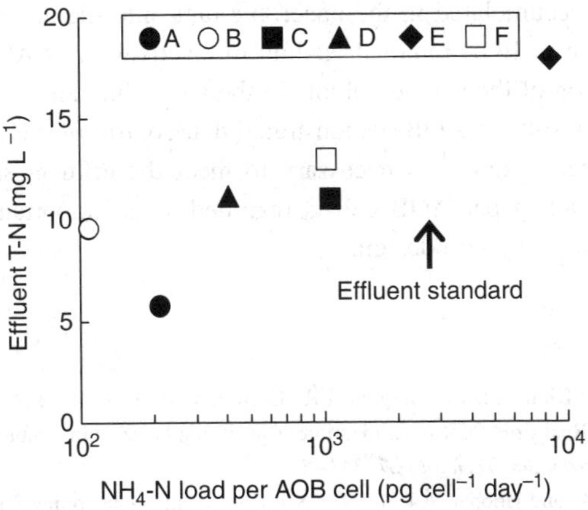

Figure 6 The effect of NH_4-N loading per AOB cell (annual mean values) on T-N removal (annual mean values)

Table 5 Mean annual characteristics of the the influent to the aeration tank, the effluent. Data are presented as means (± standard deviation). All values are in mg L^{-1}

Johkasou name	inf. BOD Mean (Std dev.)	eff. BOD Mean (Std dev.)
A	19.4 (11.3)	2.26 (2.53)
B	21.9 (14.8)	4.29 (2.38)
C	26.2 (11.0)	7.54 (2.62)
D	35.2 (26.3)	9.82 (5.30)
E	61.7 (43.4)	20.8 (22.6)
F	58.3 (44.1)	10.2 (3.73)

effluent BOD concentration of Johkasou E was much higher than those of other Johkasous (Table 5). Suspended solids (SS) concentration of the influent water flowing into the aeration tank in Johkasou E was the highest (Table 5). Therefore, influx of sludge accumulated in the anaerobic tank into the aerobic tank was thought to cause the trouble of BOD removal. Such a condition was unfavorable for AOB. This trouble might be originally caused by insufficient maintenance, such as lack of controlling of the sludge volume in the anaerobic tank. In order to avoid this problem, we should regulate the sludge volume in the anaerobic tank.

Consequently, in this type of biological nitrogen removal system equipped with porous carriers for immobilizing nitrifiers, we should regulate NH_4-N loading per AOB cell below 210 pg $cell^{-1}$ day^{-1} to meet the effluent standard of advanced Johkasou. We should regulate the sludge volume in the anaerobic tank to avoid the negative effect on AOB. Furthermore, it was found that NH_4-N loading per AOB cell is regarded as an important parameter for assessment of treatment performance and biological nitrogen removal system design.

Conclusion

Annual monitoring of six on-site decentralized systems "Johkasou" demonstrated that ammonia oxidation is a rate-limiting step in biological nitrogen removal. When the effluent NH_4-N concentration was higher than 2 mg L^{-1}, it was difficult to meet the effluent standard of advanced Johkasous (T-N 10 mg L^{-1}). In contrast, nitrogen removal efficiency was hardly affected by nitrite oxidation and denitrification in advanced Johkasous. In addition, water temperature and pH in the aeration tank seemed to have little impact on the difference in nitrification and the number AOB cells in these Johkasous.

Influx of sludge accumulated in the anaerobic tank into the aerobic tank was thought to cause deterioration of BOD removal leading to inactivation of AOB. In order to avoid this trouble, regulation of the sludge volume in the anaerobic tank is indispensable.

The quantitative results of AOB demonstrated that control of NH_4-N loading per AOB cell below 210 pg cell^{-1} day^{-1} is necessary to meet the effluent standard of advanced Johkasou. NH_4-N loading per AOB cell is regarded as an important parameter for biological nitrogen removal system design.

References

Harms, G., Layton, A.C., Dionisi, H.M., Gregory, I.R., Garrett, V.M., Hawkins, S.A., Robinson, K.G. and Sayler, G.S. (2003). Real-time PCR quantification of nitrifying bacteria in a municipal wastewater treatment plant. *Environ. Sci. Technol.*, **37**, 343–351.

McTavish, H., Fuchs, J.A. and Hooper, A.B. (1993). Sequence of the gene coding for ammonia monooxygenase in *Nitrosomonas europaea*. *J. Bacteriol.*, **175**, 2436–2444.

Okabe, S., Satoh, H. and Watanabe, Y. (1999). *In situ* analysis of nitrifying biofilms as determined by *in situ* hybridization and the use of microelectrodes. *Appl. Environ. Microbiol.*, **65**, 3128.

Rother, E. and Cornel, P. (2004). Optimizing design, operation and energy consumption of biological aerated filters (BAF) for nitrogen removal of municipal wastewater. *Wat. Sci. Tech.*, **50**(6), 131–139.

Rotthauwe, J., Witzel, K.P. and Liesack, W. (1997). The ammonia monooxygenase structural gene *amoA* as a functional marker: molecular fine-scale analysis of natural ammonia-oxidizing population. *Appl. Environ. Microbiol.*, **63**, 4704–4712.

Sinigalliano, C.D., Kuhn, D.N. and Jones, R.D. (1995). Amplification of the *amoA* gene from diverse species of ammonium oxidizing bacteria and from an indigenous bacterial population from seawater. *Appl. Environ. Microbiol.*, **61**, 2702–2706.

Welander, U., Henrysson, T. and Welander, T. (1997). Anaerobic pretreatment improves single sequencing batch reactor treatment of landfill leachates. *Wat. Res.*, **31**, 2351–2355.

Toxicity of domestic wastewater pH to key species within an innovative decentralised vermifiltration system

R.J. Hughes, J. Nair, K. Mathew and G. Ho

Environmental Technology Centre, Murdoch University, Murdoch 6150 Western Australia
(E-mail: *Robert.Hughes@murdoch.edu.au*)

Abstract The development of innovative decentralised wastewater treatment systems is pertinent to environmental sustainability. However for their appropriate use they require significant research and design to ensure they can be relied upon in the future and in most cases they have received less attention and research than large-scale sewage systems. This study assessed the likelihood of biological inhibition and disruption from pH to key species for vermifiltration. The test firstly assessed the buffer capacity of a vermicompost + manure media finding it had a relatively high buffering capacity for pH. The toxicity after the buffer capacity experiment showed that the species would survive between pH levels of 6.2 and 9.7. At the higher and lower pH levels however the survival of juveniles was impaired, probably due to their ability to uptake greater amounts of soluble salts and inability to regulate them. Overall the study showed that the tolerance of the key vermifiltration species to pH was far less than their survival as previously proposed and further research assessing their long-term population dynamics in a vermifiltration system with pH is warranted.

Keywords Biological inhibition; buffer capacity; pH; vermifiltration

Introduction

Wastewater contains a complex mixture of contaminants, including nutrients, pathogens and toxic compounds at moderate to trace concentrations e.g. endocrine disrupting compounds. The treatment of wastewater in new technologies needs appropriate research and assessment before it can be used appropriately. Decentralised and onsite treatment processes have been assessed to a limited degree when compared to centralised sewage work and although many contain similar processes to centralised systems, they may work differently in small units. The change in process efficiency may be due to the highly variable nature of effluent these systems treat. In newer processes utilised in decentralised and onsite systems, it is important they are thoroughly assessed for maintenance and operation problems beyond the development stages.

The operation and maintenance problems in wastewater treatment may be due to mechanical failures or system failures from inappropriate design. Biological inhibition is a common type of failure, which is due to inappropriate design or changes in effluent quality from disposal of toxic chemicals. Biological inhibition is commonly from disposal of heavy metals such as chromium from tanneries (Gupta and Bose, 2003) but may also occur from disposal of illegal toxic chemicals such as toluene and formaldehyde. The study of biological inhibition has rarely taken place in onsite systems, with one biological inhibition test available for septic tanks (Vaishnav and McCabe, 1996).

There is very limited knowledge on onsite wastewater toxicity. The few online studies on onsite influent have shown a highly variable level of common wastewater parameters such as pH, E_c, temperature and flow (Patterson, 2003). These changes show that the influent quality is directly related to the household use at a period of time.

Water Science & Technology **Vol 55 No 7 pp 211–218** © IWA Publishing 2007

doi: 10.2166/wst.2007.147

The pH of influent often varies considerably with household product use. The ingredients in many of the products contain caustic soda and alkaline salts, which can raise pH levels above 11 (Patterson, 2001). Influent with a high pH may cause disruption during the treatment process unless the system is able to buffer the pH. Acidity of influent may occur from putrescible wastes, such as those from a food grinder in the kitchen sink (Eriksson *et al.*, 2002).

The toxicity of pH to onsite systems has rarely been studied and its toxicological effects are largely unknown. Many traditional types (e.g. septic tanks) of onsite systems are designed with a large capacity and residence to ensure wastewater is homogenised and pH neutralised. The design of newer systems may not include such a large buffering capacity, especially where the wastewater passes through small treatment chambers or percolates over a filter bed.

The toxicity and buffering capacity of pH to vermicomposting systems has never been appropriately studied, especially in respect to wastewater. The buffering capacity of peat and pig manure compost has been shown with gas emissions where they were found to remove volatile gases and keep the filter bed within the pH range of 6.9 to 7.2 (Chan and Lu, 2005).

The toxicity of pH may pose a problem with vermifiltration units as the species involved, *Eisenia fetida* and *E. andrei*, are sensitive to environmental changes. The survival of these species is largely dependent on pH levels between pH 5 and 9 (Edwards, 1988). However it should be noted that their impairment within this range is largely unknown and it is likely that reproduction rates, growth and subsequent production of vermicompost may be reduced within this range. The lowering of the vermicomposting rate may cause a substantial problem in vermifiltration units as they rely upon the upper filter layers to be well degraded and act as a filter to remove organic material.

This research is the first study that assesses the buffering capacity of vermicompost and the effect of experimentally elevated and reduced pH levels on the two worm species *E. fetida* and *E. andrei*. These worms are the most common species in vermifiltration units in Australia and many other countries across the world.

Methods

Buffer capacity

Vermicompost for the test was produced from cow manure collected from Murdoch University's veterinary farm. The cow manure was free from herbicides and pesticides. The cow manure was vermicomposted for a 6-week period. Cow manure collected at the same time was left to mature over the same period. Both vermicomposted manure and manure were used in the test to represent the top layer of a vermifiltration unit, where both vermicompost and undegraded faecal material are present.

The initial buffer capacity test followed the same approach as Singh *et al.* (2003) but with some adaptations. The test used a 1:10 (w/v) ratio with 3 types of medium and three replicates of each medium. For each medium 5 g of vermicompost, vermicompost + manure (50:50 w/w) and manure was used. The amount added was fresh weight but was calculated on a dry weight basis using the moisture content. The initial test used acid and base concentrations of 0.01 mol l^{-1} NaOH and HNO_3 solutions, but these were changed to 0.1 mol l^{-1} in later experiments due to the high buffering capacity of the media. The pH was measured after 2 ml of the solution had been added and left on an end-over-end shaker for 1/2 hr. The pH was measured with the use of a portable Hanna pH and temperature probe. The initial tests took place over a period of approximately 6–8 hrs.

The final buffer capacity test took place in 2 L polypropylene mesocosms with 250 g of compost and 500 g of manure. The initial amount of NaOH and HNO_3 added was based upon the amount required to change the pH to a level between 4.5 and 10.5 from the initial buffer capacity test. Additions of $1\,\text{mol}\,\text{l}^{-1}$ acid/base solutions where made were the pH level was outside the appropriate level. The addition amount was calculated using a fresh weight sample (5 g dry weight equivalent based on moisture content) sample of the mesocosm medium. The sample was adjusted in a 1:10 (w/v) solution and left after each addition for over 1 hr, sometimes up to 12 hrs. The amount required to adjust the sample was used to calculate the addition required to the mesocosms. The final test took place over approximately 1 month with a number of additions occurring to the lower pH mesocosms.

The tests using $0.1\,\text{mol}\,\text{l}^{-1}$ and $1\,\text{mol}\,\text{l}^{-1}$ acid/base concentrations were used for a comparison only, as the $0.01\,\text{mol}\,\text{l}^{-1}$ test was abandoned and provided inappropriate data. The $0.1\,\text{mol}\,\text{l}^{-1}$ was transformed into $1\,\text{mol}\,\text{l}^{-1}$ equivalent to allow an accurate comparison. The addition amounts for the initial vermicompost + manure medium test were used for the comparison with the mesocosm test in the results section.

pH toxicity test

Experiment design. The toxicity experiment utilised the same 21 polypropylene containers as the final buffer capacity experiment. The test started once the pH in all the mesocosms had been adjusted to the required level. The mesocosms blocked direct sunlight and had air vents added to the top using 2 mm plastic mesh. 10 *E. fetida* and *E. andrei* were added to each container. There were seven pH levels 4.5, 5.5, 6.5, 7.5 (control), 8.5, 9.5 and 10.5. Each pH level had 4 replicates.

Monitoring. The pH and electrical conductivity (E_c) of each container was measured using a 1:10 (w/v) ratio. The method followed the standard procedure, using 1 hr on an end-over-end shaker before pH measurement (Rayment and Higginson, 1992). After the pH measurement, samples were centrifuged at 3000 rpm for 5 minutes to allow E_c measurement. The pH and E_c were measured using Hanna portable probes. The moisture content was measured weekly by weighing a sample and then drying it in a microwave until no change in weight could be determined. The moisture content was then determined by dividing the original weight against the weight lost. Visual assessments of the vermicompost were taken during and after the test. The final assessment is presented in the results.

Chemical analysis. Organic carbon was measured by loss on ignition. Samples for organic carbon were weighed and left in a muffle furnace at 550 °C for 24 hrs then left to cool in a desiccator and reweighed. The organic carbon content was then determined by dividing the original weight against the weight lost. Ammonia was measured using the standard method and extracted using $2\,\text{mol}\,\text{l}^{-1}$ KCl solutions. TKN and TP were measured by the persulphate method (APHA, 1998).

Biological endpoints. The worm weight was measured weekly using an AND 300 balance. Mortality, clitella and juvenile presence was measured weekly during the weighing of the worms. The juvenile worms were categorised into small or medium worms depending on their size, e.g. small <3 cm and medium >3 cm. The juveniles were removed from the mesocosms after they were counted.

Statistical analysis. The survival results were assessed using SPSS 11. Survival percentages and goodness of fit for the logistic regression model were calculated using the probit statistic and chi square respectively (Forbes, 1994).

Results

Buffer capacity experiment. Table 1 shows the amount of $1\,mol\,l^{-1}$ NaOH or HNO_3 required to change the pH to a specified level. The initial column represents the $0.1\,mol\,l^{-1}$ buffer capacity test and the final column represents the $1\,mol\,l^{-1}$ mesocosm test. The addition required to change the pH in the mesocosm was substantially higher than that found in the $0.1\,mol\,l^{-1}$ buffer capacity test. There is considerably more acid required for each of the tests to change the pH to the appropriate level. The final test occurred over a period of approximately 1 month and the initial test took place over a period of 6–8 hrs.

pH toxicity experiment

pH and E_c. The pH gradually increased in the test period in most of the replicates. Figure 1 below also shows a noticeable decrease in pH in the higher pH replicates (e.g. 8.5 to 10.5) after the first week of the test.

The electrical conductivity was highest in the lower and higher pH replicates. The lower pH replicates showed the highest E_c values and 10.5 also had a comparatively high E_c. Figure 2 shows that E_c tended to slightly decrease over the test period, with slight fluctuations noticed between weeks.

Toxicity endpoints. The survival of worms during the first week of the test is shown below in Figure 3. All worms died after 1 week in pH replicates of 4.5, 5.5 and 10.5. All worms in the other replicates survived over the test period. The pH values where 50% mortality occurs are pH 5.99 and 9.99. The levels where worm survival is above 95% are between 6.2 and 9.7.

Figure 4 shows the survival of juveniles that hatched from the vermicompost during the test period. It is noticed that the control replicate at pH 7.5 had the greatest number of juveniles whilst pH 6.5 and 8.5 had a low number and 9.5 had no juveniles. Only pH 7.5 had medium size worms survive whilst pH 6.5 and 8.5 had some small worms.

Discussion

The most important aspect of vermifiltration is production of vermicompost from faecal material for use as a wastewater filter medium. If the worms within the upper filter beds of a unit become inhibited then it is likely that filtration will become impaired and hydraulic conductivity will be reduced.

The initial test assessed the buffering capacity of three manure, manure + vermicompost and vermicompost mediums. A second test was used to adjust the pH to levels between 4.5 and 10.5 in a number of mesocosms filled with a vermicompost + manure medium. It was found that the actual buffering capacity of the mediums in the mesocosm was well beyond the initial buffer capacity test. Hence the use of such buffer

Table 1 The amount of 1M acid/base solutions required to change the pH in the experiments

pH	HNO₃		NaOH	
	Initial	Final	Initial	Final
4.5	5.95	229.26		
5.5	4.01	174.49		
6.5	2.21	59.67		
7.5	No additions	No additions	No additions	No additions
8.5			0.53	24.27
9.5			2.01	90.25
10.5			3.59	156.11

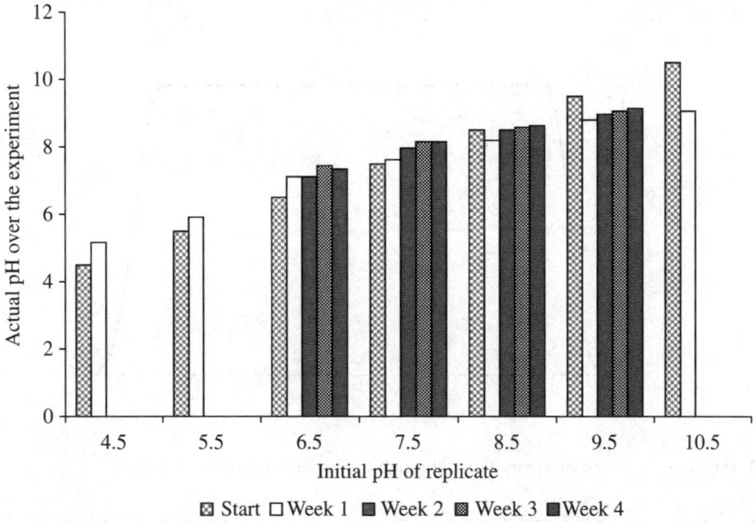

Figure 1 The pH changes over the experimental period of 4 weeks

capacity tests may lead to inaccurate buffer capacity assessments of vermicompost, compost and manures. The final buffer capacity test over a period of 1 month was more accurate and showed that the medium had a high capacity to buffer out acid and base solutions.

The buffering capacity of the medium was greater for the nitric acid additions when compared to sodium hypochlorite. It is likely that the much of the acidity (H^+) was adsorbed by free anions within the medium. There may have not been enough exchange sites to adsorb hydroxide ions to the same extent. As many wastewater constituents may contain sodium hydroxide (e.g. cloudy ammonia) further tests within reasonable pH values for wastewater should be done to assess its ability to change the pH of a vermicompost + manure medium.

The toxicity tests for pH showed the tolerance of the worms to pH and the limits where pH inhibits survival and reproduction. The mesocosms used for this test were

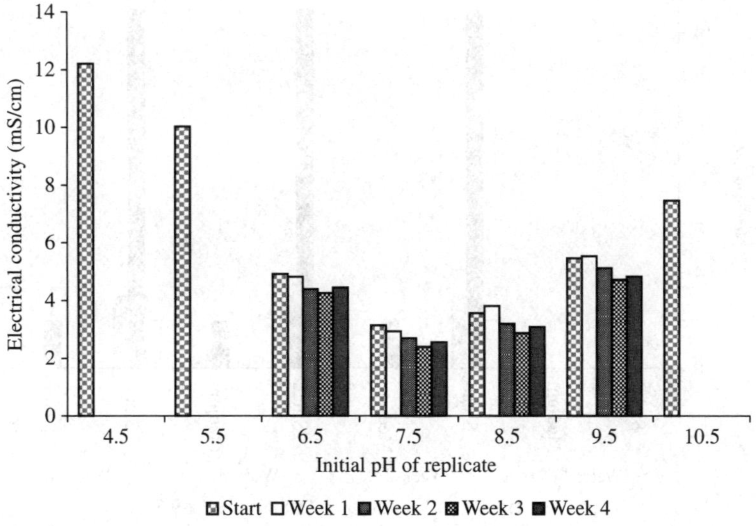

Figure 2 The electrical conductivity over the experimental period of 4 weeks

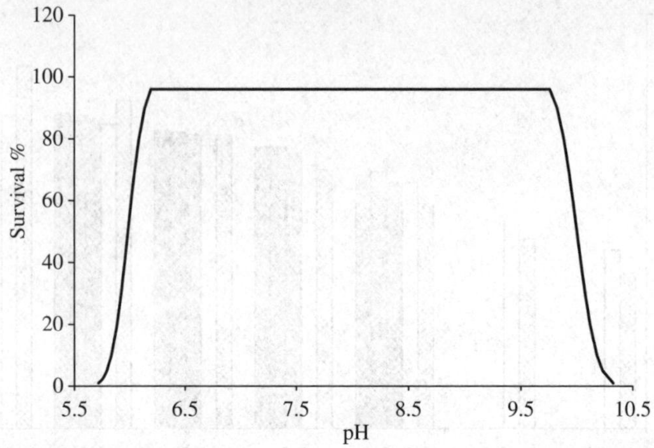

Figure 3 The survival percentages produced through the logistic regression model

adjusted to a specific pH level for accurate assessment of biological endpoints and some monitoring of pH, E_c, ammonia and moisture content took place.

The pH increased in most of the replicates over the test period, although a decrease was noticed in the higher pH replicates after the first week. This decrease is probably due to additional buffering occurring at these higher pH levels and may not be due to earthworm activity. The general increase in all of the replicates after the second week can be attributed to the action of the worms and it is common for many animal wastes to be slightly alkaline after processing (Edwards, 1998).

The E_c was affected by pH and increased at the higher and lower pH levels e.g. 4.5, 5.5, 9.5 and 10.5. This is likely to be due to the greater mobility of soluble salts and macro-elements at these pH values, causing them to easily dissolve into the extract solution.

Worm survival was effected by pH in this experiment with the pH replicates of 4.5, 5.5 and 10.5 having 100% mortality after the first week. The replicates of 6.5, 7.5 (control), 8.5 and 9.5 survived throughout the experiment.

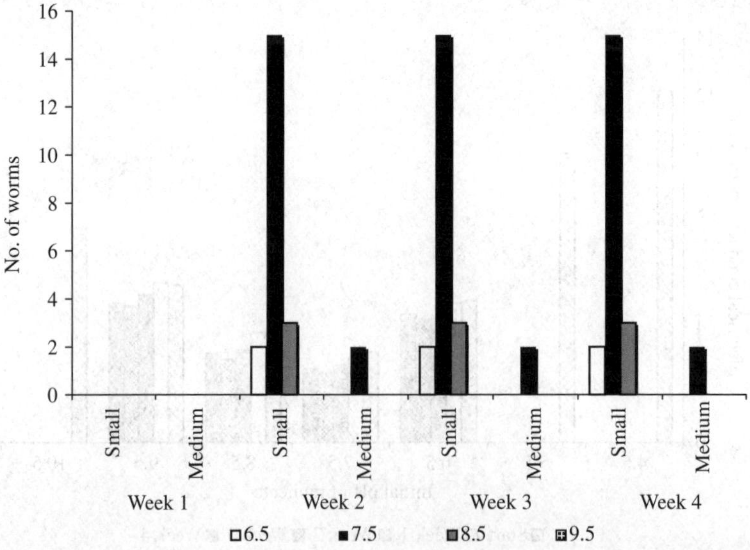

Figure 4 The numbers of small and medium worms present during each week of the test

The probit statistic was used to find the median lethal pH levels of 9.99 and 5.99. In between 6.2 and 9.7 it was found that over 95% of worms would survive suggesting that these are the tolerance limits for *E. fetida* and *E. andrei*. Edwards (1988) found that *E. fetida* would survive between >5 and < 9, implying that these values are fairly accurate. The values of Edwards (1988) show a greater tolerance in the worms nonetheless and it may suggest that other factors that work synergistically with pH (e.g. elemental solubility) have increased toxicity in this test. It has been found that in most studies pH is the factor that may affect toxicity the most, above cation exchange capacity and organic matter, and affect both the physico-chemical binding and biological uptake processes (Eijsackers, 1998).

Worm reproduction was affected by pH at 6.5, 8.5 and 9.5 when compared to the controls (pH 7.5). The reproduction of worms was greatest in the control replicate, signifying that juvenile worms tend to be more susceptible than adult worms, which is most likely due to factors such as volume to surface area ratio. There were no juveniles in the highest surviving pH replicate (9.5), implying that reproduction may be inhibited at this pH. It may also add weight to the point that the high solubility of some elements at this pH was toxic to the worms. The low number of juveniles at pH 6.5 and 8.5 may also imply that reproduction is inhibited at this point – whether the worm population can be sustained at these pH levels is unclear from this study.

The reduction in reproduction at higher and lower pH levels has never been found before in conjunction with pH and is very important where the management of worm populations is key to vermicomposting activities. The low survival of juveniles in this study, could allow a further reduction in the tolerance limits of these species closer to the control level of 7.5, e.g. 7.5 ± 05. This lower limit may provide a level where worm populations can be maintained, although further work should be used to assess this and how the reproductive rates affect vermicompost over a longer period of time.

Considering the lower tolerance of juvenile worms, it is important that the buffering capacity of wastewater is studied further with flow through mesocosms and appropriate wastewater. In this study the high level of buffering capacity with very acidic/basic solutions has shown the vermicompost has a high potential to buffer wastewater. It is also important that pH is studied with toxic compounds found in domestic wastewater that may increase or decrease in toxicity with pH, to show the tolerance limits for the worms. Hydraulic conductivity and pH should be further assessed to show if a reduction in vermicomposting with pH affects wastewater filtration of specific compounds.

Conclusion

The buffering capacity of vermicompost, vermicompost + manure and manure were found in this study. The initial test on these mediums was found to be inept and a 1 month test in a mesocosm provided suitable information on the buffering capacity of the vermicompost + manure medium.

The toxicity of pH to *E. fetida* and *E. andrei* was found in a toxicity test that assessed the affects of pH from 4.5 to 10.5. The survival figures were all similar to those provided by Edwards (1988), although the worms in this study were slightly less tolerant. Toxicity was attributed to the solubility of elements within the medium and was shown by a higher E_c at the higher and lower pH levels, e.g. 10.5, 9.5, 4.5 and 5.5.

The survival of juveniles was impaired at all pH levels when compared to the control replicates. It is likely that juveniles were unable to withstand the variable pH levels due to a greater surface to volume ratio than the adults. This may have led to diffusion of elements into their bodies above their regulation levels.

Further research assessing the buffering capacity and population dynamics of a flow through mesocosm is warranted to show if its hydraulic conductivity and population levels will change with high or low pH wastewater. The use of toxic substances that are affected by pH is also warranted.

References

APHA (1998). *Standard Methods for Examination of Water and Wastewater*, 20th edn, American Public Health Association.

Chan, W.-C. and Lu, M.C. (2005). A new synthetic biofiltration materia: poly(vinyl alcohol)/pig manure compost compostie beads. *Journal of Chemical Technology and Biotechnology*, **80**, 574–580.

Edwards, C.A. (1988). Breakdown of animal, vegetable and industrial organic wastes by earthworms. In *Earthworms in Waste and Environmental Management*, Edwards, C.A. and Neuhauser, E.P. (eds), SPB Academic Publishing, Hague, Netherlands, pp. 21–31.

Edwards, C.A. (1998). The use of earthworms in the breakdown and management of organic wastes. In *Earthworms Ecology*, Edwards, C.A. (ed.), CRC Press LLC, Boca Raton, Florida.

Eijsackers, H. (1998). Earthworms in environmental research: still a promising tool. In *Earthworm Ecologu*, Edwards, C.A. (ed.), CRC Press LLC, Boca Raton, Florida.

Eriksson, E., Auffarth, K., Henze, M. and Ledin, A. (2002). Characteristics of grey wastewater. *Urban Water*, **4**, 85–104.

Forbes, T.L. (1994). The design and analysis of concentration-response experiments. In *Ecotoxicology*, Calow, I.P. (ed.), Blackwell Scientific Publications, London, pp. 438–461.

Gupta, S. and Bose, P. (2003). Case studies on biological treatment of tannery effluent in India. *Journal of the Air and Water Management Association*, **53**(53), 8.

Patterson, R.A. (2001). Water quality relationships with reuse options. *Water Science and Technology*, **43**(10), 147–154.

Patterson, R.A. (2003). Temporal variability of septic tank effluent. In *Future Directions for on-site Systems: Best Management Practice. Proceedings of on-site*, University of New England, Armadale.

Rayment, G.E. and Higginson, F.R. (1992). *Australian laboratory handbook of soil and water chemical methods*. Inkata Press, Sydney.

Singh, B., Odeh, I.O.A. and McBratney, A.B. (2003). Acid buffering capacity and potential acidification of cotton soils in northern New South Wales. *Australian Journal of Soil Research*, **41**, 875–888.

Vaishnav, D.D. and McCabe, J.W. (1996). Development of an anaerobic sludge respiration inhibition test and its use to assess septic tank safety of consumer products. *Environmental Toxicology and Chemistry*, **15**(5), 663–669.

Tertiary filtration in small wastewater treatment plants

V. Naddeo and V. Belgiorno

Department of Civil Engineering, University of Salerno, Via Ponte don Melillo, 84084 Fisciano (SA), Italy
(E-mail: *vnaddeo@unisa.it*)

Abstract Tertiary filtration can be proposed in small wastewater treatment plants with impact on protected water bodies. Rotating disk filters may be adopted, in respect to conventional sand filters, when low availability of space and low investment costs are the prevailing conditions. The overall objective of this research was to evaluate the filtration efficiency of rotating disk filters; to compare effectiveness with traditional sand filters; to analyse thoroughly the importance of particle size distribution in wastewater tertiary filtration. In the experimental activity, conventional wastewater quality parameters were investigated and particle size distribution (PSD) was characterized to discuss the filter effectiveness. The effect of design and operation parameters of tertiary filters were discussed related to particle removal curves derived from particles counts. Analysis of particle size distribution can be very useful to help comprehension of filtration processes, design of filtration treatments and to decide the best measures to improve filter performance.

Keywords Particle size distribution; rotating disk filter; sand filter; wastewater filtration

Water Science & Technology **Vol 55 No 7 pp 219–225**

Introduction

Stringent limits on wastewater discharges imposed by new regulations in the European Union are based on territorial constraints. In tourist areas, high efficiency of treatment is required for small wastewater treatment plants, which have many difficulties related to the load variability concentrated in a few months of the year and the simplicity of the processes implemented. Wastewater reclamation and water potential for irrigation and industry may be increased by advanced treatment of secondary effluents to meet specified effluent quality objectives. The filtration as a tertiary step has a substantial effect on the efficiency of wastewater treatments as BOD_5, COD and SST, promoting an effective impact on water body quality and improving disinfection effectiveness as well as allowing the discharge in highly protected systems.

Sand filters are the most adopted option all over the world. Sand filtration needs high availability of space and often it cannot be adopted in upgrading small wastewater treatment plants (Hatziconstantinou *et al.*, 1995; Kobler and Boller, 1997; Ødegaard and Skrovseth, 1997; Hamoda *et al.*, 2004). Particularly, in some typical situations of Mediterranean coastal areas, choice of wastewater treatments is addressed toward compact technologies, due to the limitations of land availability (Dohmann *et al.*, 1996; Schwille *et al.*, 2002).

Rotating disk filters (RDF) can be interesting when low availability of space and low investment costs are the prevailing conditions. Different technologies adopted for RDF offers a means of dynamic filtration that minimizes the plugging of the filter pores with particles and the build-up of a cake layer (Ohkuma *et al.*, 1994; Serra and Wiesner, 2000; Schwille *et al.*, 2002). In the design and operation of a rotating filter, there are a large number of variables that affect the filtration performance including operational parameters that can be varied by the device operator, suspension parameters that are dependent on the physical characteristics of the suspension that is being filtered, and geometric

doi: 10.2166/wst.2008.148

parameters that are related to the physical dimensions of the device (Hatziconstantinou *et al.*, 1995; Boller, 1997; Christophe, 1998; Schwille *et al.*, 2002).

Usually the performance of the filtration systems have been based on common operational parameters such as pressure loss, cycle time or total removal of suspended solids; with these parameters, only large particles representing the major part of the suspended volume or mass were considered (Ødegaard, 1998; Adin, 1999; Marquet *et al.*, 1999).

In many studies PSD was measured with a Coulter counter or other instruments, but most of these methods are only applicable to count bigger particles and often require artificial solutes (Lawler, 1997; Kaminski *et al.*, 1997; Kobler and Boller, 1997). The PSD measuring particles down to 20 μm can reveal a lot more information than simple analysis of suspended solids. Another requirement for filtration studies is to gain information also on the number concentrations of particles in different size classes. Therefore, PSD data are best based on single particle counting. From a technical point of view this is not always an easy task. The number concentration of colloidal particles in natural systems (like treated wastewater) are often too high and, depending on the particle counter, require careful dilution of suspension samples (Kobler and Boller, 1997; Lawler, 1997; Adin, 1999).

In this study the efficiency of a pilot scale rotating disk filter (RDF) was tested and compared to a full scale conventional sand filter (SF). Their application on small wastewater treatment plants as tertiary treatment was also discussed.

Material and methods

A RDF pilot-plant was installed close to a full-scale SF unit at Mercato San Severino wastewater treatment plant; in this way all tests were carried out simultaneously with the same influent.

The treatment plant of Mercato San Severino (Salerno, Italy) is based on a conventional biological process scheme, and actually treats both domestic and industrial discharges. Pre-treatment of this plant is based on screens and grit removal, while secondary treatment is performed with activated sludge. Tertiary treatment is carried out by a conventional deep bed sand filter. Disinfection is obtained by chlorination. SF is an unstratified mono-medium unit that works in downward mode with a filtration rate of $150\,l/m^2\,min$. A batch backwash operation occurred every 45 min. The main quality parameters of treated wastewater are showed in Table 1.

In Figure 1 the RDF pilot plant is illustrated; the device is composed of two inner porous filter disks to separate particles from the fluid, and a semi-cylinder housing. The length of the diameter of the filter disks is 30 cm, and the overall surface of both filters is $1413\,cm^2$. The experimental flow rate treated in the pilot plant was fixed at 5 L/s. The filtrate was removed via ports in the bottom block at the right and left end of the test cell. Several water nozzle ports along the upper housing sides of the disks filter are necessary to purge the system in working conditions. A direct-drive stepper motor and controller were used to rotate the inner cylindrical filter. The filter medium was a porous iron disk with a specific pore size, according to the manufacturer. In this study, two

Table 1 Wastewater quality of Mercato San Severino treatment plant

	WWTP influent	Effluent (before filtration)
BOD_5 (mg/L)	150 ± 60	25 ± 10
COD (mg/L)	420 ± 75	86 ± 20
TSS (mg/L)	90 ± 30	35 ± 25
Tot P (mg/L as P)	3 ± 3	2 ± 1.5

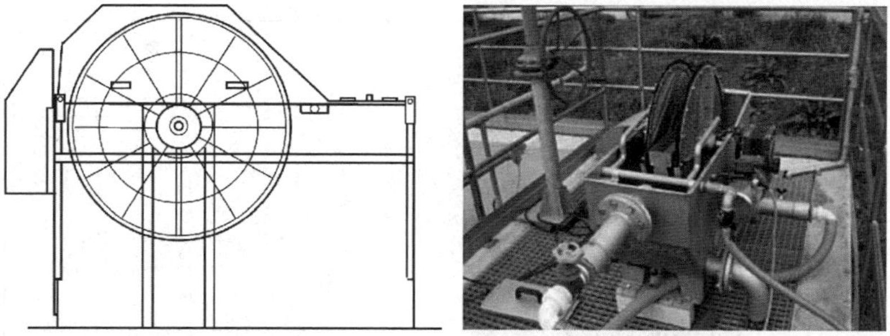

Figure 1 Rotating disk filter pilot-plant

different pore sizes of the filter medium were tested, 35 μm (RDF35) and 10 μm (RDF35) respectively.

The characterization of the main wastewater parameters were performed in the Environmental Engineering Laboratory of the University of Salerno (Fisciano, SA, Italy). All analyses were carried out at dry weather conditions on composite samples, taken with an automatic sampler system. Samples were taken, in both filtration systems, from three different points. The first sample was taken from the effluent of the secondary settling tanks immediately before the sand filter (influent of both filtration units), the second and the third samples were taken from the effluent and the backwashing. COD and TSS were detected for all collected samples following the standard methods (AWWA-APHA-WEF, 1998).

The characteristics of filtration systems and the size of filter medium were also studied with particle size distribution (PSD) analyses. Samples of 1 L stored in brown glass bottles were used for PSD analyses. The discrete PSD was determined by a SPC-device (Particles Measuring Systems Inc., Model APSS-200). It works as a combination of a laser scattering light detector for smaller particles and a laser light blockage detector for bigger particles. A Model APSS-200 was used, sampling volumes of 250 mL. In order to calibrate the instrument, thresholds of 2, 5, 10, 15, 20, 25 and 30 μm were used to establish the individual channels of the sensor over a wide range from 2 to 125 μm. These detectors are able to recognise single particles in a solution with a limited particle concentration; samples like treated wastewater had to be diluted to allow single particle counting. Therefore, the sensor was equipped with a specific dilution unit. It allows direct dilution of the suspension up to a factor of 60 without changing the dispersion medium. In this study the samples were pre-diluted with deionised and 0.1 μm filtered water and then diluted by a factor of 10 on the analysis device immediately before the particle measurement.

Results and discussion

TSS concentration in the influent ranged generally between 30 and 60 mg/L, with an average of 32 mg/L; in the effluent the concentration is drastically reduced (Figure 2). The influent COD concentration varied between 66 and 106 mg/L, with an average of 86 mg/L, while in the effluent the COD reduction over 30% has comparable for all tested systems. These results highlight how all filtration systems work well and the effluent quality was acceptable for Italian MCL (COD $<$ 125 mg/l; TSS $<$ 35). Anyway the SF unit definitely has a better performance with respect to TSS removal.

The results of the PSD and particle removal in all tested conditions are respectively shown in Figures 3 and 4. Excellent filter performance was observed in the SF system which operates at a very low filtration rate. Except for the particle size range, lower than

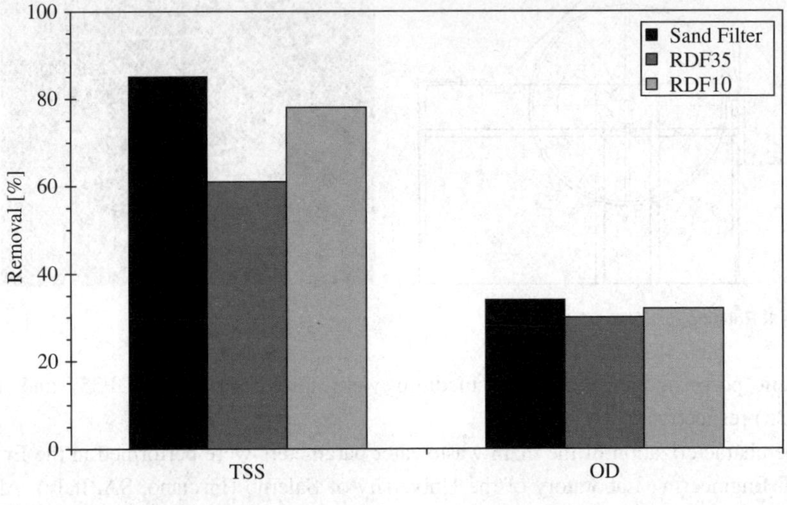

Figure 2 Comparison of TSS and COD removal

15 μm, particle removal of more than 90% was achieved and reached more than 96% in the size range above 20 μm.

The same behaviour, but with a somewhat lower efficiency, was observed in the RDF systems (Figure 4).

The results showed that in SF there is a poor colloid removal. In fact, particles below a size of 10 μm are virtually not removed in SF and may even be produced by folc disruption in the filter bed, as confirmed in the literature (Kobler and Boller, 1997; Lawler, 1997; Adin, 1999). This effect is not present in the RDF system where layer filtration occurs with respect to deep-filtration of SF.

These results are also explainable if we suppose that each particle group size is ruled by a different mechanism of removal. Particles smaller than about one micrometre are transported by diffusion. Larger particles are transported by gravity. Transport of relatively large particles may also be dominated by interception, or they may be retained by interstitial straining. Lawler (1997) showed that particles of 1–10 μm in size have minimal opportunity for removal in deep filtration systems, since transport mechanisms of these particles within the filter bed are less efficient.

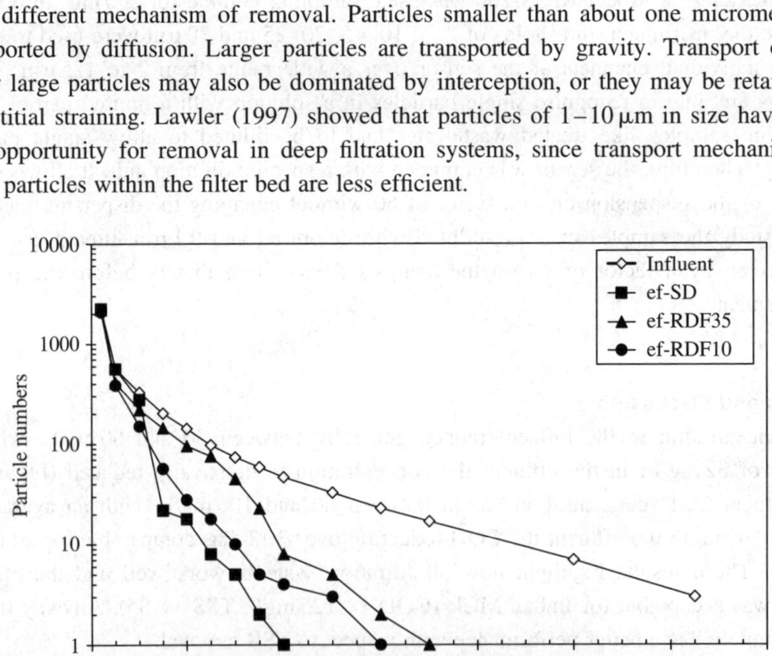

Figure 3 Comparison of particle size distribution

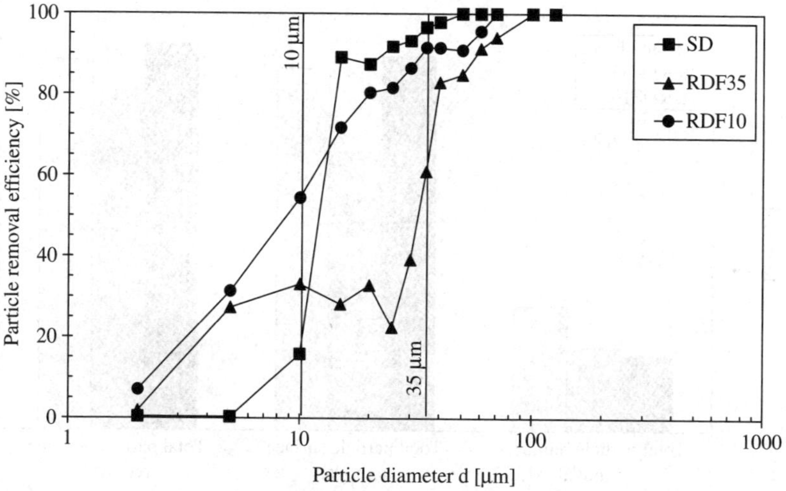

Figure 4 Comparison of particle removal efficiency

Particle overall volume distribution (PVD) can be calculated from PSD data assuming a spherical shape with a diameter equal to the logarithmic average of the particle diameter measured in each measured size range. While particle number increases inversely with particle size, PVD resembles more like a bell shaped curve. Distribution of cumulative volume of particles is plotted in Figure 5.

The clear explanation is that since particle volume is a third power of particle size, the total volume of the small particles is small in spite of their larger number, while the total volume of the larger particles is small due to their smaller number. Such a calculation may help determine solids volume entrapped within a filter bed or a dripper between predetermined particle sizes. This process theory generally assumes uniform suspended particles (Richert and Hunter, 1971; Lawler, 1997). That assumption is inaccurate for wastewater effluent suspensions, where particle size may range from several nanometres up to more than 100 micrometres. Transport equations express strong dependency of transport mechanism efficiency on particle size (Kobler and Boller, 1997; Lawler, 1997; Neis and Tiehm, 1997; Adin, 1999). In order to know on which mechanism to

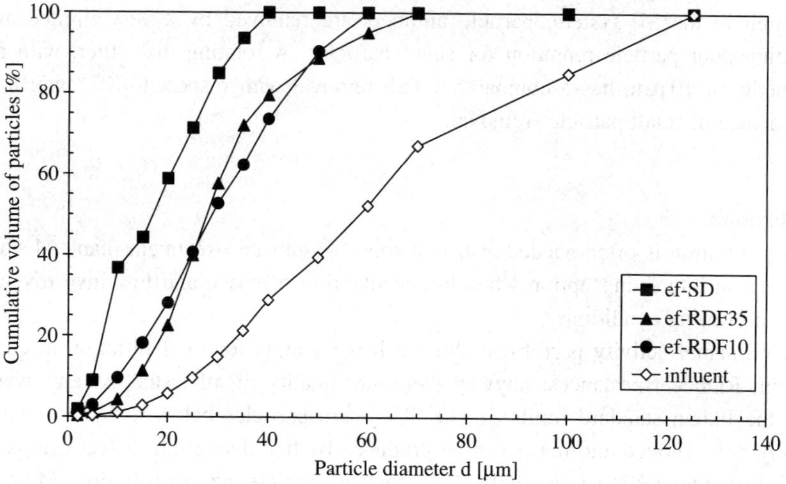

Figure 5 Comparison of cumulative volume distribution of particles

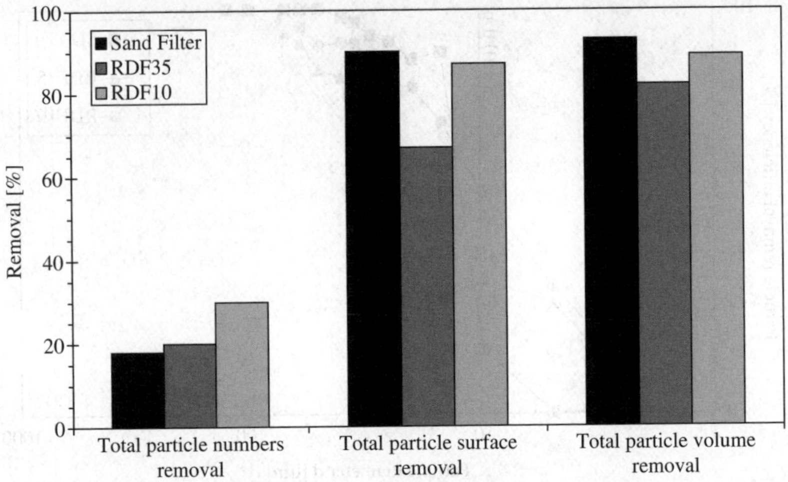

Figure 6 Comparison of total particle numbers, total surface and total volume removal

focus, it is necessary to determine the relevant size group of particles in the wastewater by PSD analysis.

The results showed that the examined filters reach particles removal rates of more than 80% if the particles volume is taken as a reference for particulate mass removal. Therefore, the particles volume which is formed mainly by big particles is not a sensitive parameter to evaluate filter performance. If particle number counts are additionally considered, the filter plants could be classified according to their different ability to remove small particles, especially in the size range lower than 20 μm.

The average TSS and COD removal measured in the all tests (Figure 2) can be compared to the removal of total particle numbers, total particle surface and total particle volume (Figure 6). From different investigations it is known that the larger particles (higher than 10–20 μm) contribute the most to the solids mass expressed as TSS concentration (Kobler and Boller, 1997; Lawler, 1997; Adin, 1999). Since the bigger particles represent also the dominant volume fraction, particle volume removal may give a good indication of TSS removal performance.

The results show that, according to the PSD, particle volume and particle numbers removal may differ enormously. Whereas solids mass and particle volume removal are very high in the SF system, particle numbers are removed by a very limited amount indicating poor particle retention for small particles. A rotating disk filter, with a pore size media of 10 μm, has a comparable TSS removal with respect to SF, against better performance in small particles removal.

Conclusions

Tertiary filtration is often needed also in a small wastewater treatment plant. Mechanical filters are an interesting option when low availability of space and low investment costs are the prevailing conditions.

Experimental activity performed showed better results for sand filtration in comparison with RDF performances; anyway the water quality of all tested systems was well under the Italian standard limits. In the SF system particles below a size of 15 μm are virtually not removed and may even be produced by floc disruption. Effectiveness of the mechanical filter (RDF) is strongly dependent on particle size distribution. More information is needed on the influence of operational conditions on RDF.

Repeated particle analysis showed that the pattern of particle removal curves for both filter systems stayed usually constant and confirmed that PSD measurement is a reproducible method for filter evaluation in practice. Particle size distribution can serve as a powerful tool for understanding filter behaviour and discrepancies in case of filtration under various conditions. It may be applicable for advanced wastewater filter design. PSD should be taken into consideration in filtration models, rather than relying on a single representative particle diameter of questionable value.

Particle removal as a function of a particle's size could be calculated and represents an excellent designer monitoring tool. Anyway more research work to standardize measures is required to introduce particles size distribution as the usual characterization parameter.

Acknowledgements

Raffaella Attianese and Rosario Casale for their contribution in the first part of the experimental activities.

References

Adin, A. (1999). Particle characteristics: a key factor in effluent treatment and reuse. *Wat. Sci. Tech.*, **40**(4–5), 67–74.

AWWA-APHA-WEF (1998). *Standard Methods for the Examination of Water and Wastewater*, 20th edn, American Public Health Association, Washington, DC.

Boller, M. (1997). Small wastewater treatment plants – a challenge to wastewater engineers. *Wat. Sci. Tech.*, **35**(6), 1–12.

Dohmann, M., Hebert, A.F., Sanz, J.P. and Zenzes, N. (1996). Operational problems of wastewater filtration plants – their analysis and possible solution. *Wat. Sci. Tech.*, **34**(3–4), 315–322.

Hamoda, M.F., Al-Ghusain, I. and AL-Mutairi, N.Z. (2004). Sand filtration of wastewater for tertiary treatment and water reuse. *Desalination*, **164**, 203–211.

Hatziconstantinou, G.J., Kalergis, C.M. and Grivas, A.P. (1995). Upgrading of a small wastewater treatment plant. design and operation aspects. *Wat. Sci. Tech.*, **32**(7), 111–117.

Kaminski, I., Vescan, N. and Adin, A. (1997). Particle size distribution and wastewater filter performance. *Wat. Sci. Tech.*, **36**(4), 217–224.

Kobler, D. and Boller, M. (1997). Particle removal in different filtration systems for tertiary wastewater treatment – a comparison. *Wat. Sci. Tech.*, **36**(4), 259–267.

Lawler, D.F. (1997). Particle size distributions in treatment processes – theory and practice. *Wat. Sci. Tech.*, **35**(4), 15–23.

Marquet, R., Mietton-Peuchot, M. and Wheatley, A.D. (1999). Characterisation of trickling filter effluent by particle size distribution and high performance size exclusion chromatography. *Wat. Res.*, **33**(6), 1415–1424.

Neis, U. and Tiehm, A. (1997). Particle sue analysis in primary and secondary waste water effluents. *Wat. Sci. Tech.*, **36**(4), 151–158.

Ødegaard, H. (1998). Optimised particle separation in the primary step of wastewater treatment. *Wat. Sci. Tech.*, **37**(10), 43–53.

Ødegaard, H. and Skrovseth, A.F. (1997). An evaluation of performance and process stability of different processes for small wastewater treatment plants. *Wat. Sci. Tech.*, **35**(6), 119–127.

Ohkuma, N., Shinoda, T., Aoi, T., Okaniwa, Y. and Magara, Y. (1994). Performance of rotating disk modules in a collected human excreta treatment plant. *Wat. Sci. Tech.*, **30**(4), 141–149.

Richert, D.A. and Hunter, J.V. (1971). General nature of soluble and particulate organics in sewage and secondary effluents. *Water Research*, **5**(7), 421.

Schwille, J.A., Mitra, D. and Lueptow, R.M. (2002). Design parameters for rotating cylindrical filtration. *Journal of Membrane Science*, **204**, 53–65.

Serra, C.A. and Wiesner, M.R. (2000). A comparison of rotating and stationary membrane disk filters using computational fluid dynamics. *Journal of Membrane Science*, **165**, 19–29.

Water Science & Technology **Vol 55 No 7 pp 227–234** © IWA Publishing 2007

Wastewater infiltration percolation for water reuse and receiving body protection: thirteen years' experience in Spain

F. Brissaud*, M. Salgot, M. Folch**, M. Auset***, E. Huertas** and A. Torrens****

*Hydrosciences, MSE, Université Montpellier II, 34095 Montpellier Cedex 05, France (E-mail: *brissaud@msem.univ-montp2.fr*)

**Institut de l'Aigua de la Universitat de Barcelona. Facultat de Farmàcia. Joan XXIII, s/n. 08028 Barcelona, Spain (E-mail: *salgot@ub.edu*; *mfolch@ub.edu*; *ehuertas@ub.edu*; *antoninatorrens@ub.edu*)

***Bren School of Environmental Science and Management, University of California, 2323 Bren Hall, Santa Barbara, CA 93106, USA (E-mail: *mauset@bren.ucsb.edu*)

Abstract Infiltration percolation (IP) is an extensive technology to treat primary or secondary effluents of small and middle size communities before reuse or disposal to sensitive receiving bodies. Thirteen years of implementation of IP in Spain has highlighted the necessity to abide by defined design and construction rules and operation conditions in order to achieve consistently the treatment objectives and guarantee a long lasting treatment capacity. From this experience, high care should be taken of (i) the characteristics of the sand constituting the filter, (ii) the drainage conditions, (iii) the influent spreading over the infiltration surface and (iv) the risks related to recurrent overloading. Simple monitoring measures are suggested in order to improve the reliability of IP plants.

Keywords Clogging; infiltration percolation; overloading, porous medium; wastewater

Introduction

Infiltration percolation (IP) is an extensive treatment technology which consists essentially of intermittently infiltrating wastewater through 1.5–2.0 m deep unsaturated sand beds (Figure 1). As the mean hydraulic load can not exceed about $0.65\,m^3\,d^{-1}$ per m^2 of sand bed, the capacity of these IP systems usually does not exceed a few thousand p.e., but may reach higher values, up to 25,000 p.e. when treating secondary effluents. Larger plants would require too much filter surface and sand volume.

More than 10 infiltration percolation plants (IPP) have been constructed in Spain to treat secondary effluents prior to water reuse or disposal in nearby river beds (Table 1). The Vall-Llobrega (Girona province) IPP has been in operation since 1992; the treated water is being used for irrigation of poplars. A second plant was recently added to Sant Lluís I (Balearic Islands) IPP, in order to face the increasing demand for reclaimed water used for alfalfa irrigation and hotel dual systems. Water treated through the two filters of Torrevieja (Alicante province) is used for urban landscape irrigation and supplying ornamental impoundments. The effluent of Piera (Barcelona province) IPP is used for market gardening irrigation. Water from Els Hostalets de Pierola (Barcelona Province) will be used for agriculture as soon as permission is acknowledged by the authorities. Elda (Alicante province) IP effluent is to be directed to the agricultural water supply system, although the plant was commissioned several years ago. IPP effluents of El Biar, El Rincón de la Bonanza in Alicante province, Ojos de Garza in Canary Islands and Els Hostalets are disposed of in usually dry stream beds.

doi: 10.2166/wst.2007.149

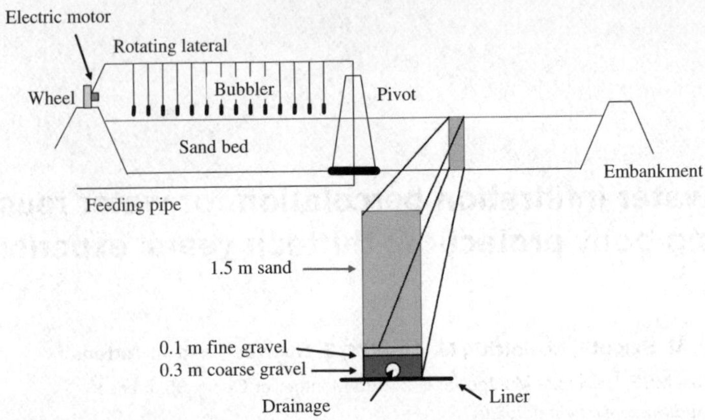

Figure 1 Typical layout of infiltration percolation systems built in Spain

This low energy consumption technology has proven to be an efficient means of reclaiming primary or secondary effluents, thus allowing water reuse and protecting sensitive receiving water bodies (Bouwer, 1996; Makni, 2001). Indeed, full size plant monitoring showed that helminth eggs are retained through pore straining in the upper layer of the filtrating beds and are totally removed (Guessab *et al.*, 1993); protozoa such as *Giardia lamblia* and *Cryptosporiidum parvum* were not detected in the filtered water of Vall Llobrega IPP by Alcalde *et al.* (2006); *Escherichia coli* content can reliably reach the less than 1000 cfu/100 mL target (Salgot *et al.*, 1996; Brissaud *et al.*, 1997; Castillo *et al.*, 2001) and monitoring performance of Piera and Vall-Llobrega IPPs exhibited several log units removal of somatic coliphages and bacteriophages RNA-F specific (Brissaud *et al.*, 1997; Alcalde *et al.*, 2006).

Laboratory investigations and field data have shown that, though IPPs are low technology systems, achieving both the designed performance, i.e. final effluent completely oxidized and efficiently disinfected, and long-lasting treatment capacity requires abiding by well defined design and construction rules and operation conditions. More than 13 years' implementation of IP systems in Spain and Mediterranean countries has brought some light on the necessity to keep to these requirements. This paper presents lessons learned from this experience.

Porous medium

The choice of the sand that will constitute the filter is a crucial step in IPP construction planning. Too high permeability due to too coarse sand would result in high percolation

Table 1 Infiltration percolation plants in Spain

Plant	Starting operation year	Useful bed surface (m^2)	Applied effluent	Capacity (m^3/day)
Vall-Llobrega	1992	575	Secondary	300
Piera	1996	428	Secondary	220
Sant Lluís I	1994	1254	Secondary	630
Sant Lluís II	2003	4955	Secondary	2300
Torrevieja I	1998	1455	Secondary	730
Torrevieja II	2002	2139	Secondary	1100
Elda	1998	4712	Secondary	2400
Els Hostalets de Pierola	1998	875	Primary	250
El Biar	1999	180	Secondary	90
El Rincón de la Bonanza	1999	1400	Secondary	700
Ojos de Garza	2001	370	Secondary	200

water velocities and, owing to oxidation and disinfection kinetics, lead to low oxidation of the treated water – unless the water is re-circulated as in trickle filters or gravel filters (Ménoret *et al.*, 2002) – and poor disinfection. On the other hand, low sand permeability together with biofilm development would reduce infiltration capacity, impede the renewal of the air phase and prevent oxygen supply, leading to anaerobic conditions which are incompatible with IP proper functioning and performance. The recommended mean grain size ranges between 200 and 800 μm with a uniformity coefficient, d_{60}/d_{10}, being less than 10. Sand may be supplied from dunes, alluvial material or quarries after crushing and sieving. Its characteristics should be closely monitored during the whole construction period.

IP is an aerobic fixed biomass process. Its treatment performances and durability rely on balanced oxygen supply and demand. Oxygen is supplied to the filter by convective and diffusive exchanges with the atmospheric air through the bed surface. Therefore, maintaining the sand bed unsaturated and the infiltration surface free of water during the main part of the operation time is crucial. Porous medium defects affect negatively the process.

Poor (or absence of) washing before supplying the sand to the construction site always results in an important deterioration of filter performance. As was observed in Torrevieja II IPP, even low fine particles content leads, when these fines are mainly clay, to a severe reduction of the treatment capabilities. The sand constituting this filter was not fully washed before the bed was laid on. Though the fraction of particles less than 20 μm did not exceed 1%, fines were put in suspension during the feeding sequences, then settled gently in the furrows modelled by the bubbler sprays on the infiltration surface, inhibiting dramatically the permeability of the upper layer of the filter, thus undermining both infiltration and oxidation capabilities (Figure 2).

Clay or silt can also enter the systems during the construction of the facilities, due to bad sand manipulation and earth contamination. When the filter starts operating, the infiltrating water flushes down the small particles, clay and silt, which migrate downward and accumulate locally and at variable depth within the sand beds, as was also observed in Torrevieja. The same process may occur when, due to operation and management deficiencies or recurrent overloading, the treatment capacity of an IP filter has been too much reduced. Indeed, the first retrofitting measure envisaged by the operator or the facility owner is often the replacement of the upper sand layers by clean new sand, as happened at Vall-Llobrega and Sant Lluís. Then, if this new sand has not been carefully chosen, it may happen that fine particles contained in the new sand are washed down to the interface between the new and the old sands, creating a low permeability layer. This phenomenon was described by Auset (2002).

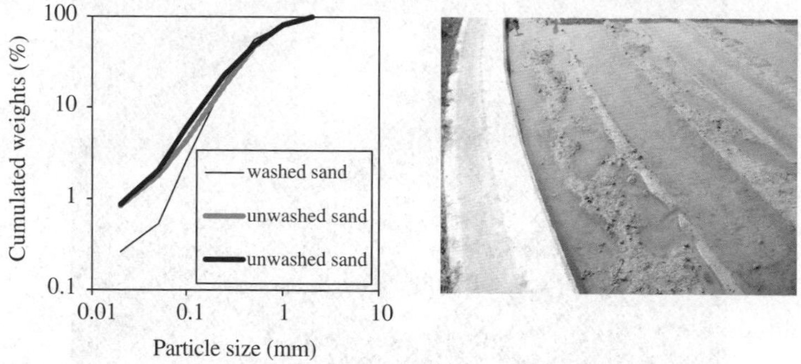

Figure 2 *Left*: particle size distribution of washed and unwashed sand of the Torrevieja II facility; *right*: flooded infiltration surface due to clay deposit

The Vall-Llobrega IP started operating in 1992, treating secondary effluent. The filter was originally constituted of dune sand with a small uniformity coefficient, $U = d_{60}/d_{10}$, equal to 2.2. After primary effluent was applied for several days in summer 1997, the filter clogged. As a significant part of the original sand had been previously lost due to different circumstances, it was decided to add about 40 cm of crushed alluvial material above the original sand. From this addition, in January 1998, the filter treated a load of $0.43\,m^3m^{-2}d^{-1}$ of activated sludge effluent, split into 10 feeding sequences a day, 5 days a week. In December 2000, a quarter of the infiltration surface was permanently flooded (Figure 3). As was also observed in other plants (Torrevieja, Sant Lluís), the persistence of impoundments at the surface of the filter had led to algae development. Degraded algae, together with organic suspended solids or clay deposits, result in a thin poorly permeable layer, which worsens the decrease of both infiltration and oxidation capacities.

The sand bed of Vall-Llobrega IPP was sampled at different depths below respectively the impounded area and the area where, conforming to the process design, water used to have completely infiltrated a short time after application. Samples were analysed for particle size distribution (Figure 4). The sand of the top 40 cm is coarser than the original dune sand and much less uniform. Below the impoundment, fines have migrated from the upper layers downward and have accumulated above the interface between the two sands. When digging into the filter, the interface layer, a few mm thin and black to dark brown, was very easily detected; its water content remained high even one week after the feeding has ceased. While most silt has been blocked at the interface, not all the clay was retained at the same depth but part of it could pass the interface and move into the dune sand. Below the well drained area, particle size distributions followed the same trend; however less fines, half the content found below the impoundments, had accumulated at the interface. When enough fine particles had accumulated at the interface, the thin impermeable layers resulting from these accumulations hindered water infiltration and gas transfer. In such circumstances, perched water saturated layers can be observed. Lasting high water saturation prevents oxygen supply, leads to local anaerobiosis, accumulation of biomass, reduction of the permeability and, eventually, permanent impoundment at the surface of the filter.

Auset (2002) could demonstrate the dramatic effect of high water saturation below the permanently flooded area. The equipment of the plant did not allow feeding of the sector containing most of the impoundment. The result was an increase from <5 to $>8\,mgL^{-1}$

Figure 3 Impoundment remaining several days after filter feeding

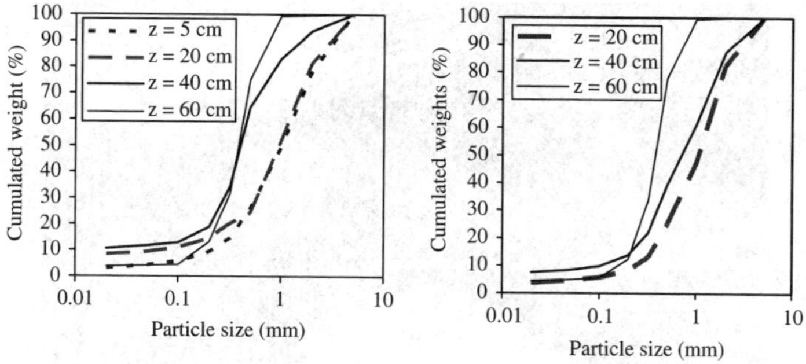

Figure 4 Particle size distribution below the impounded area (left) and the water free area (right) at Vall-Llobrega

of the dissolved oxygen content of the final effluent and an improvement from 2.4 or 2 to 4 log units of faecal coliform removal.

Design and construction

The efficiency of the treatment relies on the homogeneity of the infiltration. Uniform water spreading allows avoiding of localized overloading which would lead to clogging and performance degradation. In order to achieve uniform water distribution, all filters constructed in Spain are fed using frontal moved or centre pivot irrigation systems equipped with bubblers or special emitters designed to prevent plugging when spreading primary effluent (Figure 5).

Moreover, the infiltration surface should be maintained even; otherwise water accumulates in the depressions, leading to localized overloading, clogging risks and degraded performance. Planting turf grass on the bed surface – which is exclusively recommended for systems treating secondary effluent – helps to keep the evenness of the bed surface.

The experience is that enough care is not always taken of drainage conditions. The drainage layer, consisting of coarse gravel, should not be saturated with water and free access to the atmospheric air should be maintained. Failures of the drainage system, due to either crushing of drainage pipes during construction or to inappropriate design of drainage outlet, have resulted in the saturation of not only the bottom layer of the sand bed but, in some case, of more than half its depth; such situations were encountered at Vall-Llobrega and Els Hostalets. As not all the organic matter is fully degraded in the unsaturated part of the sand bed, the saturated bottom layers, deprived of oxygen, progressively tend to an anaerobic status. Then, over the months, anaerobic biomass develops, reducing the bed permeability. As a consequence, the water table slowly rises, worsening the situation. Digging the bed reveals sand layers with a high content of black organic matter (Figure 6). A similar situation can be observed when a soil aquifer treatment system has

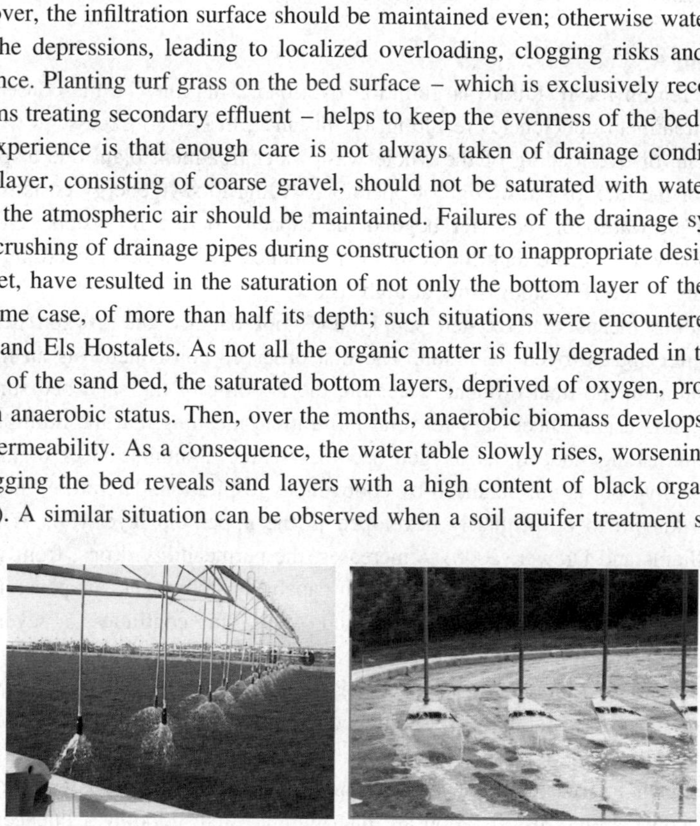

Figure 5 Emitters adopted for secondary (left) and primary (right) effluent spreading

Figure 6 Excavation in a bed clogged for misconstruction of the drainage outlet. The white line stresses the limit between the "clean" sand and the black anaerobic layer. The water of the saturated layer has invaded the bottom of the hole

been constructed upon a shallow aquifer. Then, infiltration basins may turn to leaking lagoons as was observed in the plant of Saint Gilles in La Réunion Island (France).

Hurdles to infiltration – such as manholes set in the sand bed area, as in Gabès-Tunisia pilot IPP – and every cause of preferential pathway, for instance irrigating the area above the slanting sidewall of the filter, result in lower disinfection efficiency.

Overloading

A frequent reason for IP failure is pollution overload. This may be the consequence of secondary treatment deficiencies resulting, for instance, in N-NH_4 excess, as was noticed in Torrevieja, of overloading of the whole wastewater treatment plant due to population growth or of the lack of knowledge of operators willing to save energy costs or increase the volume of water for reuse far beyond the capacity of the IP system. Overloading means not only that before long the design performance will not be reached, but also that the durability of the IP system is put at high risk.

Overloading means that oxygen supply does not balance the oxygen demand for organic matter and nitrogen oxidation. The first observed consequence is an increase of N-NH_4 content in the final effluent. Later on, the flooding of the sand bed surface lasts beyond the time usually necessary for total infiltration, meaning that the infiltration capability is diminishing. Indeed, as oxygen becomes scarce, assimilation of organic matter remains effective but is not balanced by endogenous respiration. Therefore, the biomass – which is known to be primarily developed in the upper top layer (Fox *et al.*, 2005; Rauch-Williams and Drewer, 2006) – increases, the permeability drops, from which follows the observed decrease of the infiltration capability; moreover, air phase transfer is hindered and the oxygen supply reduced. When overloading continues for several weeks, severe internal clogging occurs because the biomass evolves in anaerobic conditions, leading to organic compounds resistant to degradation. Then, and this is the worst situation for an IP system, removing the clogged sand bed may be necessary though very costly.

Less frequent internal clogging of the upper layers of the filter may result from the feeding schedule. Filters in operation are intermittently fed, flooding sequences alternating with drainage sequences, the daily number of feeding-drainage cycles, *f*, ranging between 1 and 20. It has been demonstrated that the higher the f value the higher the removal of faecal indicators (Brissaud *et al.*, 1999). However as shown by Bancolé *et al.* (2003), high fractionation of the daily load bears a major adverse effect. While, for low f values, the biofilm develops evenly over the whole depth of the bed, it accumulates in the upper layers at high f values. As the biomass increases, the hydraulic conductivity diminishes, reducing infiltration velocities and threatening the oxygen supply. Too high fractionation of the daily load threatens the process sustainability.

Conclusion

Deciphering the reasons for the failure or the defective functioning of IPPs is never easy; indeed, most of the design, construction and management defects affect the process in the same manner, i.e. performance degradation and clogging manifested through reduction of the infiltration capacity and lasting bed surface flooding. The most efficient and cost-effective approach has been to carry out autopsies of the failed filters, digging holes in order to observe bed vertical profiles. Variations of sand colour – unless the sand was dark black, as happens in a volcanic environment – consistency and water content are helpful clues to discover malfunctioning reasons. Analysing samples for particle size distribution, organic matter and water contents will provide useful complementary information.

Monitoring using as appropriate outfit should allow preventing malfunctioning and failure. The process deterioration manifests always in the filters as an increase of water content, due to permeability reduction, or of biomass as a result of oxygen need and supply imbalance. As 90% of the biomass is water, a close monitoring of water content of a few vertical profiles would provide the data required for a safe operation of the IP plants. Water content can be easily monitored through the measure of porous bed electrical conductivity. A piezometer should also be installed in the drainage layer in order to monitor the water level in the drainage layer.

Despite these above mentioned constraints, thanks to low O&M costs and high performance, IP, used alone or combined with UV (Salgot *et al.*, 2002) or another disinfection process, appears to be a very convenient wastewater reclamation technique for serving small and middle size communities or when dealing with sensitive areas, where classical secondary effluent cannot be disposed of without good treatment. In the Mediterranean, water flowing in temporary streams is usually re-used in a non-planned way, which makes it sensible to treat such water to high quality standards.

References

Alcalde, L., Folch, M., Tapias, J.C., Huertas, E., Torrens, A. and Salgot, M. (2007). Wastewater reclamation systems in small communities. *Wat. Sci. Tech.*, **55**(7), 149–154.

Auset, M. (2002). *Approche des mécanismes de décontamination en infiltration percolation* PhD Th., University Montpellier II.

Bancolé, A., Brissaud, F. and Gnagne, T. (2003). Oxidation processes and clogging in intermittent unsaturated infiltration. *Wat. Sci. Tech.*, **48**(11/12), 139–146.

Bouwer, H. (1996). Issues in artificial recharge. *Wat. Sci. Tech.*, **33**(10–11), 381–390.

Brissaud, F., Salgot, M., Foch, M., Campos, C., Blasco, A. and Gómez, D. (1997). Full scale evaluation of infiltration percolation for polishing secondary effluents. *Beneficial Reuse of Water and Biosolids, Water Environment Federation* Apr. 6th–9th, Marbella-Spain: 10/48-56.

Brissaud, F., Salgot, M., Bancolé, A., Campos, C. and Folch, M. (1999). Residence time distribution and disinfection of secondary effluents by infiltration percolation. *Wat. Sci. Tech.*, **40**(4/5), 215–222.

Castillo, G., Mena, M.P., Dibarrart, F. and Honeyman, G. (2001). Water quality improvement of treated wastewater by intermittent soil percolation. *Wat. Sci. Tech.*, **43**(12), 187–190.

Fox, P., Aboshanp, W. and Alsamadi, B. (2005). Analysis of soils to demonstrate sustained organic carbon removal during soil aquifer treatment. *J. Environ. Qual.*, **34**(1), 156–163.

Guessab, M., Bize, J., Schwartzbrod, J., Maul, A., Nivault, N. and Schwartzbrod, L. (1993). Wastewater treatment by infiltration percolation on sand: results in Ben Sergao Morocco. *Wat. Sci. Tech.*, **27**(9), 91–95.

Makni, H. (2001). Disinfection of secondary effluents by infiltration percolation. *Wat. Sci. Tech.*, **43**(12), 175–178.

Ménoret, C., Boutin, C., Lienard, A. and Brissaud, F. (2002). Use of recirculation through medium size granular filters to treat small food processing industry effluents. *Wat. Sci. Tech.*, **45**(12), 225–232.

Rauch-Williams, T. and Drewes, J.E. (2006). Using soil biomass as an indicator for the biological removal of effluent-derived organic carbon during soil infiltration. *Wat. Res.*, **40**(5), 961–968.

Salgot, M., Brissaud, F. and Campos, C. (1996). Disinfection of secondary effluents by infiltration-percolation. *Wat. Sci. Tech.*, **33**(10-11), 271–276.

Salgot, M., Folch, M., Huertas, E., Tapias, J., Avellaneda, D., Girós, G., Brissaud, F., Vergès, C., Molina, J. and Pigem, J. (2002). Comparison of different advanced disinfection systems for wastewater reclamation. *Wat. Sci. Tech: Wat. Supply*, **2**(3), 213–218.

Treatment of organic synthesis wastewater using anaerobic packed bed and aerobic suspended growth bioreactors

P. Mijaylova-Nacheva*, E. Ramírez-Camperos* and S. Cuevas-Velasco**

*Mexican Institute of Water Technology, Paseo Cuauhnáhuac 8532, Col. Progreso, 62550 Jiutepec, Morelos, Mexico (E-mail: *petiam@tlaloc.imta.mx*)

**Technological Institute of Tzacatepec, Calzada Tecnológico 27, 62780 Zacatepec, Morelos, Mexico

Water Science & Technology **Vol 55 No 7 pp 235–243** © IWA Publishing 2007

Abstract The performance of an anaerobic mesophilic packed bed reactor, with a mixture of GAC and tezontle, followed by an aerobic suspended growth system was studied for the treatment of organic chemical wastewater with a high COD concentration (22–29 g/L). The testing of the anaerobic–aerobic system was conducted in an experimental set-up for almost 2.5 years. Different operational conditions were evaluated. The anaerobic reactor showed performance stability and COD removals higher than 80% were obtained with loads up to 16.6 kg.m^{-3}.d^{-1}. The acclimation of the aerobic biomass to the substrate in the anaerobic effluent was very quick and COD removals higher than 94% were obtained even at high organic loads. The combined anaerobic–aerobic system allowed total COD removals higher than 99.5% and the accomplishment of the discharge requirements of 200 mgCOD/L when the anaerobic reactor was operated with loads of 8–11 kg.m^{-3}.d^{-1} and the aerobic reactor with 0.33 kg.kg^{-1}.d^{-1}, being the total HRT of 4.4. The average TKN removal in the anaerobic–aerobic system was 97%, the average for the anaerobic reactor being 52% and that one for the aerobic system being 94%.

Keywords Anaerobic–aerobic treatment; industrial wastewater; packed bed with GAC and tezontle

Introduction

The organic chemical industry's plants in Mexico are usually small sized. They generate small quantities of wastewater, which are however highly polluted. In addition, they are usually located far from centralized wastewater treatment facilities and on-site effluent treatment is required. Such was the case of an organic synthesis plant for fragrances and savor products with a wastewater flow of almost 25 m^3/d. The wastewater had a complex composition and high concentration of organic matter, average of 25,540 mg/L of COD. The aim was to develop a suitable system for the wastewater treatment, reaching the discharge requirements of 200 mgCOD/L. Preliminary studies indicated that both aerobic and anaerobic biodegradation with suspended growth biomass can be used for the treatment of this effluent (Nacheva *et al.*, 2002), however the anaerobic system alone did not allow the accomplishment of the discharge limits. Because of the low TSS concentration and because of the presence of VOCs in the wastewater, anaerobic biofilters with granulated activated carbon (GAC) and with tezontle (volcanic porous stone widely available in Mexico) were also tested (Nacheva *et al.*, 2003). GAC-reactors showed high effectiveness and stability, in concordance with the reported advantages of using GAC as biofilm attachment media for treatment of toxic or inhibitory wastewater (Suidan *et al.*, 1993; Kupferle *et al.*, 1995; Monteleb *et al.*, 2001). The tezontle was also shown to be a good packed bed media, but process effectiveness and stability were lower. Additional treatment was needed to reach the limits for discharge at economically reasonable organic loads in both cases. Thus a combination of anaerobic–aerobic treatment was presented, considering the successful use and the economical benefits of this system reported for

doi: 10.2166/wst.2007.150

many industrial effluents (Van Lier *et al.*, 2001; Cruz and Buitrón, 2001; Kalyuzhnyi *et al.*, 2003). The objective of this study was to determine the capacity of a combined system of an anaerobic mesophilic packed bed reactor, with a mixture of GAC and tezontle, followed by aerobic suspended growth bioreactor for the treatment of the organic synthesis effluent, allowing in this way a compact highly effective treatment plant.

Methods

Experimental set-up

The experimental set-up consisted of one anaerobic packed bed reactor, followed by an aerobic activated sludge system. Figure 1 shows the diagram of the experimental set-up. The anaerobic reactor was a jacketed acrylic column (10.3 cm diameter and 1.60 m high) with an effective volume of 10 L. Hot water was recycled through the jacket, which kept the reactor temperature at 35°C. Packed media, used in the anaerobic reactor, consisted of 50% mineral GAC (Clarimex CAGR 8 × 30) and 50% red tezontle. Both materials, GAC and ground tezontle were separately sieved and the fraction with a grain size of 2.00–2.38 mm was selected. The GAC apparent density was 0.439 t/m^3 and that of tezontle was 0.472 t/m^3. Bed media voidage of the mixture was 49.7% and bed pore volume was 1.1 ml/g. The anaerobic reactor was operated in up-flow mode, with recycling of 50% of the effluent. The biogas produced in the reactor was received in an inverted acrylic tube collector with alkaline solution. The aerobic reactor was also cylindrical but with a volume of 5 L. Air was provided using a porous air diffuser. The mixed liquor from the reactor was directed to an external settler with a volume of 1 L. The sludge recycling was 50%. The aerobic system was operated at ambient temperature.

Wastewater and nutrients

The wastewater under study resulted from an organic synthesis plant for fragrances and savor products. The effluent from the storage tank was used for feeding of the experimental reactors. During the study the COD concentration varied in the range of 22,120–29,540 mg/L. Basic organic constituents were: alcohols, aldehydes, ketones and

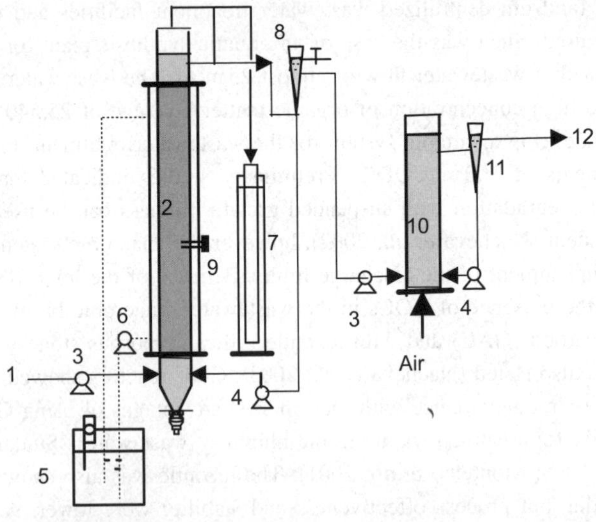

Figure 1 Schematic diagram of the experimental set-up for anaerobic–aerobic treatment: 1 – influent; 2 – anaerobic packed bed reactor; 3 – peristaltic pump; 4 – recycle pump; 5 – heating water bath with thermostat; 6 – hot water recycling pump; 7 – gas collector; 8 – controller of the effluent from the anaerobic reactor; 9 – sampler; 10 – aerobic suspended growth reactor; 11 – settler of the aerobic system; 12 – aerobic effluent

amines. Some toxic organics were determined to be present, such as toluene (407–696 mg/L) and phenol (102–172 mg/L). Nitrogen concentration was high (TKN of 950–1900 mg/L), most of it being organic nitrogen (640–1370 mg/L). Phosphates ranged from 0.1 to 4.0 mg/L. Total suspended solids were between 150 and 180 mg/L. The pH was neutral and the conductivity was 1.3–2.2 mS/cm. Phosphorus and micronutrients were added during all the experimentation. The ratio COD:N:P was controlled to be 450:7:1. The addition of Fe, Ni, Co and Mo was made based on the following rates: 1, 0.15, 0.1 and 0.002 mg/gCOD respectively. The micronutrients Zn, Mn and Cu were added at 2, 0.7 and 0.3 μg/gCOD. Addition of Na, K, Ca and Mg was based on 5, 20, 13 and 10 mg/gCOD respectively.

Experimental procedures and analytical methods

Inoculation of the anaerobic reactor was done with sludge taken from an UASB reactor for domestic wastewater treatment. The inoculums were added to the experimental reactor together with the mixed support materials. The feeding began at a volumetric organic load (VOL) of 0.2 kg $COD.m^{-3}.d^{-1}$, practicing dilution with household wastewater (80%) and yeast addition (0.01 $g.m^{-3}.d^{-1}$) to enhance the biofilm development. Gradual VOL increasing was carried out, evaluating the process performance, the biomass transformation and the biofilm development. During the start-up, dilution and yeast addition were progressively reduced and finally eliminated. The duration of the start up was almost 2 months. After this, the biofilter performance was studied over a wide range of organic loading conditions (Table 1). The aerobic system was connected after the stabilization of the anaerobic process. The inocula were taken from an activated sludge municipal treatment plant. They were submitted to acclimation to the anaerobic effluent applying low organic load (F/M), with a semi-continuous operation and using only 1 L of the reactor volume. The biomass was allowed to grow without sludge purges. During the next stages, the effective volume, the biomass concentration and F/M were gradually increased. The operational parameters for all experimental stages are indicated in Table 1. The flow rate to the aerobic system ($Q_{i\,aer}$) was reduced to 50% of the flow rate to the anaerobic reactor ($Q_{i\,an}$) during stages 7 and 8, and to 20% of the $Q_{i\,an}$ during stage 9. This was done to avoid organic overloads in the aerobic system.

The COD concentrations were determined in the influent, in the anaerobic and in the aerobic effluents daily. Temperature, pH and biogas generation were also monitored in both reactors daily. Biogas composition was determined periodically by gas chromatography using a thermal conductivity detector. Volatile fatty acids (VFA) were controlled weekly based on alkalinity determination at pH of 4.3 and 5.75. Gas chromatography analysis for VFA, using a flame ionization detector, were performed periodically. SST and SSV in the effluent were determined once a week. The removals of TKN and of NH_4-N were determined for each operational phase. Biomass quantification in the fixed bed was conducted at the end of each experimental phase. Samples of 30–50 ml were taken from the middle of each column; they were first added to distilled water in order to remove unattached particles and then transferred to a salinity solution, where biomass was detached mechanically. Then VSS were determined in both solutions and the biomass was calculated as mg VSS per litre of liquid phase. The biomass in the aerobic reactor was determined weekly. All analytical measurements were done according to *Standard Methods for Examination of Water and Wastewater* (1998). Micrographs of grains from the anaerobic reactor were done using a scanning electron microscope, JSM-6400 Noran.

Table 1 Operational parameters of the anaerobic–aerobic treatment system

Stage	Period, months	Anaerobic system				Aerobic system					
		$Q_{i\,an}$, L/d	Average COD_{ian}, g/L	Average VOL, kg.m^{-3}.d^{-1}	HRT_{an}, days	$Q_{i\,aer}$, L/d	V_{aer}, L	Average MLVSS, mg/L	Average F/M, kg.kg^{-1}.d^{-1}	HRT_{aer}, days	Average T_{aer}, °C
1	3 to 6	0.5	23,510	1.18	20.0	0.5	1	1182	0.149	2.0	25.3
2	7 to 9	0.7	23,503	1.64	14.3	0.7	2	1532	0.15	2.9	26.1
3	10 to 12	1.1	23,528	2.59	9.1	1.1	2	1519	0.247	1.8	26.8
4	13 to 15	1.5	25,055	3.76	6.7	1.5	2	2544	0.318	1.3	28.1
5	16 to 18	2.2	24,721	5.44	4.6	2.2	5	2087	0.312	2.3	27.9
6	19 to 21	3.4	23,515	8.00	2.9	3.4	5	4099	0.324	1.5	27.8
7	22 to 24	4.5	25,238	11.36	2.2	2.25	5	4176	0.326	2.2	27.9
8	25 to 27	5.7	29,136	16.61	1.8	2.85	5	4207	0.75	1.8	26.5
9	28 to 30	10.0	24,216	24.22	1.0	2.0	5	4224	0.848	2.5	27.9

Results and discussion

Start-up of the anaerobic reactor

The anaerobic packed bed reactor reached COD removal of 80%, biodegradation rate of 0.83 kgCOD.m^{-3}.d^{-1} and biogas production of 0.31 m^3/kg COD_{rem} after 56 days of operation (Figure 2). The increases of the organic load produced process destabilizations which were less marked over time. The most sensible parameter was the biogas production which was affected at each load increase and 5–10 days were needed for its recovery. Methane content in the biogas was determined of 69% at day 55 from the start-up. The alkalinity increased from 540 mg $CaCO_3$/L to 1680 mg $CaCO_3$/L and the pH from 6.9 to 7.5. The concentrations of the VFA were 50–100 mg/L.

Important changes in the biomass structure were observed during the start-up period. The inoculum's biograins were destroyed, CAG and tezontle grains were covered with a bio-film (Figure 3) and some small not attached bioflocs were formed in the void spaces of the packed bed. A large proportion of thin flocks were going out of the reactor with the effluent, causing an increase of the effluent TSS from 114 mg/L (at the start-up) to 330 mg/L (day 45). Nevertheless the biomass quantity in the reactor increased from 1.75 kg/m^3 (at the start-up) to 2.53 kg/m^3 (day 60).

Analysis of the process performance

During the first experimentation stage (days 63–182 from the start-up) the average VOL was 1.18 kgCOD.m^{-3}.d^{-1} in the anaerobic reactor. More than 98.5% COD removal was reached on day 150 and maintained until day 182. The biogas production increased to

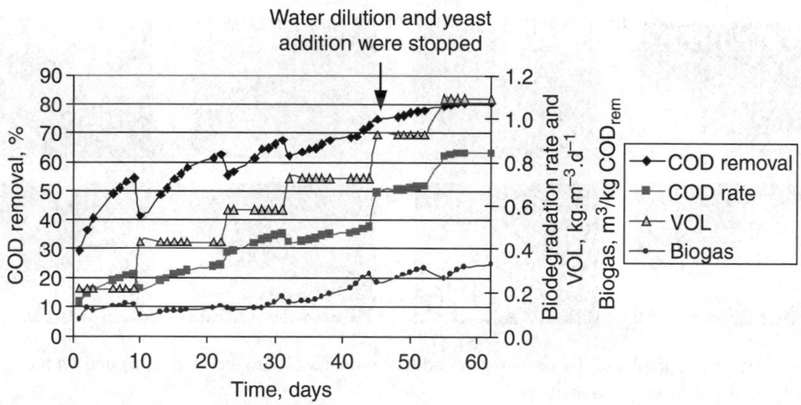

Figure 2 Biofilter performances during the start-up

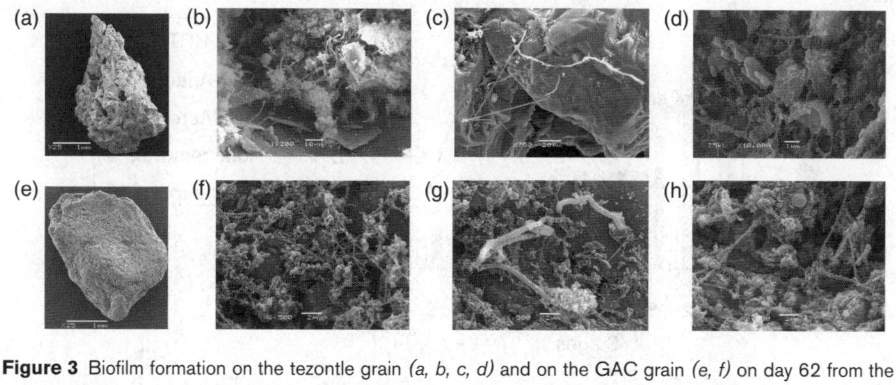

Figure 3 Biofilm formation on the tezontle grain *(a, b, c, d)* and on the GAC grain *(e, f)* on day 62 from the start-up

$0.48\,m^3/kg\ COD_{rem}$, methane was 66–68%. The average of the alkalinity was 1680 mg $CaCO_3/L$ and the pH was in the range 7.4–7.6. The concentrations of the VFA were between 168 and 214 mg/L. Biomass concentration in the reactor reached $5.58\,kg/m^3$ at day 180 from the start-up, the not attached VS being almost 30% of the total biomass in the biofilter. Biofilm was already dense and recovered completely the grain surfaces (Figure 4). The TSS concentration in the effluent was 305 mg/L on average.

The aerobic reactor was connected on day 126 from the start-up of the anaerobic biofilter, when the anaerobic process was already stable, with 91% COD removal and with high biogas production. The initial MLVSS was 1000 mg/L. The acclimation was relatively quick, effluent COD of 78 mg/L was obtained in one month of operation. The load increases done all through the experimentation did not affect significantly the process performance. Only slow efficiency reductions were observed, which have been overcome in 1–3 days. The average COD removals per experimental stage are presented on Figure 5, indicating as well the hydraulic residence time (HRT) applied in both systems. The remaining results that indicate the performance of the anaerobic packed bed reactor are summarized in Table 2. Figure 6 shows the correlations between organic loads, removal efficiencies and effluent concentrations in the anaerobic reactor and in the aerobic system. As can be seen, the anaerobic–aerobic treatment system allowed COD removal efficiency higher than 98% during all the experimentation (Figure 5). The increasing of the organic load caused efficiency reduction in the anaerobic reactor (Figure 6a), which was successfully compensated by the aerobic system (Figures 5 and 6b). The anaerobic reactor did not allow the accomplishment of the requirements for discharging the wastewater

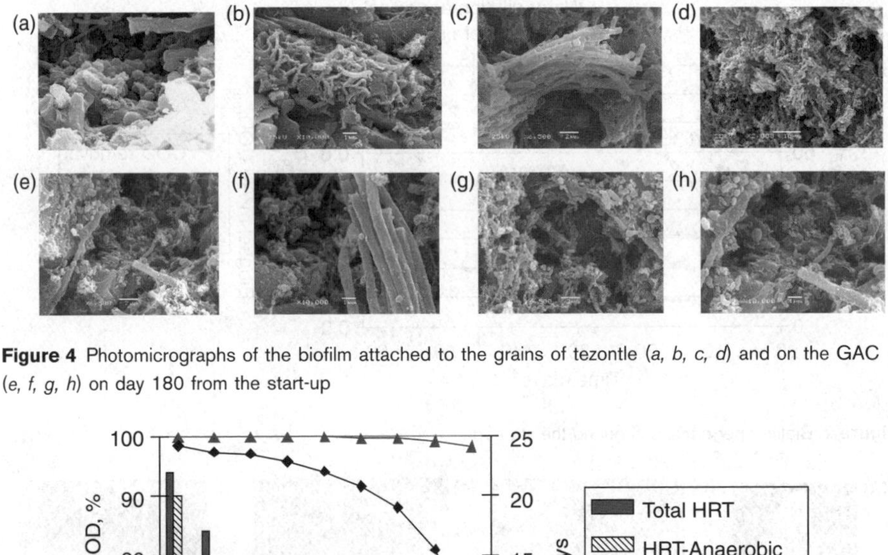

Figure 4 Photomicrographs of the biofilm attached to the grains of tezontle (*a, b, c, d*) and on the GAC (*e, f, g, h*) on day 180 from the start-up

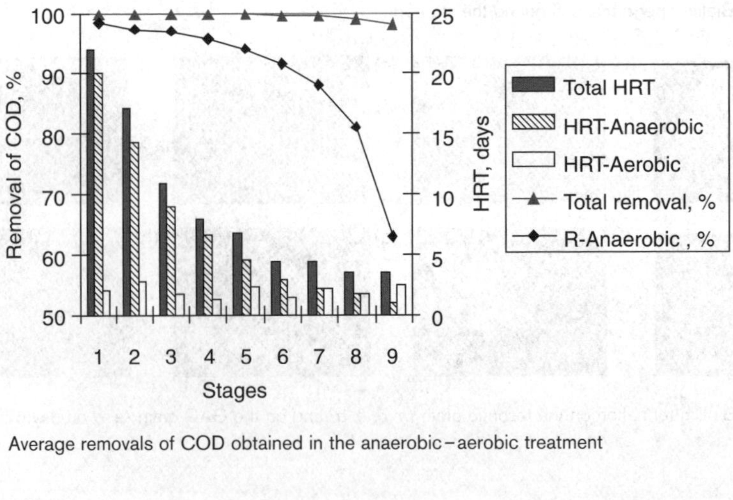

Figure 5 Average removals of COD obtained in the anaerobic–aerobic treatment

Table 2 Average results per experimental stage obtained in the anaerobic packed bed reactor

Stage	Alkalinity, $mgCaCO_3/L$	VFA, mg/L	pH	Biogas, $m^3/kgCOD_{rem}$	Methane, %	Biomass, $kgVS/m^3$	Biomass retention time, d
1	1680	193	7.5	0.44	66.7	4.12	279
2	2130	95	7.6	0.47	67.2	7.51	292
3	2703	108	7.7	0.48	67.9	11.29	106
4	2175	115	7.6	0.48	68.2	11.39	64
5	1988	117	7.5	0.49	69.1	11.59	32
6	1725	138	7.3	0.52	69.5	11.62	29
7	1507	175	7.2	0.54	70.0	11.87	27
8	1350	223	7.2	0.44	66.2	11.97	24
9	1078	290	7.0	0.39	64.5	12.22	22

(200 mg/L) even at the lowest organic loads (Figure 6a). The connection of the aerobic system allowed the accomplishment until stage 7 (F/M of 0.33 $kg.kg^{-1}.d^{-1}$). During the last two stages, despite the good performance of the aerobic system (94–95% COD removal), the total removal decreased and COD concentrations higher than 200 mg/L were determined in the aerobic effluent (Figure 6b).

Nitrogen removal is illustrated on Figure 7. Most of the organic nitrogen was removed in the anaerobic reactor (Figure 7a); the efficiencies varied between 86 and 94% with an average of 91%. Additional removal was obtained in the aerobic system (Figure 7b), thus the average organic nitrogen removal in the anaerobic–aerobic system was 97.6%. Most of the NH_3-N was removed in the aerobic system (average of 97%). Ammonia concentration in the anaerobic effluent was almost 60% higher than that in the raw wastewater due to the ammonia release during the biodegradation of nitrogenated organic compounds. The average ammonia removal for the anaerobic–aerobic system was almost 95%. The average TKN removal of the anaerobic–aerobic system was 97% (averages of 52% in the anaerobic reactor and 94% in the aerobic system).

Anaerobic reactor. The anaerobic reactor showed stable performance and good biogas production (0.33–0.57 m^3/kg COD_{rem}), at pH of 6.9–7.8, VFA of 70–390 mg/L and methane content in the biogas of 62–73%. COD removals higher than 90% were obtained with loads up to 8.0 $kg.m^{-3}.d^{-1}$. Toluene and phenol were not detected in the anaerobic effluent. The methane production yield was 0.326 $m^3CH_4/kgCOD_{rem.}$ during the operation at VOL up to 8.0 $kg.m^{-3}.d^{-1}$. Inhibition of the methane production was observed at VOL of 16.6 $kg.m^{-3}.d^{-1}$ (stage 8). In spite of this, the anaerobic reactor allowed 80% COD removals with a load of 16.6 $kg.m^{-3}.d^{-1}$, while this result was possible with VOL up to 4.45 $kg.m^{-3}.d^{-1}$ in biofilters with tezontle alone. TSS in the anaerobic effluent were 235–395 mg/L. The biomass growth was relatively low, 0.028–0.032 $kgVS/kgCOD_{rem}$. Biomass extractions were not practiced until day 330 (stage 3), when the accumulated solids began to cause clogging effects and gas bubble trapping in the bed. TS and VS were 17.4 kg/m^3 and 13.91 kg/m^3 respectively at that time. The proportion of not attached biomass trapped in the media was almost 41%. Bed washing was practiced and the biomass was adjusted to 9.8 $kgVS/m^3$, reducing this way the proportion of not attached biomass to almost 27%. The washing procedure did not affect the process performance. The second washing was applied at the end of the stage 3 and the solids were reduced from 12.8 to 10.1 $kgVS/m^3$. Subsequent washings were done each 20 d during stage 4, each 10 d during stages 5, 6 and 7, and each 8 d during the last two stages. Treated water was used, injecting it in the bottom of the reactor for 5–7 min at a pressure which allows 15% bed expansion. During stages 5 to 9 the biomass

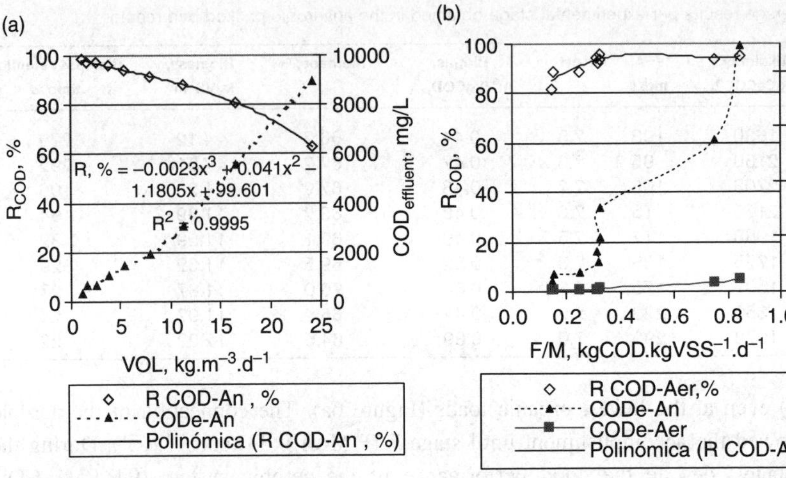

Figure 6 Correlations between the organic loads and the COD removal efficiencies in the anaerobic packed bed reactor *(a)* and in the aerobic suspended growth system *(b)*

retention times were 22–32 d. The optimum biomass concentration in the anaerobic reactor was determined to be between 10 and 13 $kgVS/m^3$.

Aerobic system. The aerobic reactor began to work at its full volume and in continuous regime from the beginning of stage 5; with an influent COD concentration of 1320 mg/L, having MLVSS of 2 kg/m^3, at F/M of 0.31 $kgCOD.kgVSS^{-1}.d^{-1}$ and HRT of 2.27 d. Effluent COD concentration (CODe) of 82 mg/L was obtained. From the beginning of stage 6, the biomass concentration was progressively increased up to 4 kg/m^3, maintaining almost the same F/M as in the previous stage, but with lower HRT (1.47 d). The average COD removal was 94%. During stage 7, the influent COD concentration was 50% higher, the same biomass concentration was maintained, the HRT was increased to 2.22 d to have the same F/M as in the previous stage. The removal increased, the average CODe was 121 mg/L. The anaerobic effluent had high COD concentrations during the last two stages, thus high organic loads were also tested in the aerobic system. The average removal efficiencies were 94 and 95% for stages 8 and 9 respectively. The biodegradation rates increased 2.2–2.6 times with respect to stages 5, 6 and 7, but the CODe rose, having averages of 332 and 448 mg/L for stages 8 and 9 respectively. The kinetic coefficients for the aerobic degradation were determined assuming a first order model and Monod relationship and considering the obtained data during the experimental stages 4 to 8: biodegradation rate coefficient (k) of 0.932 kg

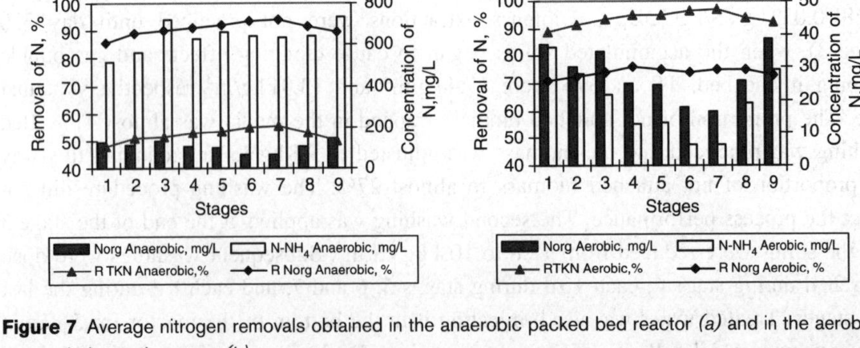

Figure 7 Average nitrogen removals obtained in the anaerobic packed bed reactor *(a)* and in the aerobic suspended growth system *(b)*

COD.kgVSS^{-1}.d^{-1}, half-velocity coefficient (K_s) of 199 mg/L, cell yield coefficient (Y) of 0.257 g VSS/g COD_{rem} and endogenous decay coefficient (k_d) of 0.010 d^{-1}. Compared with the kinetic coefficients obtained in the aerobic treatment when used as first treatment of the wastewater (Nacheva *et al.*, 2002), the values determined in this study of k, K_s and k_d are lower (26, 49 and 64% respectively). The value of Y, however, was double which means that the anaerobic pre-treatment reduced the inhibition of the aerobic biomass growth caused by the toxic compounds present in the wastewater and allowed operation at higher organic loads and shorter hydraulic and cell retention times.

Conclusions

The combined system of mesophilic anaerobic packed bed reactor, with a mixture of GAC and tezontle, followed by aerobic suspended growth biomass reactor is an effective technology for the treatment of the organic synthesis wastewaters with high COD, of 22–29 g/L, which contains xenobiotic compounds. The combined GAC-tezontle packed bed makes possible 3.7 times higher degradation rates compared with tezontle alone, allowing in this way a more compact and effective treatment plant. Removals of COD higher than 99.5% can be obtained operating the anaerobic bio-filter with VOL of 8–11 kg.m^{-3}.d^{-1} and the aerobic reactor with F/M of 0.33 kg.kg^{-1}.d^{-1}, at in total HRT of 4.4 days.

References

Cruz, A. and Buitrón, G. (2001). Biodegradation of disperse blue 79 using sequenced anaerobic/aerobic biofilters. *Wat. Sci. Technol.*, **44**(4), 159–166.

Kalyuzhnyi, S., Gladchenko, M. and Epov, A. (2003). Combined anaerobic-aerobic treatment of landfill leachates under mesophilic, submesophilic and psychrophilic conditions. *Wat. Sci. Technol.*, **48**(6), 311–318.

Kupferle, M.J., Tsaichu, C., Gallardo, V.J. and Lindberg, D.E. (1995). Anaerobic pretreatment of hazardous waste leachates in publicly owned treatment works. *Water Environment Research*, **67**(6), 221–230.

Monteleb, M.A., Cuidan, M.T., Kim, J., Davel, J.L. and Adrian, N.R. (2001). Anaerobic degradation of 2, 4, 6-trinitrotoluene in GAC fluidized bed and batch reactors. *Wat. Sci. Technol.*, **43**(1), 66–75.

Nacheva P.M., Peña-Loera B. and Cuevas-Velasco S. (2003). Anaerobic treatment of organic synthesis wastewater using bio-filters. *Preprints IWA International Conference on Sustainable Development*, Cape Town, South Africa.

Nacheva P.M., Ramírez-Camperos E., Mendoza-Roque J. and Rocha M.I. (2002). Comparison of Aerobic and Anaerobic Systems for the Biotreatment of Synthetic Perfume Industry Wastewater. *Preprints of the 3rd IWA World Water Congress*, Melbourne, Australia.

Standard Methods for the Examination of Water and Wastewater (1998). 20th edition, American Public Health Association/American Water Works Association/Water Environment Federation, Washington DC, USA.

Suidan, M.T., Schoeder, A.T., Nath, R., Krishnan, E.R. and Brenner, R.C. (1993). Treatment of leachates by carbon-assisted anaerobic fluidized beds. *Water Sci. Technol.*, **27**(2), 273–282.

Van Lier, J.B., Tilche, A., Ahring, B.K., Macarie, H., Moletta, R., Dohanyos, M., Hulshoff Pol, L.W., Lens, P. and Verstraete, W. (2001). New perspectives in anaerobic digestión. *Wat. Sci. Technol.*, **43**(1), 1–18.

Biological nitrogen removal with three different SBBR

A. Valdivia*, S. González-Martínez* and P.A. Wilderer**

*Institute of Engineering, National University of Mexico, Ciudad Universitaria, 04510 Mexico DF, Mexico (E-mail: *sgm@pumas.iingen.unam.mx*)

**Institute of Water Quality Control and Waste Management, Technical University of Munich, Am Coulombwall, 85748 Garching, Germany (E-mail: *peter@wilderer.de*)

Abstract In order to compare the performance of biofilms growing on different support media, three reactors were fed with municipal wastewater from the city of Garching, Germany, and operated under the sequencing batch procedure. The support media tested have the commercial names of Kaldnes, polyethylene special support for moving bed reactors with approximate diameter of 12 mm; Liapor, ceramic spheres with diameters between 4 and 6 mm; Linpor, plastic foam cut in cubes of 15 mm. The bench-top reactors were tested for COD, TSS and ammonia nitrogen removal. During 452 days runs with organic loads between 0.5 and 8.0 gCOD/m^2·d were tested. Thin biofilms (Kaldnes and Liapor) perform better for COD and ammonia removal under lower organic loading values (< 2.5 gCOD/m^2·d). For organic loads over 3.0 gCOD/m^2·d, the reactor packed with Linpor (thick biofilm) showed a better COD and ammonia nitrogen removal than the other two. Linpor achieved the highest NOx-N production reaching values between 15 and 20 mg/l. For low organic loading rates Linpor and Liapor present similar average NOx-N concentrations. Kaldnes shows the lowest concentrations throughout the whole experimental period. The difference between ammonia nitrogen removal and NOx-N generation is simultaneous denitrification inside the deep biofilms. The average mean cellular retention times were 5.4 days for Liapor, 10.0 days for Kaldnes and 22.9 days for Linpor. This is the reason why Linpor achieved complete nitrification even with higher organic loads.

Keywords Biofilm; denitrification; diffusion; nitrification; packed-bed reactors

Water Science & Technology **Vol 55 No 7 pp 245–254** © IWA Publishing 2007

Introduction

It has been shown that simultaneous nitrification and denitrification occurs during aeration due to anoxic zones in biofilms (Garzón-Zúñiga and González Martínez, 1996; Arnz *et al.*, 2001; Gieseke *et al.*, 2002). Some authors mention the probable formation of micro niches with anoxic conditions in the deep layers of biofilms where denitrifying bacteria take advantage of nitrates coming from the superior layers where nitrifying activity is observed (Pastorelli *et al.*, 1997; Helness and Odegaard, 1999). To study the activity of thin and thick biofilms during the simultaneous removal of organic substances and inorganic nitrogen compounds from municipal wastewater, the performance of three laboratory sequencing batch biofilm reactors (SBBR) was analysed.

Method

Three different commercial biofilm support media were selected and used to pack three sequencing fixed-bed batch reactors. These biofilm support media have been used for the treatment of different wastewater types with positive results. Every one of the three reactors was packed with different support material. Table 1 presents the main characteristics of the biofilm support media. Three pilot reactors, fed with the municipal wastewater of the city of Garching, Germany, were operated for a period of 452 days. The reactors were made of Plexiglas 19 cm in diameter and 110 cm height (Figure 1). Its dimensions and influent conditions are given in Table 2 and 3, respectively. They were operated under the sequencing batch procedure with 8-hour cycles: 30 min fill, 120 min anaerobic,

doi: 10.2166/wst.2007.151

Table 1 Characteristics of the support media

Parameter	Kaldnes®	Liapor®	Linpor®
Size (mm)	10 × 7	4–8	15 × 15 × 15
Specific surface area (m^2/m^3)	500	480	270
Density (g/cm^3)	0.17	1.4	0.018
Bed height (m)	0.17–0.33	0.16–0.33	0.32–0.52
Apparent porosity (%)	44	38	58
Total surface area (m^2)	2.7–8.3	2.7–8.1	2.7–3.9

300 min aerobic, 30 min draw. A heat exchanger was necessary to keep the temperature in the reactor approximately at 25 °C. The reactors were equipped with four sample ports to take samples of the biofilm support material. Backwashing was performed every 48 hours. Every cycle 100% of the water was replaced with fresh wastewater.

During the 452 days of the experimental procedures the organic load was varied between 0.5 and 8.0 gCOD/m^2·d. Periodical sampling of influent and effluent was made for COD, suspended and volatile solids, inorganic nitrogen compounds and Kjeldahl-Nitrogen. Table 4 shows the overall operation conditions of the three reactors.

Total nitrogen balance

For the balance nitrogen is considered in both organic and inorganic (ammonia, nitrites and nitrates) forms. The balance is represented by Equation 1 (Nikolavcic *et al.*, 2000; Zwerger *et al.*, 2000).

$$(Q \times TN)_{\text{influent}} - (Q \times TN)_{\text{effluent}} = (Q \times TN)_{\text{backwashing}} + N_2 \uparrow \quad (1)$$

where

TN Total nitrogen, g/m^3 (sum of nitrate, ammonia and organic nitrogen in dissolved and suspended forms)

Q Wastewater flow, m^3/day

$N_2 \uparrow$ Molecular nitrogen produced by denitrification, gTN/day

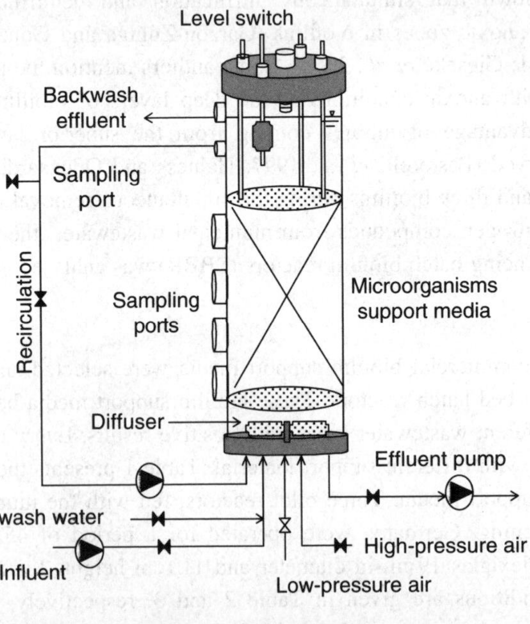

Figure 1 Schematic representation of the laboratory scale SBBR

Table 2 Reactor main characteristics

Diameter	0.19 m
Sectional area	0.028 m^2
Total volume	0.031 m^3
Water exchange volume	100%
Recirculation	0.2 m^3/h

Table 3 Mean influent concentration values

Parameter	mg/L
Total COD	214 ± 96
Dissolved COD	76 ± 26
NH_4-N	34 ± 8
Total Nitrogen	50 ± 1
PO_4-P	4 ± 1

Cycles profiles

Two representative cycles of operation were analyzed. The first one corresponds to organic loads under 2.5 gCOD/m^2·d, and the second one to organic loads over 3.0 gCOD/m^2·d.

Results

SBBR performance under an average organic load of 0.8 gDQO/m^2·d

Figure 2 shows the COD and dissolved oxygen profiles for the three reactors under an organic load of 0.8 gCOD/m^2·d. The time zero corresponds to influent concentrations. The vertical dotted line represents the end of the anaerobic reaction. The last values correspond to effluent concentrations.

COD removal. All three reactors present the same behavior. The COD decrease during the two hours of anaerobic reaction (first 120 minutes) is attributed to adsorption and biochemical transformations. Bouwer (1987) states that particles have to be adsorbed to the biofilm surface by diffusion, settling or entrapment before biochemical reactions can take place. After the anaerobic reaction is finished the COD continues decreasing until, after minute 300, a stable value is reached. In terms of COD removal it is considered that the cycle duration can be reduced to 6 hours.

The reactor packed with Kaldnes (Figure 2–A) showed total COD effluent values of 56 mg/L and 31 mg/L for dissolved COD. The reactors packed with Liapor and Linpor (Figures 2–B and 2–C, respectively) showed a similar behavior reaching total COD values in the effluent between 17 and 20 mg/L, and 15 and 16 mg/L for dissolved COD.

Table 4 Operational characteristics of the three reactors

Operation characteristics	Organic load					
	0.5–3.0 gCOD/m^2· or ·d			3.5–8.0 gCOD/m^2· or ·d		
	Liapor	Kaldnes	Linpor	Liapor	Kaldnes	Linpor
Wastewater flow, m^3/d	0.066	0.063	0.072	0.069	0.069	0.063
Bed height, m	0.33	0.33	0.52	0.16	0.17	0.32
Bed volume, m^3	0.009	0.0093	0.0148	0.005	0.0055	0.010
Total exposed area, m^2	8.3	8.1	3.9	2.7	2.7	2.7

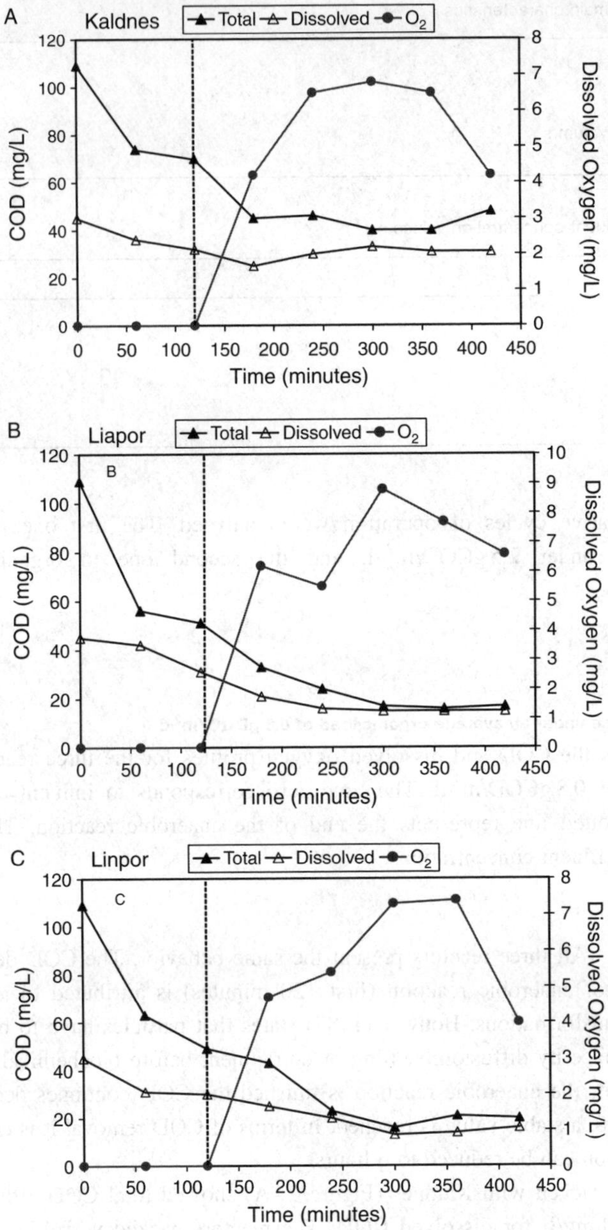

Figure 2 COD and dissolved oxygen profiles under an organic load of 0.8 gCOD/m^2·d. A) Kaldnes; B) Liapor; C) Linpor

Nitrogen removal. The pattern of the ammonia nitrogen profiles during the anaerobic phase was the same for the three reactors (Figure 3). The influent contained 22 mgNH$_4$-N/L (time zero). During the first two hours of anaerobic reaction the ammonia concentration decreased approximately to 14–27%. In previous studies it was shown that the biofilm strongly adsorbed ammonium explaining the initial reduction after filling (Nielsen, 1996; Gieseke *et al.*, 1999, 2002). Immediately after the aeration begins the activity of the nitrifiers was observed through decreasing ammonia concentration completing nitrification.

With Liapor and Linpor the nitrate concentration increased and eventually decreased at the end of the cycle. With Kaldnes the nitrite concentration increased almost till the end of the cycle. As nitrite is produced at the biofilm surface, a high concentration

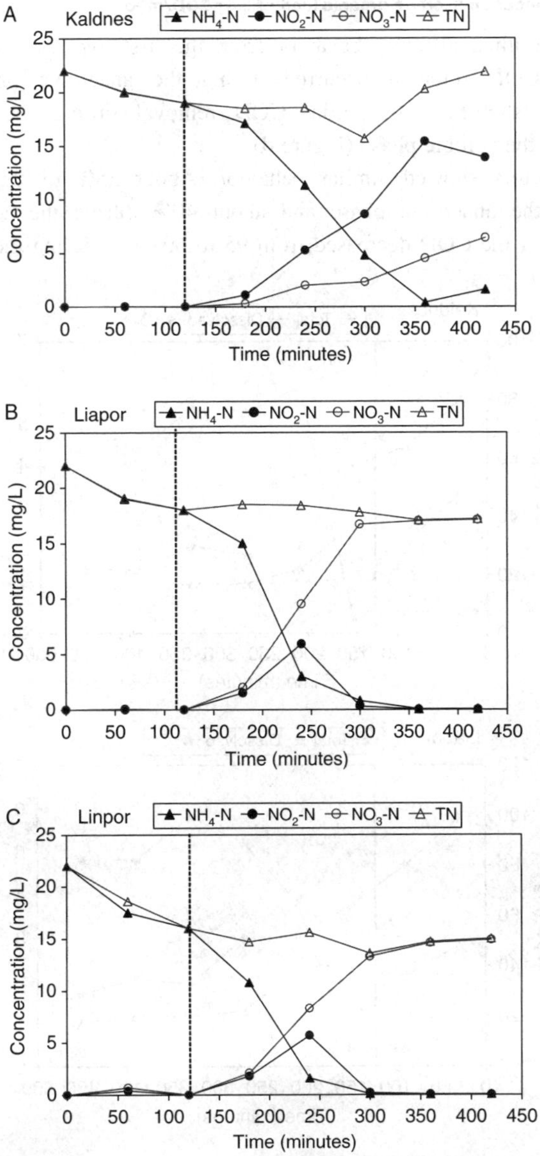

Figure 3 Ammonia, nitrite and nitrate nitrogen in the bulk liquid at an organic load of 0.8 gCOD/m^2·d. A) Kaldnes; B) Liapor; C) Linpor

gradient with an initially low nitrite concentration in the bulk fluid is induced during nitrification (Nijhof and Klapwijk, 1995). With the exception of Kaldness, nitrite was transformed in to nitrate at the end of the cycle nitrate being the sole nitrogen form detected. This situation indicates that the growing biofilm in Kaldnes presents characteristics that make difficult the diffusion into the deeper layers. The biofilm structure affects the transport of nutrients into and out of the biofilm (Beyenal and Lewandoski, 2000). When the biofilm density increases, the resistance to the diffusion increases (Cassey *et al.*, 2000).

The reactors packed with Liapor and Linpor remove the ammonia during the first three hours of aerobic reaction (300 min), indicating a strong activity of nitrifiers with the consequent presence of nitrate in the bulk liquid. The complete transformation of ammonia with the reactor packed with Kaldnes occurred after the fourth hour of aeration.

SBBR performance under an average organic load of 3.6 gCOD/m^2·d

COD removal. With a different behavior from the first experimental run, where the most important COD removal occurred during the anaerobic phase, this second experimental run showed only partial COD removal during the anaerobic phase continuing during the aerobic phase (Figure 4).

The three reactors showed similar behavior. About 35% of the total COD was removed during the anaerobic phase and about 40% during the aerobic phase. For Kaldnes and Linpor the COD decreased from 95 to 60 mg/L during the anaerobic phase.

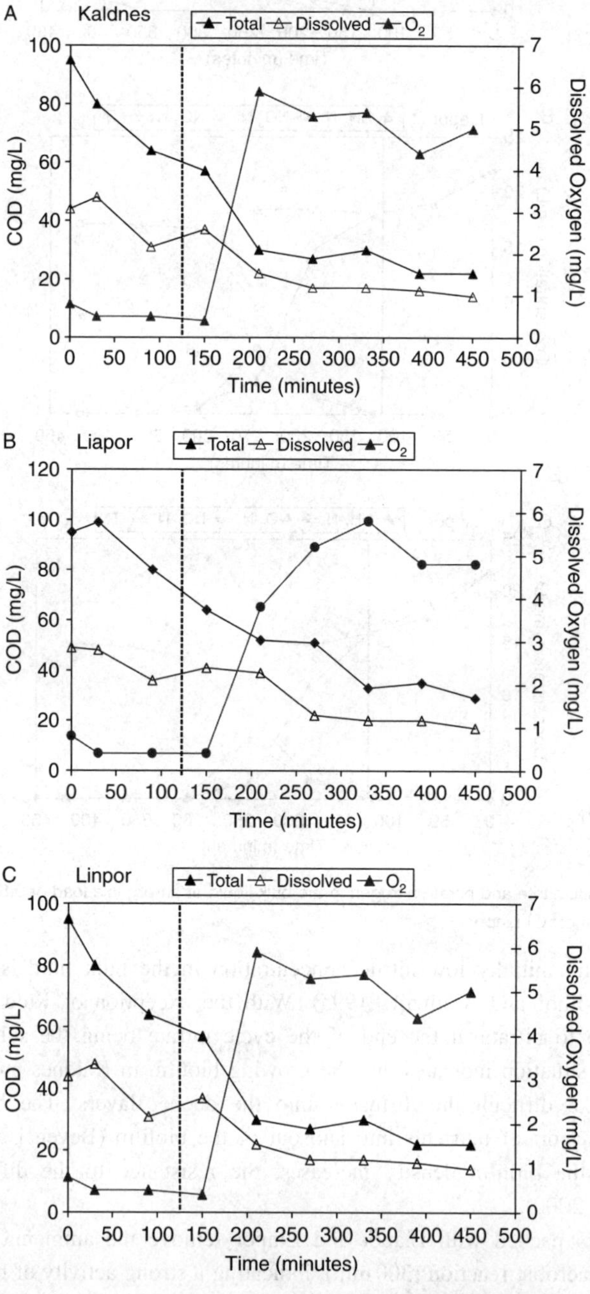

Figure 4 COD and dissolved oxygen profiles under an organic load of 3.6 gCOD/m^2·d. A) Kaldnes; B) Liapor; C) Linpor

These two reactors achieved a final total COD of 20 mg/L, while Liapor reached 30 mg/L at the end of the cycle. For the three reactors, at the end of the cycle the dissolved COD reached values slightly under 20 mg/L with Liapor showing the higher values.

Nitrogen removal. Ammonia concentration in the influent was 39 mgNH_4-N/L (Figure 5). At the beginning of the aerobic phase the ammonia oxidation did not begin until the biofilm was fully penetrated with oxygen (Castillo *et al.*, 1999). The oxygen requires time to diffuse through the biofilm and then the nitrifiers start their metabolism.

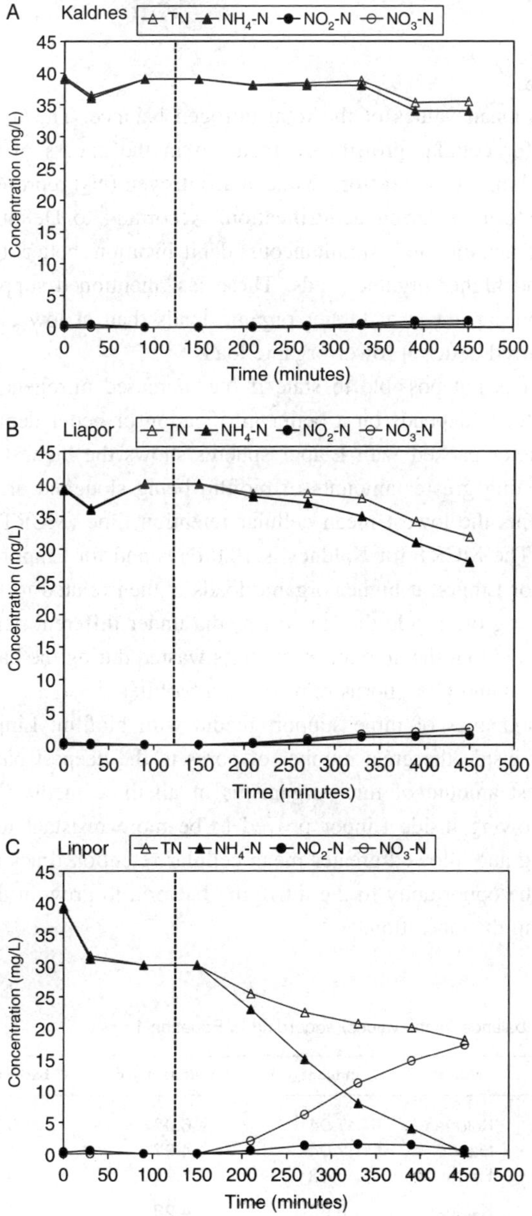

Figure 5 Ammonia, nitrite and nitrate nitrogen in the bulk liquid at an organic load of 3.6 gCOD/m^2·d.
A) Kaldnes; B) Liapor; C) Linpor

The reactors packed with Kaldnes and Liapor showed a minimal nitrifying activity; the ammonia concentration showed values near the total nitrogen concentration and nitrites and nitrates showed very low concentrations (Figures 5A and B).

The reactor packed with Linpor had a better response for nitrification when the organic load increased, allowing total nitrification at the end of the cycle (Figure 5–C). In the case of Linpor, considering the ammonia concentration at the beginning of the aerobic phase (22 $mgNH_4$-N), the reduction in total nitrogen can only be explained when a simultaneous denitrification in the deeper layers of the biofilm is considered (Figure 5–C). Previous studies with 'deep' biofilms growing on Linpor and Kaldnes allow the assumption that nitrates diffuse to the regions with the lower oxygen concentration (Gieseke *et al.*, 1999; Arnz *et al.*, 2001; Falkentoff *et al.*, 2001).

Total nitrogen balance

Table 5 shows the mean values of the total nitrogen balance. The nitrogen compounds that can be used for cellular growth are included in the excess of biomass detached during the backwashing of the reactors. Molecular nitrogen (N_2) represents the amount of total nitrogen transformed during denitrification. According to Default 1 and Table 5, Linpor allows nitrification and simultaneous denitrification better than Kaldnes and Liapor at lower and higher organic loads. These last mentioned support media present better denitrification capacities at higher organic loads than at lower ones. For Linpor, denitrification occurred better at lower organic loads.

From Table 5 it is not possible to state if the increased nitrogen production during the lower organic loads depends on a better nitrification or better denitrification or both combined. The reactor packed with Liapor spheres shows the highest values of TN for backwashing indicating greater amounts of biofilm being sloughed off during this operation. Liapor presents the lowest mean cellular retention time (MCRT) with an average value of 5.4 days. The MCRT for Kaldnes is 10.0 days and for Linpor is 22.9 days. The better nitrification of Linpor at higher organic loads is then related to the higher MCRT, allowing nitrifiers to grow inside the support media under different conditions. Park and Chang (2000) reported that the amount of biomass wasted during the backwash procedure also affected nitrogen and phosphorus removal in a biofilter.

Figure 6 shows images of three support media with biofilm. Linpor allows microorganisms to grow inside the great number of pores to the deepest parts of the material allowing the highest amount of microorganisms of all three media (Figure 6–C). The microorganisms growing inside Linpor proved to be more resistant to the stress forces during backwashing and allowed greater mean cellular retention times (23 days) than the other two, giving the opportunity to the nitrifying bacteria to grow in the system without being washed during this operation.

Table 5 Total nitrogen balance (mean values) according to Equation 1

Organic load $gCOD/m^2 \cdot d$	Media	NT influent g/d	NT effluent g/d	NT backwashing g/d	$N_2\uparrow$ g/d
0.5–2.5	Kaldnes	7.04	6.07	0.33	0.6
	Liapor	7.04	4.97	1.22	0.8
	Linpor	6.43	1.91	0.46	4.1
3.0–8.0	Kaldnes	6.90	4.28	0.72	1.9
	Liapor	6.90	4.24	1.23	1.4
	Linpor	6.30	2.34	0.48	3.5

A Kaldnes B Liapor C Linpor

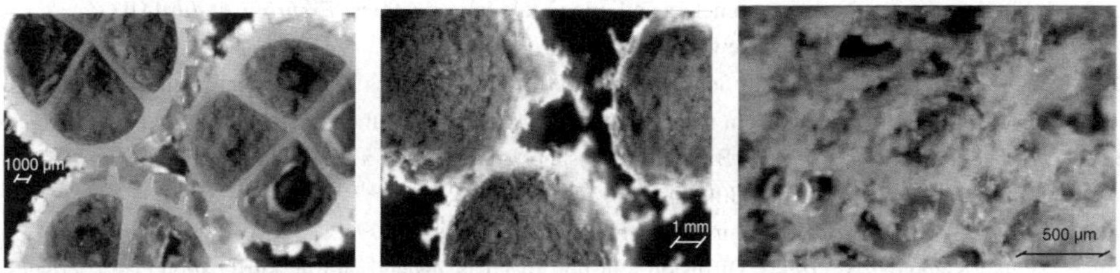

Figure 6 Media support with biofilm. Images correspond to the higher tested of organic load

Conclusions

1. All three reactors showed a more stable operation at organic loading rates under 2.0 gCOD/m^2·d. Linpor achieved higher COD removal rates at higher organic loads. Liapor showed the best performance at lower organic loads. Kaldnes showed a stable performance at medium organic load values.
2. In 95% of the experiments Linpor achieved NH_4-N removal values of over 50%. For organic loads under 2.5 gCOD/m^2·d Liapor and Kaldnes achieved high nitrification rates near 100%. For organic loads over 2.5 gCOD/m^2·d, the NH_4-N removal remained mostly under 40% without any tendency.
3. Linpor achieved the highest NOx-N production reaching values between 15 and 20 mg/l. For low organic loading rates Linpor and Liapor present similar average NOx-N concentrations. Kaldnes shows the lowest concentrations throughout the whole experimental period.
4. The explanation given to the differences between ammonia nitrogen removal and NOx-N generation is simultaneous denitrification inside the deep biofilms.
5. The average mean cellular retention times were 5.4 days for Liapor, 10.0 days for Kaldnes and 22.9 days for Linpor. This is the reason why Linpor achieved complete nitrification even with higher organic loads.

Acknowledgements

The authors wish to thank the German Academic Exchange Service (DAAD) and the Directorate General for Academic Affairs (DGAPA) of the National University of Mexico for their support.

References

Arnz, P., Arnold, E. and Wilderer, P.A. (2001). Enhanced biological phosphorus removal in a semi full-scale SBBR. *Wat. Sci. Tech.*, **43**(3), 167–174.

Beyenal, H. and Lewandowski, Z. (2000). Combined effect of substrate concentration and flow velocity on effective diffusivity in biofilms. *Wat. Res.*, **34**(2), 528–538.

Bouwer, E.J. (1987). Theoretical investigation of particle deposition in biofilm systems. *Wat. Res.*, **21**(12), 1489–1498.

Cassey, E., Glennon, B. and Hamer, G. (2000). Biofilm development in a membrane-aerated biofilm reactor: Effect of flow velocity on performance. *Biotechnol. Bioeng.*, **64**(4), 476–486.

Castillo, P.A., González-Martínez, S. and Tejero, I. (1999). Biological phosphorus removal using a biofilm membrana reactor: Operation at high organic loading rates. *Wat. Sci. Tech.*, **40**(4–5), 321–329.

Falkentoft, C., Harremoës, P., Mosbaek, H. and Wilderer, P. (2001). Stability of a lab-scale biofilm for simultaneous removal of phosphorus and nitrate. *Wat. Sci. Tech.*, **43**(1), 335–342.

Garzón-Zúñiga, M. and González-Martínez, S. (1996). Biological phosphate and nitrogen removal in a biofilm sequencing batch reactor. *Wat. Sci. Tech.*, **34**(1–2), 293–301.

Gieseke, A., Arnz, P., Schramm, A. Amman, R., Wilderer, P.A. (1999) Nutrient removal with a sequencing batch biofilm reactor. Process parameters and microscale investigations. *Preprints of 4th IAWQ Conference on Biofilm Systems*. New York, USA.

Gieseke, A., Arnz, P., Amann, R. and Schramm, A. (2002). Simultaneous P and N removal in a sequencing batch biofilm reactor: Insights from reactor- and microscale investigations. *Water Research*, **36**, 501–509.

Helness, H. and Odegaard, H. (1999). Biological phosphorus removal in a sequencing batch moving bed biofilm reactor. *Wat. Sci. Tech.*, **40**(4–5), 161–168.

Nielsen, P.H. (1996). Adsorption of ammonium to activated sludge. *Wat. Res.*, **30**(3), 762–764.

Nijhof, M. and Klapwijk, A. (1995). Diffusion transport mechanisms and biofilm nitrification characteristics influencing nitrite levels in nitrifying trickling filter effluents. *Wat. Res.*, **29**(10), 2287–2292.

Nikolavcic, B., Schweighofer, H., and Kroiss, H. (2000) COD balance and nitrogen removal in a biofilter pilot plant. *Wat. Sci. Tech.*, **41**(4–5), 69–76.

Park, D. and Chang, W. (2000). Simultaneous removal of nitrogen and phosphorus in a two-biofilter system. *Wat. Sci. Tech.*, **41**(12), 101–106.

Pastorelli, G., Andreottola, G. and Canziani, R. (1997). Pilot plant experiments with moving bed biofilm reactors. *Wat. Sci. Tech.*, **36**(1), 43–50.

Zwerger, B., Arnold, E., and Wilderer, P.A. (2000) Nutrient balances for combined nitrification and denitrification in biofilters. *Wat. Sci. Tech.*, **41**(4–5), 91–95.

Biological aerated filtration of municipal wastewater using a low-cost filtration media

S. González-Martínez, T. Millán and O. González-Barceló

Institute of Engineering, National University of Mexico, Ciudad Universitaria, 04510 México DF, Mexico
(E-mail: *sgm@pumas.iingen.unam.mx*)

Abstract The main objective of this research was to demonstrate that selected natural lava stones can be successfully used for low-cost aerobic biofiltration of municipal wastewater. To demonstrate the procedure a pilot filter was built using 6 mm lava stones as support material. The filter depth was 3.0 m. Provided with sampling ports at different depths analysis of the wastewater could be made for COD, TSS, ammonia and nitrates nitrogen, pH, temperature and Kjeldahl nitrogen. Backwashing was performed every 72 hours. Total and dissolved COD and TSS behaved similarly with the organic load: The highest removal rates were observed with the lowest organic load of 0.8 kgCOD/m^3 d. These removal rates decreased to a minimum value at organic loading rates of 1.5 kgCOD/m^3 d and then remained without noticeable changes to the highest value of 3.5 kgCOD/m^3 d. The highest total and dissolved COD removal values were 81 and 84%, respectively. For TSS the best removal value was 95%. Up to 75% ammonia removal was achieved at the lowest organic load of 0.8 kgCOD/m^3 d. Ammonia removal decreased to 36% with a higher organic load of 1.6 kgCOD/m^3 d. The Mean Cellular Retention (MCRT) time varied from 1 to 6 days with an average of 3.2 days. This fact proves that the MCRT depends on the backwashing frequency more than of any other factor involved. The bed volume decreased in about 5% after 300 days of operation. Microscopic observations showed that the small stones were rounder after 300 days and that the volume losses were caused when the edges of the stones were cut by the abrasion caused by backwashing.

Keywords Backwashing; biofilm; biological filtration; support media

Water Science & Technology **Vol 55 No 7 pp 255–262**

Introduction

The Biofor® process manages surface velocities of 10 to 33 m^3/m^2 h with relatively high organic loading rates of 2 to 3 kgDBO/m^3 d for organic substances and 1.5 kgN-NH$_4$/m^3 d for ammonia nitrogen achieving removal rates of 70% for BOD and between 80 and 90% for ammonia nitrogen. Biofor filters usually have a depth of 3.0 metres (Pujol *et al.*, 1994; Pujol, 2000). Actually there are reports of efficient treatment of municipal wastewater with organic loads of 18 kg COD/m^3 d and with backwashing every 24 hours (Bédard, 1999). The best backwashing performance was through the fluidisation of the filter bed with a combination of high-pressurized air and water from the bottom of the filter (Humby and Fitzpatrick, 1996).

The size of the media particles is selected according to the use of the filter. For example, particles with sizes above 6 mm are recommended for COD removal and nitrification. For tertiary treatment and nutrient removal, particle sizes between 2 and 3 mm are recommended (Moore *et al.*, 2001). Among different synthetic materials, Ojeda and Buitrón (2001) tested the growth speed of biofilms and concluded that lava stones have the fastest colonisation rates. Rough surfaces produce the more stable biofilms and improve solid retention (Mendoza-Espinosa and Stephenson, 1999).

The main objective of this research was to demonstrate that selected natural lava stones can be used successfully for low-cost biofiltration of municipal wastewater.

doi: 10.2166/wst.2007.152

Method

A pilot filter was built according to Figure 1 and Table 1. The filter is a PVC pipe, with an internal diameter of 19 cm, in a vertical position, filled with a 3.0-metre lava-stone-bed. The lava stones were sieved to a size of 6 mm. The filter was fed with municipal wastewater from a neighboring settlement near the wastewater treatment plant of the National University of Mexico. Up-flow operation was selected. Every 72 hours the filter bed was backwashed with high pressurized air and tap water for 2 minutes. The back-wash water was collected to an approximate volume of 120 litres. Analyses of the back-wash water were performed for total suspended solids (TSS) and Kjeldahl-Nitrogen of the TSS.

Table 2 shows the characteristics of the filter material. The filtration material consisted of small lava stones easily found in Central Mexico with an approximate cost of US$ 150 for a 6 m^3 load already sieved to the desired size. Figure 2 shows an electron-micrograph of the lava stones used during this research. The surface is highly porous making the material ideal to immobilise microorganisms. The pores observed in the figure present diameters between 50 and 150 μm.

Twenty four-hour samples were collected from influent and effluent. Analyses were made for total and dissolved COD, ammonia nitrogen, nitrates, and TSS. On another set of experiments, using the sampling ports, grab samples were taken along the filter height in order to make concentration profiles of different parameters. The distance between

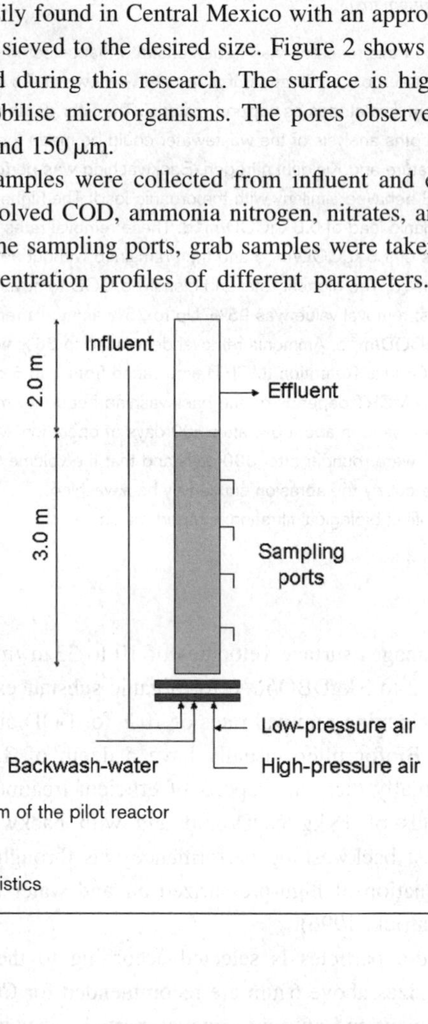

Figure 1 Schematic diagram of the pilot reactor

Table 1 Pilot filter characteristics

Total height	5.0 m
Internal diameter	0.19 m
Filter bed depth	3.0 m
Media volume	0.039 m^3
Total surface area	33 m^2

Table 2 Filter media characteristics

Specific surface area	390 m^2/m^3
Average particle size	6.0 mm
Specific number of particles	1,842 stones/l
Apparent density	1.51 kg/l
Packing material porosity	46%

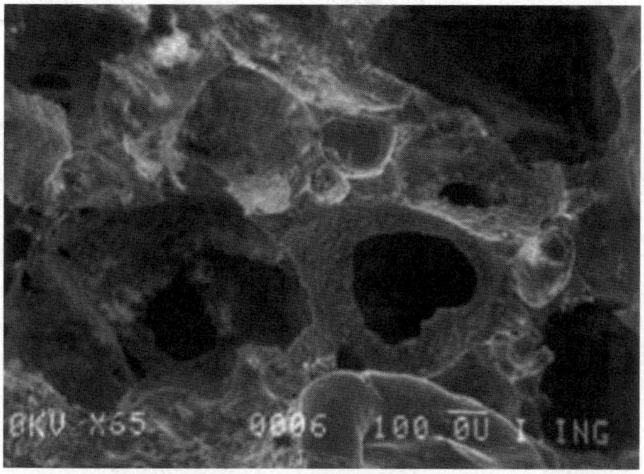

Figure 2 Electronic micrograph of the surface of a lava stone

sampling points is 60 cm. The operation procedure consisted in controlling the influent flow to adjust the desired organic load to 1.0, 1.5 and 2.5 kgCOD/m^3 d.

Results

After approximately 350 days of operation, due to abrasion caused by backwashing, the media lost 5% of its volume. Regularly, the lost volume was replaced with raw material.

The dissolved oxygen was kept above 4.0 mg/l during the whole experiment. The pH values fluctuated between 7.6 and 7.9 for the influent and between 7.2 and 7.5 for the effluent. The average temperature varied between 19 and 21 °C. According to regular measurements, the water temperature oscillated between a minimum of 16 °C at night and 24 °C during the day.

The experimentation was divided into four parts: start-up and three different organic loading adjustments. During the first 100 days the reactor was operated under an average organic load of 1.5 kgCOD/m^3 d and samples were taken to analyse only TSS and total COD to obtain average TSS removal values of 84% and COD removal of 72%. After one week of operation, the reactor showed TSS and total COD removal values of 70 to 90% and 30 to 60%, respectively.

Figure 3 shows the organic load from the beginning to the end of the experimentation (400 days). Four periods representing different stages of the experimental work resulted in four average organic loading rates.

Because of the constant variations of the wastewater quality, the achieved organic loads are slightly different from the originally programmed ones. The start-up of the system took 100 days where the organic load was controlled over the influent flow to guarantee a hydraulic retention time of 1.25 h. Instead of the originally planned organic load values of 1.0, 1.5 and 2.5 kgCOD/m^3 d, the achieved values were 0.8, 1.2, and 2.6 kgCOD/m^3 d.

COD and TSS removal

Figure 4 shows that dissolved COD removal decreased with the organic load. A similar figure was observed for total COD but showing values slightly lower than the ones for dissolved COD. The highest COD removal rates of 80–90% were observed with the lowest organic loading rates. By increasing the organic load up to 3.25 kgCOD/m^3 d, the COD removal rates tend to reach a stable value of 65% for dissolved COD and 60% for

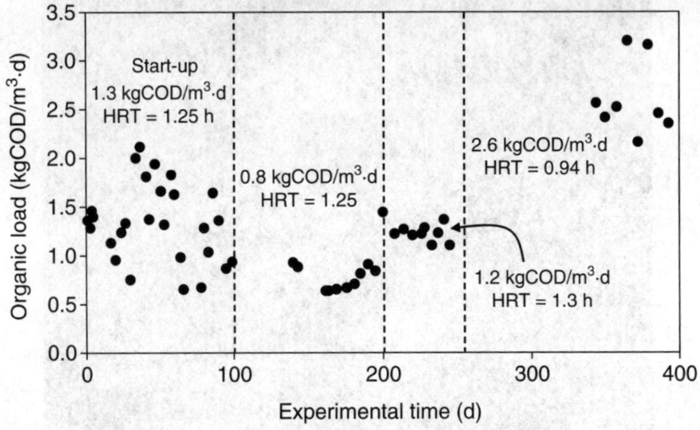

Figure 3 Organic loading rates during the reactor operation

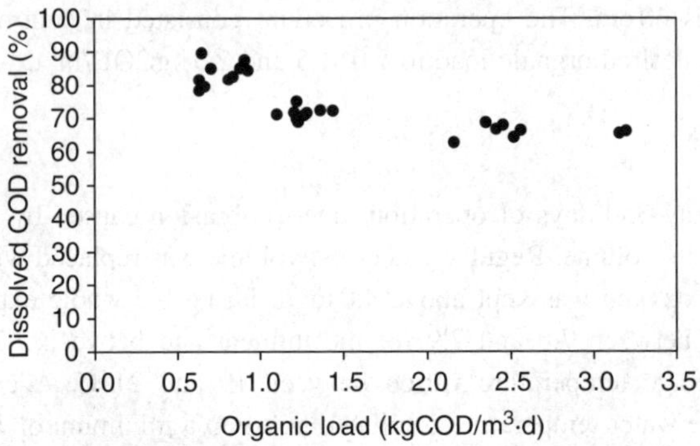

Figure 4 Dissolved COD removal and organic load

total COD (Table 3). Figure 4 shows that an active biofilm is responsible for the removal of dissolved organic substances.

Table 3 presents the overall results for COD removal according to the periods indicated in Figure 3. The COD removal rates described in Figure 5 and Table 3 show a linear relationship between removed COD and organic load. These results indicate also that the COD removal rates are independent of the hydraulic retention time.

Figure 6 shows the total COD profiles along the filter depth. The triangles (dates 10/08/2002, 03/09/02 and 10/09/02) indicate sampling days for the lowest organic loading rate of 0.8 kgCOD/m^3 d; the diamonds indicate samples during the organic load of 1.2 kgCOD/m^3 d (05/10/02, 11/10/02 and 15/10/02) and the squares (11/04/03, 19/04/03 and 05/05/03) indicate the highest organic load of 2.6 kgCOD/m^3 d. The lower and middle organic load curves (triangles and diamonds) show a similar pattern: having the

Table 3 Average total and dissolved COD values

	Organic load (kgCOD/m^3 d)	HRT (h)	Influent (mg/l)	Effluent (mg/l)	Removal (%)	Removal (gCOD/d)
Total COD	0.8	1.25	87	17	81	52
	1.2	1.30	147	48	67	71
	2.6	0.94	221	88	60	133
Disolved COD	0.8	1.25	60	10	84	38
	1.2	1.30	110	31	71	57
	2.6	0.94	148	50	66	98

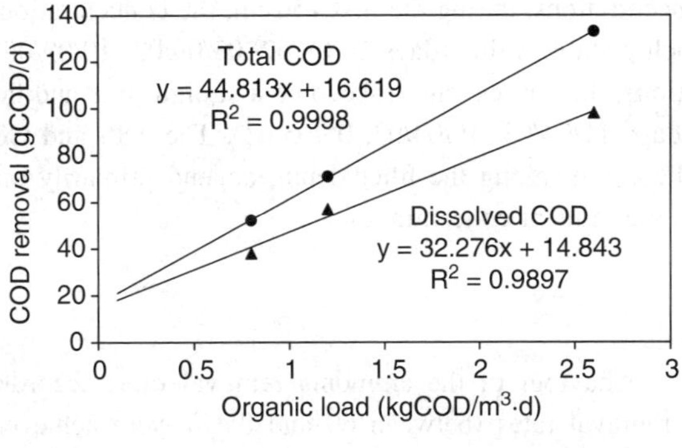

Figure 5 COD removal rates and organic load

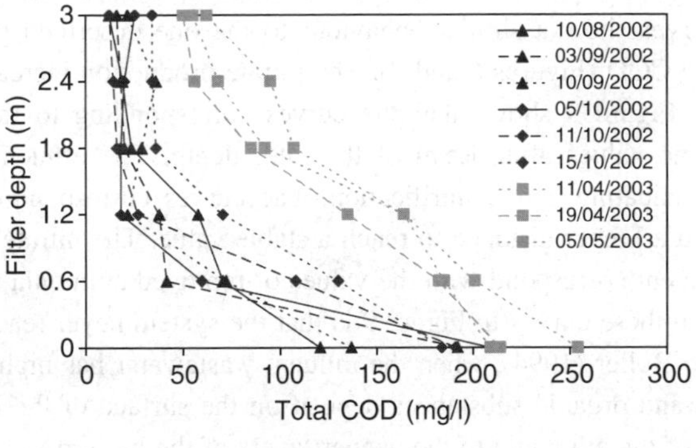

Figure 6 COD profiles for different days and organic loads

filter in up-flow operation, the highest COD values were measured at the bottom (influent) and rapidly decreased during the first 1.2 m until a relatively stable value was achieved, continuing unchanged in the last 1.2 m before exiting the filter. For the COD measured during the highest organic load (squares) the behaviour is different: the COD values decrease with an almost constant slope along the filter depth, reaching higher effluent values than the curves with lower organic loads.

The TSS removal showed similar behaviour to COD: with the lowest organic loads, removal values between 85 and 98% were achieved. With organic loads above 1.0 kgCOD/m^3 d the TSS removal rates were kept between 65 and 75% with no recognisable tendencies. Figure 7 shows the behaviour of TSS along the filter depth. The curves in this figure show the highest solids concentration at the influent (depth zero) and the lowest at the effluent (depth 3.0 m). TSS and COD present the same behaviour: with the

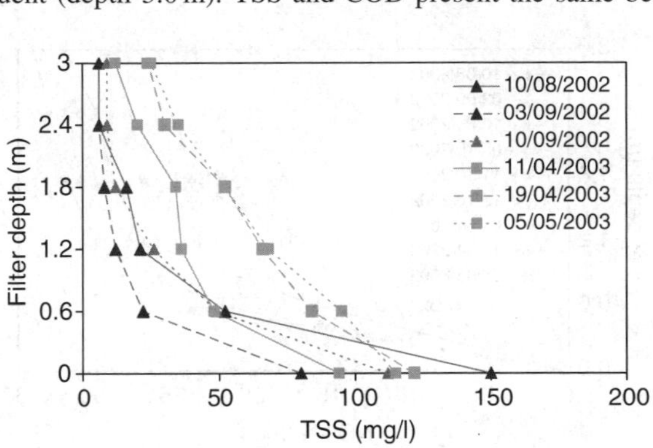

Figure 7 Suspended solids along the filter depth

lowest influent concentrations, during the first 150 cm, the concentration decreases rapidly and the curves reach a stable value (days 10/08/02, 03/09/02, 10/09/05). With increasing influent concentrations, the curves tend to become a straight line and the effluent concentrations increase (days 11/04/03, 19/04/03, 05/05/03). The TSS and COD removal rates, and the form of the curves along the filter depth, depend primarily on the organic load and secondarily on the influent concentration.

Ammonia nitrogen removal

Figure 8 shows the behaviour of the ammonia removal rates according to the organic load. The highest removal rates (between 60 and 85%) were achieved with the lowest organic loading rates of 0.5 to 1.0 kgCOD/m^3 d. The removal decreases consistently with increasing organic loads. This behaviour can be compared with the curves in Figure 9. In this figure the curves show a similar behaviour to the one described previously for the curves of TSS and COD (Figures 6 and 7). The nitrate production increases with decreasing organic load. Figure 9 shows that the curves corresponding to the lowest organic loads tend to stable values after 1.5 m of the filter depth. The effluent values fluctuate around 30 mg/l, indicating active nitrification. The curves corresponding to the highest organic loads need a longer distance to reach a stable value. The nitrate nitrogen concentrations in the effluent correspond with the values of removed ammonia nitrogen.

Interesting from these curves in Figure 9 is that the system never reached full nitrification. According to Boller (1994), when the influent wastewater has higher COD and TSS values, the solids and organic substances deposit on the surface of the biofilm inhibiting the mass transfer of the ammonia to the deeper layers of the biofilm.

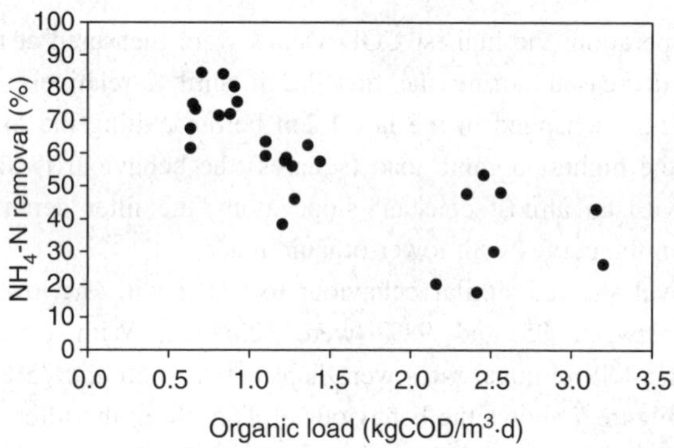

Figure 8 Ammonia nitrogen removal depending on the organic load

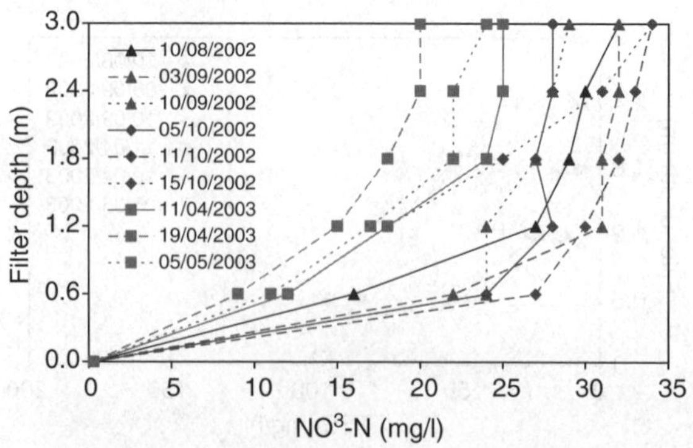

Figure 9 Nitrate production along the filter depth for different organic loads

Table 4 shows the average values obtained for nitrification. These values show that the sum of the ammonia nitrogen and nitrates nitrogen at the effluent coincide with the ammonia nitrogen at the influent, indicating a low nitrogen removal caused by cellular growth and eventual simultaneous denitrification in the deeper biofilm layers.

Mean cellular retention times (MCRT)

The MCRT was calculated through the determination of Kjeldahl-Nitrogen in the stones, of the suspended solids in the effluent and in the water from backwashing and performing a mass balance.

According to Figure 10, the MCRT values vary from 1.5 to 4.5 days with an average of 3.2 days. With these results it is shown that the MCRT depends on the frequency of backwashing (every 72 hours) and not on the organic load.

Backwashing and pressure drop

Figure 11 indicates that the backwashing frequency of 72 hours can be increased to higher values. The maximal pressure drop of 13 cm was achieved with the highest

Table 4 Average values of ammonia nitrogen and nitrate production

Organic Load (kgCOD/m^3 d)	HRT (h)	NH_4-N			NO_3-N	
		Influent (mg/l)	Effluent (mg/l)	Removal (%)	Influent (mg/l)	Effluent (mg/l)
0.77	1.25	42	11	75	<0.3	31
1.24	1.30	33	15	56	<0.3	19
1.60	0.94	51	33	36	<0.3	17

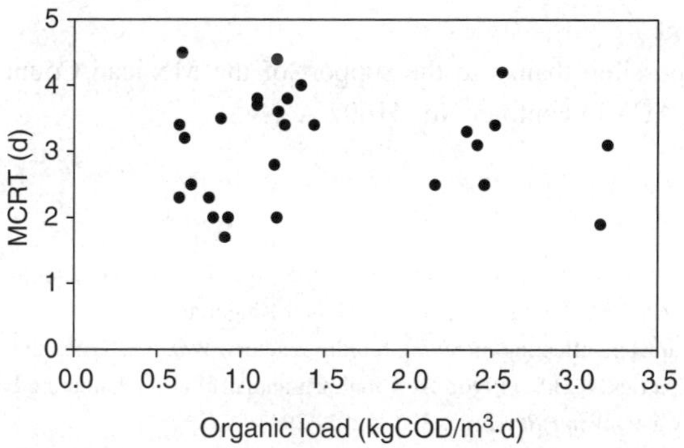

Figure 10 Mean cellular retention time

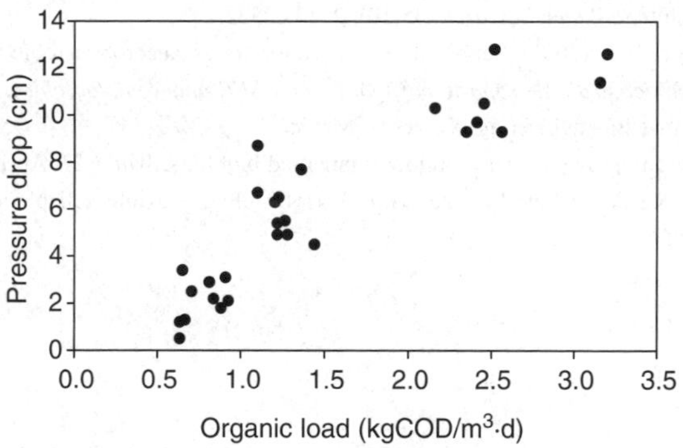

Figure 11 Pressure drop and applied organic load

organic load. For practical purposes a larger operational run of the filter is attractive because of the reduction of manpower for operation and maintenance.

Conclusions

- The filter showed a rapid biological adaptation, showing removal values of 80% for TSS and 50% for COD after 10 days of operation.
- Total and dissolved COD and TSS behaved similarly with the organic load. The best removal rates were observed with the lowest organic loads of 0.8 $kgCOD/m^3$ d. These removal rates decreased to a minimum value at organic loading rates of 1.5 $kgCOD/m^3$ d and then remained without noticeable changes to the highest load value of 3.5 $kgCOD/m^3$ d. The highest total and dissolved COD removal values were 81 and 84%, respectively. For TSS, the best removal value of 95% was achieved with an organic load of 1.5 $kgCOD/m^3$ d.
- Nitrification was achieved up to 75% ammonia removal at the lowest organic load of 0.8 $kgCOD/m^3$ d. Ammonia removal decreased to 36% with a higher organic load of 1.6 $kgCOD/m^3$ d.
- The Mean Cellular Retention (MCRT) time varied from 1 to 6 days with an average of 3.2 days. This fact proves that the MCRT depends on the backwashing frequency more than any other factor involved.
- The bed volume decreased by about 5% after 300 days of operation. Microscopic observations showed that the small stones were rounder after 300 days and that the volume losses were caused when the edges of the stones were cut by the abrasion caused by backwashing.

Acknowledgement

This project was possible thanks to the support of the Mexican Council for Science and Technology (CONACYT) contract No. 41697/A-1.

References

Bédard, P (1999). High rate biofiltration. Proceedings of the Third International Symposium on Biological Aerated Biofilters (*BAF-3*). Cranfield University, United Kingdom.

Boller, M. (1994). Parameters affecting nitrifying biofilm reactors. *Wat. Sci. Tech.*, **29**(10–11), 1–11.

Humby, M.S. and Fitzpatrick, C.B.S. (1996). Attrition of granular filter media during backwashing with combined air and water. *Water Research*, **30**(2), 291–294.

Mendoza-Espinosa, L.G. and Stephenson, T. (1999). A review of biological aerated filters (BAFs) for wastewater treatment. *Environmental Engineering Science*, **16**(3), 201–216.

Moore, R., Quarmby, J. and Stephenson, T. (2001). The effects of media size on the performance of biological aerated filters. *Water Research*, **35**(10), 2514–2522.

Ojeda, R.L. and Buitrón, M.G (2001). Support media selection for an anaerobi/aerobic SBR (original in Spanish) (*Selección del medio de soporte para un reactor SBR anaerobio/aerobio*). IX National Congress on Biotechnology and Bioengineering. Veracruz, Mexico.

Pujol, R (2000). Process improvements for upflow submerged biofilters. *Water 21*. April, pp. 25–29.

Pujol, R., Hamon, M., Kandel, X. and Lemmel, H. (1994). Biofilters: flexible, reliable biological reactors. *Wat. Sci. Tech.*, **29**(10–11), 33–38.

Keyword Index